GALILEO UNBOUND

GALILEO UNBOUND

A Path Across Life, the Universe and Everything

David D. Nolte

Purdue University

Great Clarendon Street, Oxford, OX2 6DP,
United Kingdom

Oxford University Press is a department of the University of Oxford.
It furthers the University's objective of excellence in research, scholarship,
and education by publishing worldwide. Oxford is a registered trade mark of
Oxford University Press in the UK and in certain other countries

First Edition published in 2018

Impression: 1

Published in the United States of America by Oxford University press
198 Madison Avenue, New York, NY 10016, United States of America

British Library Cataloguing in Publication Data
Data available

Library of Congress Control Number: 2017962119

ISBN 978–0–19–880584–7
DOI: 10.1093/oso/9780198805847.001.0001

Printed and bound by
CPI Group (UK) Ltd, Croydon, CR0 4YY

In June of 1633 Galileo was found guilty of heresy by the Holy Inquisition and sentenced to house arrest for what remained of his life. He was a renaissance Prometheus, bound for giving knowledge to humanity. With little to do, and allowed few visitors, he at last had the uninterrupted time to finish his life's labor. When *Two New Sciences* was published in 1638 it contained the seeds of the science of motion that would mature into a grand and abstract vision that permeates all science today. In this way, Galileo was unbound, not by Hercules, but by his own hand as he penned the introduction to his work:

> *. . . what I consider more important, there have been opened up to this vast and most excellent science, of which my work is merely the beginning, ways and means by which other minds more acute than mine will explore its remote corners.*
>
> Galileo Galilei (1638) *Two New Sciences*

Preface

Beware! You are traveling at alarming speed, flying eastward at over a thousand kilometers per hour as the Earth spins on its axis propelling you along. In the years between Copernicus and Galileo this speed was considered absurd and blasphemous, which is why Copernicus never published until near his death, and why Galileo spent the last years of his life under house arrest. Yet this speed pales in comparison with the Earth's year-long rush around the sun, traveling at over one hundred thousand kilometers per hour. To this must be added to the speed of the Sun orbiting the center of the Milky Way galaxy. It may take the Sun 200 million years to orbit once, but it does so at nearly a million kilometers per hour. Something as massive as our Milky Way might seem to be an anchor to which we can tether ourselves in safety, but it too is flying through space, pulled towards a region in the Universe known as the Great Attractor at over 3 million kilometers per hour (0.3% of the speed of light). Thus, your personal trajectory through the Universe is a collection of motions stacked upon motions in a relative Galilean cascade that would have amazed Galileo himself.

Galileo Unbound: A Path Across Life, the Universe and Everything explores the continuous thread that took us from Galileo's discovery of relative motion, and the parabolic trajectory, to today's modern dynamics and complex systems. It is a history of expanding dimension and increasing abstraction, until today we speak of entangled quantum particles moving among many worlds, and we envision our lives as trajectories through a space of a thousand dimensions spanned by genes and proteins that define our essence. Remarkably, common themes persist that predict the evolution of species as readily as the orbits of planets. Galileo laid the foundation upon which Newton built a theory of dynamics that could capture the trajectory of the moon through space using the same physics that controlled the flight of a cannon ball. Late in the nineteenth-century, concepts of motion expanded into multiple dimensions, and in the 20th century geometry became the cause of motion rather than the result when Einstein envisioned the fabric of space-time warped by mass and energy, causing light rays to bend past the Sun. Possibly more radical was Feynman's dilemma of quantum particles taking all paths at once—setting the stage for the modern fields of quantum field theory and quantum computing. Yet as concepts of motion have evolved, one thing has remained constant—the need to track ever more complex changes and to capture their essence, to find patterns in the chaos as we try to predict and control our world. Today's ideas of motion go far beyond the parabolic trajectory, but even Galileo might recognize the common thread that winds through all these motions, drawing them together into a unified view that gives us the power to see, at least a little, through the mists shrouding the future.

Galileo Unbound is the trade nonfiction companion to the college textbook *Introduction to Modern Dynamics: Chaos, Networks, Space and Time* (Oxford, 2015). The textbook uses ideas of abstract spaces to unify a broad range of topics in dynamics. Whereas the *Introduction to Modern Dynamics* is aimed at advanced undergraduates, *Galileo Unbound* is written for anyone who has an interest in the history of modern thought. Many memes of scientific literacy have found their way into cultural literacy through popular TV shows like *The Big Bang Theory* and through popular science writers like Brian Greene and Neil Degrass Tyson. *Galileo Unbound* shows where these ideas came from, how they evolved and how they shape our daily lives.

The seed from which this book grew was planted in the first years of the new millennium when I became enthralled by the beauty and the power of multidimensional spaces with their potential to capture behavior as complex as the trajectory of life. My own personal journey writing this book was helped along its way by many whom I would like to thank. Foremost among these is my mother Nancy Nolte, who went to New York City as a young woman in her early 20's to begin a career as an editor in the book publishing industry of the 1940's. She was my writing coach and my role model from my earliest days through the early stages of writing this book. I would like to thank my long-time colleague Anant Ramdas in the Physics Department at Purdue University, with whom I have shared many enjoyable hours talking about fascinating episodes in the history of physics. Helpful insights and corrections in the manuscript stage were provided by Andrew Hirsch and Ephraim Fischback at Purdue, and by Leon Glass and Jeremy Gray on selected chapters, for which I am grateful. The editors at Oxford, Sonke Adelung and Ania Wronski, provided constant encouragement and keen oversight during the process, helping to bring this story to life. Finally, I thank my wife Laura and my son Nicholas for loving support and inspiration. They are the force center of my own personal trajectory through this complex universe.

Contents

1. Flight of the Swallows 1
    Trajectories at Dusk 2
    Galileo's Parabola 4
    The Wealth of Motions 5
2. A New Scientist 11
    Introducing Galileo 12
    The Starry Messenger 15
    Pulp Friction 20
    Cometary Prelude to a Dialog 29
    Orbit Out of Control: Galileo's Tragedy 34
3. Galileo's Trajectory 39
    Form is the Function 39
    The Arc of War 45
    Inclined to Move 52
    Saved from the Ashes 59
4. On the Shoulders of Giants 64
    The Plague on You 65
    Pride and Prejudice 70
    The Principle of Least Action: Opening Act 74
    Of Friends and Forgers 79
    Lagrange's View 86
    The Fourth Dimension and Beyond 93
5. Geometry on my Mind 95
    Manifolds from Heaven 96
    The Cat that Walks Through Walls 104
    Fractional Worlds 110

6. The Tangled Tale of Phase Space 121
Liouville's Theorem 122
Clausius' Transformation 125
Boltzmann's Phase 129
The Three-Body Problem 132
Boltzmann's Atoms 139
Eherenfest's Legacy 140
New Abstractions 143
7. The Lens of Gravity 145
What Stays the Same if Everything is Relative? 146
Warping Space-Time 150
Schwarzschild's Radius 161
Black Hole Trajectories 166
Galileo Bound 170
8. On the Quantum Footpath 172
Matrix Mechanics 173
Wave Mechanics 179
Uncertain Physics 182
Entanglement 187
Drums in Brazil 194
Feynman's Diagrams 198
Galileo in Quantumland 203
9. From Butterflies to Hurricanes 205
(KAM) and the Fate of the Earth 206
Lorenz's Butterfly 215
Feigenbaum's Ratio 221
Galileo Poised 230
10. Darwin in the Clockworks 232
The Origin of Species 233
The Limits of Growth 238
The Wealth of Nations 244
Genetics in the Mix 247
Landscapes of Life 253

Pauling's Molecular Clock 258
Dawkin's Meme 262

11. The Measure of Life 267
Firefly Waves 268
Autonomous Oscillations 271
Unruly Heart 277
Ghost in the Shell 282
Seven Bridges of Königsberg 289

12. Epilogue: Galileo at Home in the Multiverse 301

Endnotes 303
Bibliography 319
Index 327

1

Flight of the Swallows

Without Aristotle's Physics there would have been no Galileo.
MARTIN HEIDEGGER, The Principle of Reason[1]

You are a projectile flying through space. You are a worker or investor improving your earnings. You are a mother or father or child in a social network that grows throughout your life. You are a biological machine, with molecular motors driving countless motions inside your cells, propelling you down a hallway or a racetrack. These motions, these trajectories, these changes in your life are defined, followed and measured by physicists, mathematicians and engineers as thin silver traces winding their way through spaces of unimaginable dimensions. The path across life, the universe and so many hyperdimensional worlds is being captured by new disciplines within new sciences like *chaos theory*, *entanglement*, *network science*, *econophysics* and *evolutionary dynamics*. The foundations of these new disciplines can be traced back to a single man with a singular obsession to capture the permanent qualities that define change—to Galileo and his discovery of the parabolic trajectory of a rock thrown from a clifftop.

Nothing less is at stake than our lives. Our health changes and evolves in time on its path through a space defined by all the molecules in our bodies and by all their permutations as they combine to do their jobs to keep us healthy and alive. There are regions of this health space where our trajectories are free to traverse in good health. But when disease strikes, we are knocked off course. One of the challenges of the new century will be to find the best ways to measure the path to health, and to detect early signs of disease to give doctors the best chance to get us back on track. Our fate, then, will not be in Aristotle's perfect stars, but will be in Galileo's trajectory, albeit generalized into abstract spaces of immense dimension, but maybe still recognizable to the old master who launched us on this path.

Galileo Unbound. David D. Nolte, Oxford University Press (2018).
 DOI: 10.1093/oso/9780198805847.001.0001

Trajectories at Dusk

In rural Indiana, on the river bottom of Wildcat Creek, lies an old hay field. The Wildcat is tamer than its name, meandering across central Indiana in lazy loops and ox bows. As it approaches the city of Lafayette, it has carved a high bluff out of the Indiana sandstone where it picks up speed as it passes by the hay field. Sometimes at dusk, in the summertime as the heat of the day yields to the cool breath of moist air creeping from the trees by the river, you can stand in the tall dry grass in the fading light. This is when the swallows, who nest in the steep bluffs, come out to fly. They dart. They dive and swerve. You crane your neck to follow their flight as they flit past. They streak to the heavens, rising from the dark backdrop of white sycamores into the bright clarity of sunburnt sky, their silhouettes black against the last brightness of day. The logic of their flight defies interpretation. The abrupt turns, the quick stops, repeating, but never exactly. They are three-dimensional beings living in a three-dimensional space. Their trajectories fill their space, but not fully. How would you define them? What mathematics could you employ to capture their essence?

This problem of capturing motion is a feat at odds with itself, because it seeks to find what is constant in the impermanent quality of motion. How do you describe a path, the path of the Wildcat, or the path of a dozen swallows? What permanent qualities capture the rush across rocky rapids or the swooping dives? These questions have challenged every generation of scientist since the ancient Greeks first tried to apprehend the nature of change. Some of them went so far as to say that motion is impossible, that change is an illusion.

Zeno's paradox is the classic misdirection that tried to stop motion in its tracks. Motion is impossible, Zeno argued, because to get anywhere, you first must go half way. Let's say you take that step. But then half the distance remains, and you must step again to the next half, so that a quarter of the distance remains. By extension, you must step, in succession, to 1/8, then 1/16 and then 1/32 and so on. The fraction of the journey remaining may get progressively smaller, but the fractions never stop. They are infinite in number. The time it takes to step on an infinite number of fractions is itself infinite—you need an infinite amount of time to go anywhere, so you cannot go anywhere. Motion is impossible! Of course, no one bought into this argument, even at the time. But no one could prove it wrong either, except by demonstration.

When the philosopher Diogenes of Sinope (circa 300 BCE), also known as Diogenes the Cynic, first heard Zeno's paradox, he said nothing, but simply arose and walked.

Aristotle knew instinctively why Zeno was wrong, noting that the times to cover the successively smaller distances became successively smaller themselves. He did not have the mathematical tools at hand to turn his instinct into rigorous proof, but he was acutely aware that things *did* move and change in time. The nature of change was one of the primary interests in his *Physics*. He sought to define why things moved and what moved them. More importantly, he sought to define what all motion had in common, what universal forms and ideas encompassed all types of motion, regardless of their idiosyncratic details. In other words, Aristotle was searching for invariants.

Aristotle's search for universals (invariants) of motion is the same endeavor physicists engage in today. The theory of Special Relativity is first and foremost a study of invariants—the things that remain true no matter what the relative location or speed of an observer may be. This aspect of relativity theory, first developed by the German mathematician Hermann Minkowksi, and initially received coldly by Einstein, is often overlooked. After all, postmodern relativism, which took its inspiration, in part, from relativity theory, holds that there are no absolutes. However, the central quantities of relativity theory are combinations of properties of trajectories that change *individually*, but that remain absolutely constant *collectively* no matter by whom they are observed, no matter how they are moving. These constants of motion become the anchors from which the many counter-intuitive conclusions of relativity theory can be derived with certainty (moving clocks slow down, moving meter sticks shrink, and twins don't stay twins when one of the pair flies off in a spaceship). Therefore, Aristotle's search for invariants of motion is as modern today as it was then.

Nevertheless, Aristotle was not a modern, and he constructed a theory of motion that did *not* stand the test of time. This is forgivable, because he fought to define a new field of study where none existed before—to create something out of a vacuum, so to speak. What Aristotle did succeed in doing was to erect a theoretical edifice that would challenge future generations of philosophers (and early scientists) to demolish it. A thousand years passed before Galileo could finally meet that challenge—even at the peril of his own freedom.

Galileo's Parabola

Galileo's scientific career was spent tearing down Aristotle's edifice. He worked hard to spread his discoveries that supported the Copernican system of the world, with the Sun at the center, publishing at his own expense and writing in Italian instead of Latin for wider dissemination to the professional classes in Italy. But Galileo was not antagonistic to Aristotle. He went so far as to suggest that Aristotle, if made acquainted with his new findings on the moons of Jupiter and the phases of Venus, would have embraced these as new particulars and would have arrived inductively at a Heliocentric Universe. Rather, Galileo's ire and energy were directed at what he considered to be the false-Aristotelians, those who had abandoned Aristotle's inductive-deductive system by starting instead with Aristotle's conclusions as if they were immutable truths. This is what got him into trouble with the Inquisition and led to his house arrest for the remainder of his life.

One of the ironies of history is that Galileo's incarceration at the hands of his enemies was the catalyst that gave us Galileo's greatest and most influential discourse on modern science. It was while he was bound by house arrest, at his home in Arcetri, that Galileo finished his work on the science of motion. Much of his experimental work had been performed thirty years previously when he was a young professor at Padua, but his thinking had continued to evolve and finally crystallized in the form of his *Two New Sciences* (1638).

In this grand synthesis, published at the beginning of the seventeenth century, Galileo combined the principle of constantly accelerated vertical motion with the principle of steady (force-free) horizontal motion to arrive at the parabolic trajectory of projected objects. This law of parabolic motion was not a simple joining of two ideas. Until Galileo, there was no reason to believe that orthogonal motions were independent, being easier to believe that the faster an object goes horizontally, the slower it falls. This was the prevailing wisdom of the times. But Galileo combined exhaustive experimental studies and meticulous measurements with rigorous numerical analysis to verify his ideas. His methods (which would be recognized by any scientist today) were to: 1) control the conditions of the experiment; 2) perform quantitative measurements (e. g., time and distance) repeated many times (e. g., statistical significance); 3) find relationships between variables (e. g., mathematical analysis); and 4) uncover common principles

(e. g., distance proportional to time squared). Galileo's work on motion was revolutionary, not because of his discoveries and conclusions, but because of his *methods.*

After Galileo had freed the Earth to move through the heavens around the Sun, the problem of *why* it moved remained, a problem that could not be solved even with Galileo's new science of motion. Galileo's science was the science of kinematics—how things moved. What was needed was a science of dynamics—why things moved. A dynamical theory is a theory of cause and effect in which objects gain impetus from other sources. Therefore, a robust dynamical theory should treat all objects equally through a universal law that applies to fired cannon shots as well as the imposing motion of the planets. Another genius in another century would make this happen, but as Galileo wrote in the introduction to his *Two New Sciences,*

> . . . what I consider more important, there have been opened up to this vast and most excellent science, of which my work is merely the beginning, ways and means by which other minds more acute than mine will explore its remote corners.
>
> Galileo, 1638

When his ultimate work was published in the free Netherlands—an act forbidden by Holy decree—Galileo's ideas were freed from bondage. In the four centuries since Galileo penned this prophesy, his familiar trajectory has wandered like a thread into spaces of increasing dimension and increasing abstraction. Today we talk of quantum particles moving among a countless number of worlds, and we envision our lives as a path spanning a space of a thousand dimensions whose axes are the genes and proteins that define our state of health. Modern concepts of motion and trajectories may seem intangible and daunting, but at their heart is Galileo's trajectory—Galileo's echo heard in motion.

The Wealth of Motions

Concepts of space are not tied as closely to our familiar directions as we might think. We have become complacent living in our three dimensional world, finding it nearly impossible to visualize an alternate reality with even a single extra dimension. Yet, a space is just a set of variables contained in an *n-tuple*, and adding a new dimension is child's play. For instance, "real" three-dimensional space is composed of a

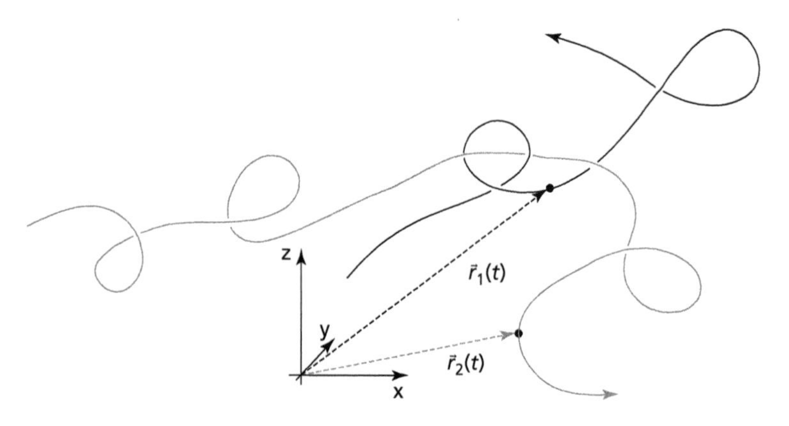

Fig. 1.1 The flights of two swallows in 3-space

3-tuple (x, y, z), and adding a fourth dimension is as simple as (x, y, z, w). If one thinks broadly, then coordinates can be anything—not just directions in space—they can be arc lengths or temperatures or even colors. One can have 5-tuples, and 6-tuples and beyond, describing generalized spaces including all the movements inside them.

Imagine again the swallows of the Wildcat, flitting in large numbers above the tall grass of the hay field. Each bird has access to three freedoms of movement. If there are only two birds, there are two sets of 3-tuples (x_1, y_1, z_1) and (x_2, y_2, z_2). In their steep dives and swooping turns, each position variable changes in time, each bird tracing out its own three-dimensional path that curves and winds past and through each other (as in Fig. 1.1), although swallows never collide. If the birds left behind thin luminescent traces, they would be beautiful, like the curves etched on fresh ice by figure skaters, though spread across three dimensions. But if there are many birds, the beauty degenerates into untraceable chaff that only a Zamboni machine could renew.

To tame the complexity of too many birds, there is a trick that we learned as children in grade school back in the 60s when the New Math was being taught to second graders. Take a union of the two sets of 3-tuples describing the flights of two swallows to create a 6-tuple (x_1, y_1, z_1, x_2, y_2, z_2). Each coordinate of the 6-tuple changes in time, so this new construction traces out a single continuous path. It does not matter that three of the coordinates relate to one bird and the other

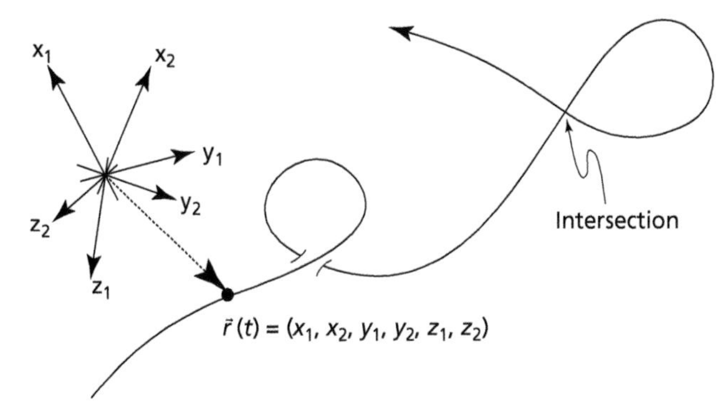

Fig. 1.2 The flight of two swallows in 6-dimensional configuration space is a single trajectory. The continuous curve in this example intersects itself at a point

three to the second bird, it is still a single curve, shown in Fig. 1.2. The price that is paid for this simplification is the promotion of the curve to a 6-dimensional space that cannot be visualized easily. Nonetheless, each point on the curve represents the instantaneous configuration of the two-bird system—this 6-dimensional space is called the *configuration space* of the system.

Despite the new simplicity, a problem with this picture occurs at the second loop in Fig. 1.2 where the 6-dimensional path intersects itself. The configuration of the system at an earlier time recurs at a later time. Therefore, a point in configuration space cannot uniquely define the path—additional information is needed to lift the ambiguity. To solve this problem, one can imagine tacking on a momentum vector to every point on the trajectory in configuration space. At the crossing point, the momentum vectors at the different times would point in different directions, but this procedure obscures the simplicity of the curve. Therefore, a further compromise is required: to the 6-tuple of the configuration attach the 6-tuple of momenta to create a 12-tuple. The new 12-tuple now describes a continuous curve that never crosses itself (as in Fig. 1.3) in a 12-dimensional space. This space is called *phase space* (we will see in a later chapter how this space got its name), and alternatively *state space* because every point on a curve in state space uniquely defines the future (and past) state of the curve. In his popular

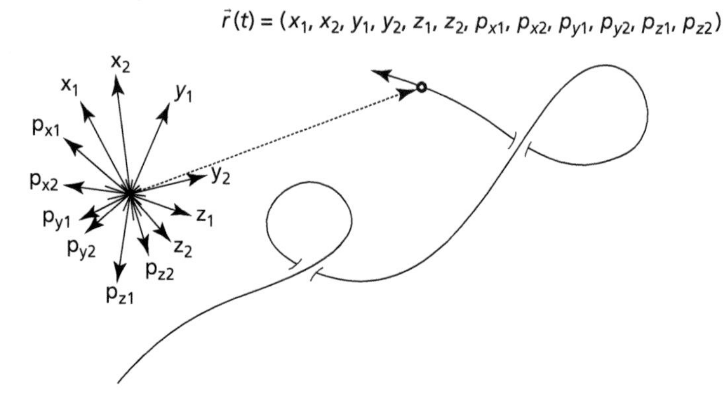

Fig. 1.3 Two birds in 12-dimensional phase space. The single continuous curve describing the collective motion of both birds is unique and can never cross itself

book *Chaos* on the history and science of chaos theory, James Gleick called phase space "one of the most powerful inventions of modern science."[2] Why? What is the power?

The power of this construction comes when there are N birds, especially when N is a large number in the hundreds, or thousands and beyond. The phase space describing the complete system then has 6N dimensions. A system of a hundred birds above the Wildcat (which happens on rare occasions) would require a phase space of 600 dimensions to define it uniquely. We seem to have paid a dear penalty, creating a complex and abstract space which we cannot hope to visualize. But there is a beauty and simplification that comes with the abstraction, and it is this—a single continuous curve, that can never cross itself, captures in its one-dimensional spider's silk the incomprehensible complexity of motion of all hundred birds. The intertwined paths of a hundred swallows becomes a single flight, a single geometric curve, through phase space. And there is more.

A single strand in a 6N-dimensional phase space arises from a single initial condition—its initial 6N-tuple. But there are an infinite number of possible initial conditions in a space, defining an infinite number of trajectories. Rather than define each curve separately, there is a final step—define a field of paths that pervades the space. The trajectories lose their identity to become infinitesimal filaments that combine into threads, and the threads into strings, and the strings into cables, until all of phase space is filled by a continuous dense flow of lines like the flow of water. The infinitesimal filaments never cross, but can converge on special locations where the system gyrations quiet down and reach steady values, or steady oscillations, or steady chaotic patterns. These special locations and regions are called attractors or limit cycles, and they exert an influence over all nearby flow lines, defining the geometry of those lines and defining the geometry of the entire dynamical space.

An example of a dynamic flow field in a two-dimensional phase space is shown in Fig. 1.4 for a special type of dynamical system that has been used to represent the flow of money in an economic system.[3] The figure displays selected streamlines and tangent vectors and is called a "phase portrait," capturing in a single glimpse a global representation of all possible system behaviors. There are critical points in the plot

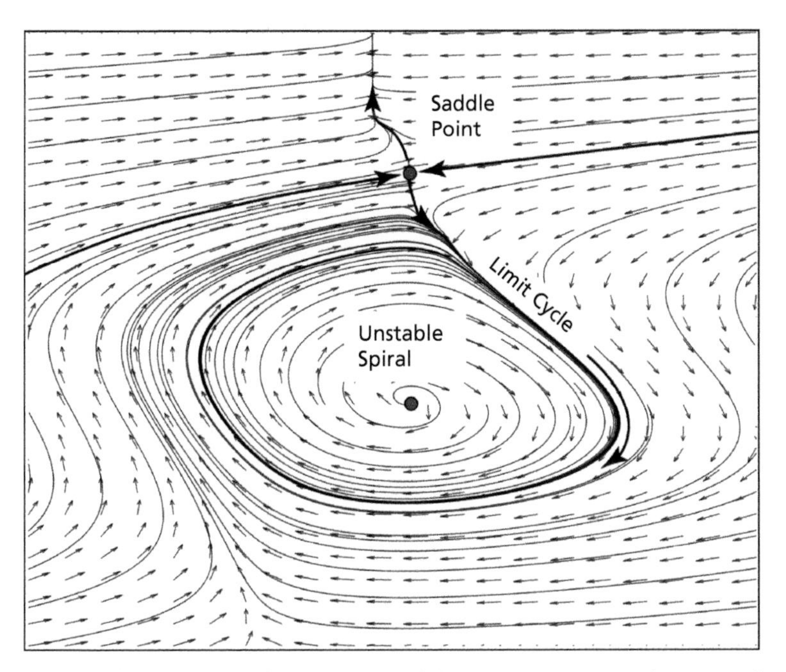

Fig. 1.4 Phase portrait of streamlines and flow vectors in a two-dimensional phase space. There are two fixed points: the source of an unstable spiral, and a saddle point. There is one limit cycle that attracts nearby flow lines. The vectors represent a field of tangent vectors that fill space. Each flow line is a system trajectory, and the flow field represents every possible system trajectory[4]

where unstable equilibrium occurs. These special points are called fixed points of the dynamics. There are two in the figure: one is the source of an unstable spiral; the other is a saddle point. The spiral streamlines spiral out until they approach another special feature known as a limit cycle. The limit cycle is a limiting streamline that attracts all paths in its neighborhood, onto which all nearby trajectories eventually relax as the system repeats itself periodically, endlessly. Limit cycles are dynamic equilibria that lie at the heart of autonomous oscillators in control systems as well as in living systems, sustaining your beating heart and the neural pulses of your mind.

As we track the expanding concept of Galileo's parabola through the centuries, we will learn that we are immersed in a sea of trajectories. From subatomic strings vibrating with quantum uncertainty, to numberless frenetic molecules performing their random thermal dance. From the planets in their stately orbits to the paths of light rays bent by the gravity of the Sun. From the emergence and evolution of life on Earth, to the chaotic plunge of asteroids threatening that life with extinction-level cataclysms. Everything moves, everything changes, and everything needs a language that can capture and define and describe motion. If ideas of geometry and space have increased in abstraction through the centuries, so surely has our view of motion. The history of physics follows the fate of Galileo's trajectory. He launched an idea that has taken on a life of its own, threading its way into diverse fields of math and physics and biology and economics, becoming ever more strange and powerful with the ability to describe the evolution of the universe as well as the evolution of life, the infinite paths of quantum particles as well as the flow of wealth across nations, the flutter of an aircraft wing as well as the pulsations of thought traveling down neural fibers. These topics are his legacy.

2

A New Scientist

... I see many organists wearing themselves out
trying vainly to get the whole thing into perfect harmony.
GALILEO, Sunspot Letters (1613)[1]

The creative act is not a monopoly held by artists, musicians, and authors. The scientist, too, is engaged in creative endeavor. While an artist creates new works or new objects that have never existed before, a scientist creates new ideas or observes new phenomena that have never been thought or seen before. Artists and scientists deal with reality at some basic level—even abstract art must resonate with the corporeal viewer. The creative artist looks at the same thing as millions of others, and sees something different. The same is true for the creative scientist, who can look at a confusion of phenomena and suddenly apprehend simple principles that govern the seemingly complex behavior.

History offers us a handful of individuals, artists and scientists, whose lives are seen as a blaze of intense creation. Leonardo da Vinci comes to mind, filling hundreds of notebooks with a deluge of thoughts and observations. Nothing was too small or too big to catch his attention—he captured the minutest details of bird feathers and the grandest gyres of water, impressing them on paper. He invented the wildest war machines and flying vehicles, while rendering exquisite studies for ambitious public paintings exploring the new science of perspective.

A hundred years after Leonardo, another guiding light of intense mind, another gifted Italian, entered the world stage. Galileo Galilei inherited a worldview based on Aristotle, but his mind ran free of the prejudices of the past and rose above the particular to see ideal behavior as a simplifying principle. Ideal systems were simple systems, governed by only a few rules (not yet truly laws—that would be Newton's contribution). The most important consequence of this new way of thinking was that many manifestations, behaving differently, can arise from an underlying system. Galileo provided this unifying

Galileo Unbound. David D. Nolte, Oxford University Press (2018).
 DOI: 10.1093/oso/9780198805847.001.0001

view, one of the first unification principles in theoretical physics, at the same time that he was guided by the results of experimental physics. Galileo invented the modern experimental method by uncovering ideal behavior as the limiting action of real systems, and he bequeathed the method to following generations of scientists down to this day.[2]

Introducing Galileo

Among the influences that shaped Galileo's character were his noble birth into a minor patrician family of Tuscany, and his curious polemical father, Vincenzo Galilei, who explored musical theory, performing experiments and engaging in bitter academic feuds. Galileo's patrician status gave him access throughout his career to the highest levels of society, while his father was a role model as a seeker and defender of truth. The Galilei were never rich, having no inherited wealth, and Vincenzo was a lutinist who played the lute wherever he could, which is no good way to earn a living. However, it was enough to send the eldest child, an inquisitive and eager Galileo, to the University at Pisa in 1581 with the hope that he would become a doctor. Galileo, on the other hand, had no interest in medicine, but when he took physics, one of the required courses for doctors (as it still is today for premeds) he was captured by its beauty.

Galileo learned Aristotelian physics from two professors at Pisa who came at the subject from opposite directions. Francesco Buonamici was an Aristotelian of the classical Greek school, fully versed in Ptolemaic theory. Girolamo Borro, on the other hand, was an Aristotelian of the Averro School, following the Andalusian Arabic scholar as a secular influence. For this predilection, and his obstinacy, the Inquisition incarcerated Borro more than once. Both Buonamici and Borro had an interest in the science of motion, and each produced works they titled *De Motu* (On Motion). Buonamici assembled a thick tome of over 1000 folio leaves on all that was known about motion and its role in natural philosophy. Borro's smaller *De Motu* focused more tightly on the properties of motion, while he published a second book, in the form of a dialogue, on his theory of the tides. Galileo had all three books in his personal library, and he referred to them fifty years later in his *Dialogo sopra i due massimi sistemi del mondo* (*Dialogue Concerning the Two Chief World Systems*).

Despite Galileo's interests and talents, he dropped out of school after four years without a degree and spent his time among richer aristocrats, gambling with success and relieving them of a little of their wealth.[3] Fortunately, Galileo was not a natural wastrel, and he was driven to continue learning on his own. He became infatuated by Euclid and was privately tutored by the mathematician Ostilio Ricci, a friend of his father's and likely a pupil of Nicolo Tartaglia (1500–1557), the Venetian mathematician famous for solving cubic equations. Galileo also took up the study of Archimedes, who had recently been reintroduced into Renaissance Italy through Tartaglia's Italian translation.

Around this time, Galileo's father became embroiled in a public dispute over the "correct" tuning of musical instruments. Because he was a gifted musician with a natural ear for pitch, he knew from experience that music theory of the day, which he learned while studying under Gioseffo Zarlino in Venice, was incorrect. Zarlino used an approach developed by Ptolemy that extended the Pythagorean ratios of the numbers 2, 3 and 4 to include the numbers 5 and 6 relying on superparticular ratios (in which the numerator is one unit more than the denominator) of 3/2, 4/3 that were extended to include 5/4 and 6/5 as the basis of consonance. However, practicing lutenists knew that tuning on these ratios prevented continuous modulation across scales. Therefore, empirically, Vincenzo settled on a superparticular ratio of 18/17 as a closer approximation to even temperament that allowed the player to transition smoothly among scales. He published his modern theory of music intonation in a book in 1581 (the same year that his son began attending classes at the University in Pisa).

Zarlino was stung by this attack from a former pupil, and he campaigned to have Vincenzo's book suppressed in Venice, though it was published in Florence and some years later came to be known by a handful of polymaths across Europe. One was Simon Stevin (1548–1620), the Flemish scientist and engineer, who recognized in Vincenzo's empirical ratio $18/17 = 1.0588$ a close approximation (within a tenth of a percent) to the twelfth root of two $2^{(1/12)} = 1.0595$, which Stevin proposed as the basis of equal temperament intonation (used nearly universally for tuning today). Over several years following the publication of Vincenzo's book, the open debate with Zarlino intensified, dividing Venetian and Tuscan musicians into camps, amplifying the polemics launched by either side against the other. To further support his modern tuning, Vincenzo engaged in a series of detailed experiments performed on strings of different lengths, diameters and tensions. Vincenzo published the results in a diatribe against Zarlino's numerology in 1589, proposing that physical reality and experimental verification took precedence over abstract ratios. This was a manifesto against Pythagorean philosophy, as well as Aristotelian, where pure ideas trumped the vagaries and imperfections of the real world.

The character of Vincenzo's measurements, subjecting well-controlled systems to systematic numerical study, holds much in common with the character of Galileo's studies of balls on inclined planes carried out 20 years later. Biographers believe it likely that Galileo assisted his father in these experiments, and that either the care and character of the musical studies reflect Galileo's fine experimental touch, or perhaps Galileo picked this up from his father, or more likely, they developed their scientific approach working together. In any case, these musical studies with his father had a profound formative effect on Galileo's own later experimental approach to his science of motion. Unfortunately, Galileo also seems to have inherited his father's tendency towards diatribes and open disputes with colleagues, a character trait that would not benefit him in later life, especially when the opponent was a Pope in Rome rather than a music theorist in Venice.

Around the time that Galileo was helping his father, he began to develop ideas of his own in engineering fields. One concerned his observation of the natural motion of the lantern in the Duomo in Pisa that swayed with perfectly even timing regardless of its amplitude. He realized that this presented a natural time-keeping device that, if miniaturized, could be helpful for timing the pulse of a patient (he was after all thinking like a doctor). Another idea concerned the construction of a hydraulic balance that improved on Archimedes famous approach to measuring the alloy content of a gold crown. As Galileo pursued a mathematical demonstration of his balance, he stumbled upon a mathematical proof of centers of gravity that created a stir of interest among local mathematicians. Within a year, in 1588, he was offered a lectureship in mathematics at the same university where he had never obtained a degree.

Once established as a junior lecturer at the Pisan university, he continued to be influenced by Borro and Buonamici and their singular interest in the problem of motion. During the three years that Galileo taught at Pisa, he began compiling his own *De Motu*, though he never published it, because he was never convinced that he had gotten to the core of the problem. It was a series of notes and observations, building on Borro and Buonamici, but presented more critically, bolstered by observations and supported by mathematics. During this time, he is supposed to have dropped two lead balls of different weights from the Leaning Tower of Pisa, observing them to fall at the same rate, in direct contradiction with Aristotle. However, he could not yet recognize the critical new element in these results—the independence of the rate of fall from mass—because he had not yet broken free of all Aristotelian ideas. This would come in time, after he had moved from Pisa in Tuscany to the University at Padua in the Republic of Venice in 1592. But even as he explored in greater depth the science of earthbound motion, he became distracted when he turned his eyes to the Heavens, aided by the newly invented telescope.

The Starry Messenger

Galileo's move to the Venetian Republic was a boon to his career and personal life and began "the best eighteen years of my life."[4] He was now a faculty member at a university, no longer just a lecturer, with leave and encouragement to explore new ideas and launch new projects. Chief among these was an increasingly detailed study of falling objects, to which we will return in Chapter 3. In the romantic city of Venice, Galileo had a full, if possibly too fast, social life that reveled in wine, women and gambling, yielding three offspring as byproducts of this lifestyle, two daughters and a boy, whom he readily acknowledged and made provisions for. Most significantly, Galileo fell in naturally with a group of open-minded literati and kindred spirits, some of whom became his great friends, like Paolo Sarpi and Giovanni Sagredo. This group, called the *Morisimi*, also counted Bruno Giordano as a peripheral member. Giordano, a few years later, would be burned at the stake for heresy by the Inquisition in Rome—an evil portent that Galileo should have paid closer attention to. The *Morisimi* were also virulently anti-Jesuit, with a history of verbal attacks and abuse that the Jesuits never forgot.[5] Although Galileo was not opposed to the Jesuits, his association with the *Morisimi* would not work in his favor when he himself was before the Inquisition thirty years later.

Paolo Sarpi (1552–1623), one of Galileo's closest friends, was a monk of the Servite order serving as theological as well as scientific advisor to the

Venetian Senate. In 1606, he was appointed state theologian, and a year later, when Pope Paul V threatened the entire Venetian Republic with excommunication (in part because of the strong anti-Jesuit feelings held generally in Venice), Sarpi advised that the state resist the edict. He was then called up to appear before the Holy Inquisition, but he refused to go to Rome and was duly excommunicated in January 1607. This did not seem to be a hardship for him, but later that year he narrowly survived an assassination attempt[6] that many suspected originated within the Roman Curia (the administrative arm of the Holy See). Sarpi resented the assassination attempt far more than the excommunication, and he became a staunch and outspoken critic of the Church.

Sarpi was well educated and widely admired as one of the finest minds of the Republic. When news arrived in Venice in November of 1608 of a new Dutch invention that made things far away appear close (the telescope), Sarpi was the first to hear of it. A few months later, when merchants offering to sell the new devices approached the Senate, they asked Sarpi to evaluate it. Sarpi knew basic optical theory, and he quickly understood the working principles of the primitive telescopes. The device consisted of two lenses: a positive objective lens with a long focal length located at the far side of a tube that captured light rays, and a negative ocular lens at the near side that intercepted the rays from the objective to create an upright virtual image for viewing by the eye. The instrument Sarpi inspected was crude, and he immediately thought of his friend Galileo, who had established a profitable side business in instrument making.

The salary of a faculty member at a University was then, as always, not lucrative. Furthermore, when Galileo's father died, the responsibility for family expenses fell to him. The sizes of dowries were experiencing rapid inflation during this time in Italy, and Galileo inherited a sizable debt for his sister's weddings. He also had, by now, children of his own who needed support. In short, Galileo was strapped for cash, and he became a small businessman to supplement his income. Although he styled himself a mathematician and a philosopher, Galileo had a talent for working with his hands, and he had invented several useful devices, like a military compass, which he manufactured and sold. Part of the manufacturing process included glass lens grinding. When Sarpi approached him with the proposition to improve the performance of the instrument, Galileo saw a business opportunity and took it.

There were several problems with the original Dutch telescope design. Foremost, the magnification power was small, only about 3x, because the curvatures of the glass lenses were too small. However, lenses with stronger curvature were difficult to grind with accuracy, leading to severe optical aberrations. This was where Galileo had an edge. He was methodical and patient, willing to test and retry something many times until things were just right. He had a knack for nuances, and a knack for exploring permutations in his trial-and-error approach. Galileo also had an edge in glass quality, being literally at the source of the Venetian glass that remains famous to this day. With these two advantages, Galileo finally constructed, by August 1609, a high-quality telescope with an unprecedented magnification of 9x. He was preparing to sell this telescope on the open market, with a clear competitive edge over the poor Dutch telescopes that were by now commonly for sale, but Sarpi shrewdly advised that instead he should give it as a gift to the Senate.

The unveiling of Galileo's telescope was orchestrated as a grand media event, not unlike Steve Jobs in Apple's heyday unveiling the next great consumer electronics product. Along with Galileo's knack for research and development came market savvy, and he was never shy about self-promotion. He placed the telescope at the top of the bell tower of St. Mark's square, aimed out to sea where the senators took turns spying distant ship's sails, the ship's hulls hidden below the horizon. The telescope could spot an arriving ship two hours before the sails became visible to the naked eye. Such a device had obvious commercial as well as military utility for the seafaring nation of the Republic of Venice. Several old senators climbed up the steep stairways more than once to look through this wonder. The event was a great success, showering adulation upon Galileo and upon the Republic for having such an ingenious servant. For his services, Galileo was awarded a lifetime position at the University of Padua and a large raise in salary, which he welcomed and needed.

At the peak of his success in Venice, his mother arrived for a long visit, like a dark cloud from Florence.[7] Galileo's mother, Giulia, was dissatisfied with her son, with his lifestyle, with his profession, with his lack of money, and certainly dissatisfied with Galileo's paramour Marina, the unsuitable mother of her grandchildren. Part of her design was to wrench Virginia, Galileo's eldest daughter, from Marina's grasp. When finally she returned to Florence with Virginia in tow, Galileo likely felt a great weight lift from his shoulders, and he lifted his eyes to the heavens, eager to forget the visit and absorb himself once more in his work. On 1 December 1609, distractedly pointing his newest telescope, now with a 20x magnification, on this and that, he turned it on the moon and must have gasped at what he saw.

The moon was waxing, only four days old, and the terminator between sunshine and shadow on the face of the moon stood out in jagged contrast between white and black. The new telescope had a small field of view, covering only about a quarter of the moon's disc. As Galileo swept his field of view over the terminator, he recognized shadows cast in the light of the sun by individual structures—mountains and crater rims. The novelty and importance of this realization is easy to overlook today when we all know what the moon is. But in Galileo's day, celestial objects were known to be perfectly smooth spheres. The dim forms of brighter and darker portions of the moon, in other words the "man in the moon," were thought to be regions of higher or lower reflectance of the smooth surface. Even other learned observers, like Thomas Harriott in England who was training his own crude telescope at the moon about the same time as Galileo, did not recognize what they saw.

Galileo, in contrast, realized almost at once that the moon had a rugged landscape. He quickly applied his mathematical skills to his observations, noting the size of the shadows relative to the size of the moon, and the inclination angle of the promontories to the incident rays of the Sun, and calculated that the mountains on the moon were about four or five miles high. The height of Mt. Everest was not known at the time, and most of the mountains in the Alps were only about two miles high, so Galileo realized that mountains on the Moon were about twice as high as the known mountains on the Earth.[8] This was an astounding discovery, and he knew it. The celestial spheres were not perfect, made of some unknown magical stuff—they were just like the Earth. And if the Earth and the Moon are alike, then there was no longer a reason why the Earth should be the center of motion any more than the Moon might be, or the Sun. He must have been almost giddy at this realization. Galileo had worked diligently over the past twenty years to find compelling evidence in favor of the Copernican system, yet always encountered obstinate resistance by the established dogma of the day. As an indefatigable proponent of post-Aristotelian science, he recognized that here was a smoking gun aimed right at Aristotle's head. And there was more. Jupiter was rising.

As Christmas of 1609 came and went, and a year of propitious portents began, Galileo turned his attention to the bright planet rising in the east at a perfect angle for viewing by the awkwardly long telescope tube that had to be kept extremely still to be usable. On 7 January, he noticed three little stars near Jupiter that formed a perfect straight line, two lying to the east and one to the west of the planet. When he looked again on 8 January all three were to the west, which was puzzling, because it meant that Jupiter must have moved east overnight, even though the astronomy tables clearly had Jupiter moving retrograde at that moment. Two nights later two of the little stars were now to the east, as if Jupiter had suddenly jumped west, which made no sense. Then on 13 January Galileo observed yet another little star, so that now there were four all in a line with the planet. It dawned on him that it was not Jupiter moving erratically against a permanent backdrop of fixed stars, but these little stars were moving around Jupiter! Furthermore, he had noticed that Jupiter was a small disc in the field of his telescope, with no trace of the usual twinkling effect of brightening and dimming seen for point-like stars, and the four little stars moving around Jupiter did not twinkle either.

Galileo came to the astounding conclusion that these must be moons of Jupiter, orbiting the planet with surprisingly short periods.[9] Jupiter had its own little solar system. It was a center of motion, meaning that there were at least two different centers of motion in the known solar system: the Moon went around the Earth as always, but the new moons of Jupiter went around Jupiter. If there were two centers, then the Earth no longer had a monopoly as a special place, and there was no reason why the Earth and Jupiter couldn't go around the Sun as a third center. In other words, there was a plethora of centers unlike anything Aristotle could explain, yet perfectly compatible with the Copernican system.

Galileo knew that these discoveries were earthshaking, world-changing, and that he was in grave danger of losing priority if anyone else were to turn the increasingly available telescopes (although of inferior quality than his) to the heavens. He was forty-five years old, already a ripe age at that time, and though always exploring and discovering, had never finished or published anything. Now he raced to get his discoveries into print—writing, revising, proofing and printing *Sidereus Nuncius* (The Starry Messenger[10]) within two months. In early March 1610, exactly 515 copies came off the press and sold out within a few weeks. *Sidereus Nuncius* was a sensation, a blockbuster. It went viral, and Galileo became a rock star in great demand as invitations poured in for personal demonstrations of his telescopic discoveries, giving him the perfect chance to get the academic post he most coveted. When he published *Sidereus Nuncius*, he took a calculated risk by naming the new moons after the Florentine Medici of his hometown. Although shocking and angering his republican friends and supporters in Venice, the gambit paid off as he was offered a lifetime appointment to the University of Pisa, exempt from teaching, free to pursue his own interests, in the personal service of a prince, his former student Cosimo II, the Grand Duke of Tuscany.

Pulp Friction

In 1610, there were approximately five alternative systems of the world. The first and foremost was the Ptolemaic system with the Earth at the center and the Sun and the Moon and all other planets revolving about it. The second was the Copernican system proposed by the Prussian-Polish astronomer Nicolaus Copernicus (1473–1543) in which all planets revolved in circles about the Sun, except the Moon that revolved around the Earth. The third was the Tychonic system, named after the Danish nobleman Tycho Brahe (1546—1601), in which the Moon and the Sun revolved around the Earth, but all the other planets revolved around the Sun. The fourth was a hybrid Tychonic-Ptolemaic system in which Mercury and Venus revolved around the Sun, but the Sun, Moon and other planets revolved around the Earth. Finally, the fifth was the Keplerian system, newly proposed in 1609 by Johannes Kepler (1571–1630) as a modification to the Copernican system, but in which all the planets moved in elliptical orbits with the Sun at one focus of the ellipses. The systems of Ptolemy, Tycho and Copernicus are shown in Fig. 2.1 illustrating the three main systems: Earth-centered (Ptolemy), Sun-centered (Copernicus) and hybrid (Tycho).

The Ptolemaic system was a compendium compiled by Ptolemy, a Greco-Egyptian philosopher writing in Alexandria around 160 AD, of numerous Greek contributions plus his own. The Moon, Sun, planets, and the celestial sphere carrying the stars, moved around a stationary Earth. Already, in the second century BC, astronomy was accurate enough to recognize that circular orbits of the planets around the Earth required adjustments to agree with observation. Most notably,

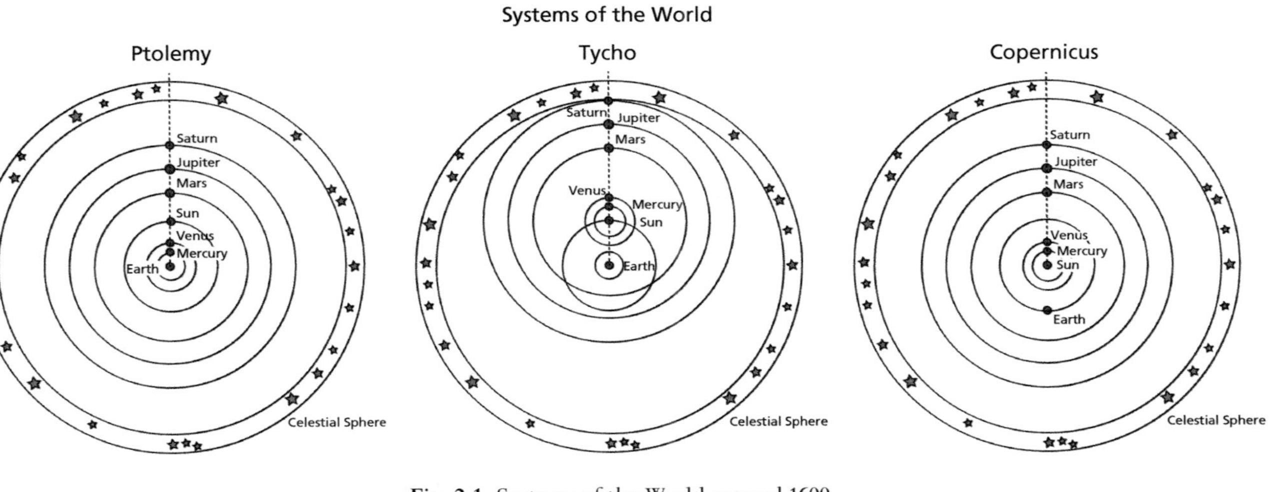

Fig. 2.1 Systems of the World around 1600

there were times of the year when planets moved retrograde, backward, westward, relative to their usual eastward motion through the rest of the year. Apollonius of Perga (c. 262–c. 190 BC) introduced a circle (called a deferent) centered on a point offset a little from the Earth (called an eccentric) to partially explain these observed motions, and Ptolemy added the idea of an equant (a point around which angular speeds were constant) to the deferent to explain variations in the speeds of planets. This combined invention was later called an epicycle, which produced curly orbits of the planets around the Earth. Each planet needed its own epicycle. Mercury and Venus required the most dramatic epicycles, while the Moon and Sun required only minimum epicycles. As astronomy advanced into the Middle Ages, new epicycles were added to the old to give calculations the kind of accuracy that was needed for astrological purposes. When St. Thomas Aquinas reconciled Aristotelian philosophy with church doctrine in the thirteenth century, the Ptolemaic system tagged along and became the *de facto* system of the world. By the time of Galileo there were nearly 40 epicycles in use by the Ptolemaic experts, despite there being only seven celestial objects. An example of a single epicycle is shown in Fig, 2.2 for the orbit of Mars. In the figure, when the Sun is at location *A*, the center of the epicycle is at *a*', and the planet Mars is at *a*. When the Sun moves to location *B*, the planet Mars is at *b* with the center of the epicycle at *b*', and so on.

In contrast to the Ptolemaic system, the Copernican system is similar to our system today, with the Earth and other planets revolving around the Sun. However, Copernicus was still enough of an Aristotelian to require the planetary orbits to be circular. This caused inaccuracies (because the orbits are elliptical) that forced him to concoct his own set of epicycles to bring theory into agreement with observation. The weakest point in Copernicus' system was the lack of stellar parallax that would have proven the motion of the Earth around the Sun. To fix this problem, Copernicus moved the stars to a great distance. The other drastic feature of his system required the Earth to rotate once per day at great speed. The vast distances of the stars and the speed of the Earth violated common sense, and even Copernicus was reluctant to publish his work.

A half century later, Tycho Brahe (1546–1601), a Danish aristocratic astronomer who had his own island observatory, produced the most accurate astronomical measurements of planetary motion. He

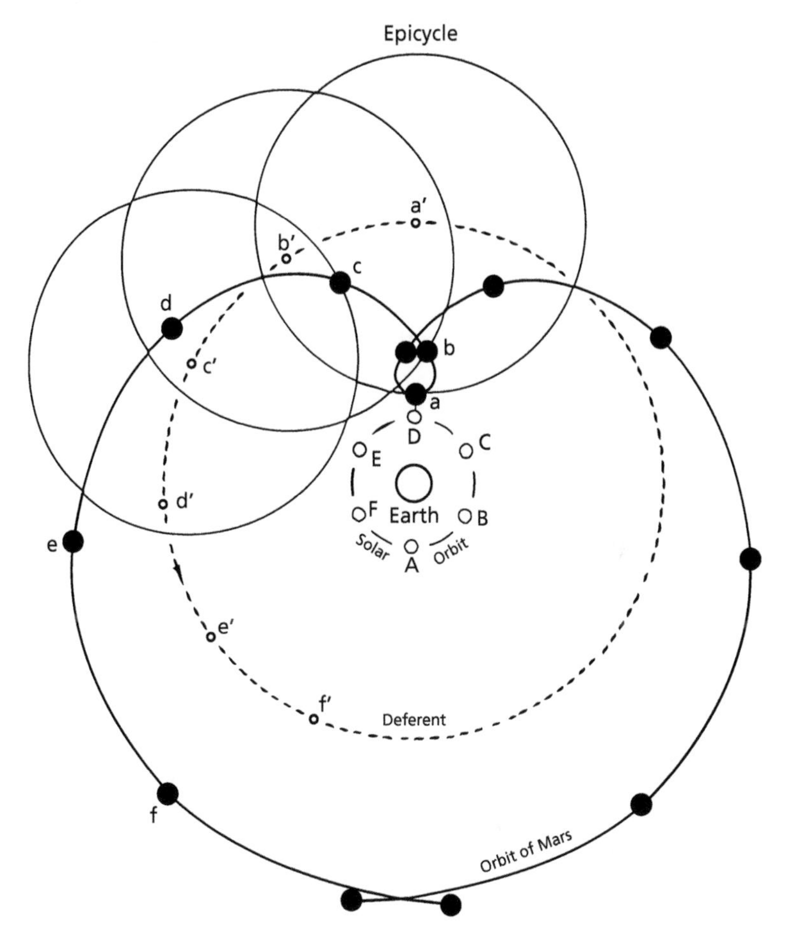

Fig. 2.2 Epicycle for Mars

assiduously looked for stellar parallax, but could not find it, prompting him to reject Copernicus' moving Earth. On the other hand, his accurate measurements only made sense if all the other planets revolved around the Sun. This Tychonic model, with the Sun revolving around the Earth, and all other planets revolving around the Sun, satisfied no one, in part because the celestial spheres would inter-penetrate. As a compromise, some astronomers, notably the astronomers at the Jesuit Roman College, adopted the hybrid model with Mercury and Venus revolving around the Sun, but with the other planets revolving around the Earth. This bastardized system worked well qualitatively, but was never accurate. As for Kepler's elliptical orbits, which were extremely quantitative, almost no one knew what to make of them. Galileo ignored them completely, and he ignored the epicycle modifications of the Copernican system that brought them accuracy. Galileo's Copernicanism was qualitative and heuristic. He did not bother himself with details.

In his new position at the court of the Medici's, Galileo's job was to provide an occasional diversion for court life by discovering an interesting thing or two. He also was there to reflect the light of greatness upon his ruler and his people. On both scores, he got to work right away. By late summer, Saturn was easily visible in the night sky, and Galileo immediately discovered that it carried two appendages that never varied. He assumed they were two massive moons that somehow never moved, although why was a great mystery that he could not solve. Galileo also hoped to discover additional moons, like those that revolved around Jupiter, but he did not find any. Christiaan Huygens solved the mystery of the appendages nearly fifty years later with a better telescope, understanding that these appendages were rings. Huygens also discovered Saturn's largest moon Titan.

When Galileo made his observations of the Moon and Jupiter, the planet Venus had been on the far side of the Sun, but now it was easily visible in the evening sky.[11] There was already considerable astronomical evidence that Venus and Mercury circled the Sun, which would mean that Venus and Mercury would display phases, just like the Moon. Therefore, by the time Galileo turned his telescope on Venus and observed that it displayed crescent-shaped phases, he may not have been surprised, although it added new evidence that the Sun acted as a center of motion, further weakening arguments for the special place of Earth in the scheme of the Universe.

Even in his early days teaching at Pisa, although he presented the Ptolemaic system in his public lectures, Galileo privately argued the merits of Copernicus against the old Aristotelian model. He openly began supporting Copernicus in 1597 after his move to Padua, and the Copernican system was the implied backdrop to his discussions and demonstrations in *Sidereus Nuncius*. Yet his definitive transformation, committing himself to the promotion of the Copernican world system, came after the detection of sunspots. Late in 1611, using his telescope and taking care to protect his eyes, Galileo observed darkened flecks moving slowly across the face of the Sun. Independently, in Ingolstadt Bavaria, a Jesuit priest named Christoph Scheiner likewise observed sunspots using a telescope of his own construction. Scheiner believed that the dark flecks were numerous small planets, or starlets, orbiting the Sun similarly as Mercury and Venus. He published his observations and his theory in a booklet with the help of his friend Mark Welser in Augsburg, using the pseudonym Apelles. Welser forwarded one of these booklets to Galileo, asking for his opinion. Galileo was not generous, and replied with unkind words about the so-called Apelles and his mistaken ideas.

Galileo's response to Welser and Scheiner began a pattern of behavior that would persist, and worsen, through the remainder of his life, and would be the cause of his considerable later troubles. Whether it was fame and adulation going to his head after his discoveries published in the *Nuncius*, or a deep-seated insecurity instilled by his moneyless patrician origins and his chronic lack of funds, or the example of his father's diatribe against Zarlino and musical temperament, Galileo's responses to others who tread upon his prized subjects were cutting and discourteous. Even worse, they were arrogant and insulting. Galileo seems to have come to believe in his own infallible genius, and in the imbecility of others. Once he had a clear idea in his head, he would consider all other ideas as rubbish and all those presenting such ideas as unworthy.

In a series of three letters to Welser, each one more detailed than the last, Galileo denounced Scheiner's starlets as nonsense, and proposed instead that they were imperfections on the face of the Sun, like clouds on Earth. He correctly deduced, by making meticulous measurements combined with geometric calculations that were his trademark, that the sunspots rotated on the surface of the Sun with a period around a lunar month, and that some of those that disappeared on the far side would reappear about a month later, changed in shape, while other spots would not survive the month-long journey. Supporting his ideas, Galileo made numerous detailed drawings of the sunspots by projecting images of the Sun on white paper and tracing the shapes of the spots with charcoal, later recreating the images with ink and washes. Welser was so impressed with the detailed letters and drawings, despite their condescending tone towards his friend Scheiner, that he consented to let Federico Cesi (1585–1630), the head of the Lyncean Academy, a literary and scientific academy in Rome, publish the letters as a book. Galileo's letters on sunspots were the Academy's first publication, appearing in 1613 and presenting to the world his theory of a rotating Sun.

A rotating Sun, Galileo knew, was the most decisive element yet discovered supporting the Copernican system. The greatest difficulty that intelligent and open-minded scholars had with the Copernican system was the requirement that the ponderous Earth must spin on its axis once per day. This required a speed at the Equator, easily calculated even in Galileo's time, of one thousand miles per hour, an incomprehensively large speed well beyond human experience. Surely, everything on the face of the Earth would be stripped away by such a terrible speed. A spinning Earth simply defied common sense. Yet Galileo now had proven a spinning Sun, an object much larger than the Earth. If the Sun can spin on its axis, why not the Earth? And just as the sunspots lazily followed the rotation of the Sun, so clouds on Earth amble along, impervious to the dizzying speed of rotation. The clouds simply are carried along by the Earth's rotation—the first definitive example of what has today become known as Galilean relativity. Galileo's sunspot letters were his first and strongest declaration of the Copernican system as *the* system of the world—not a hypothesis or mathematical model to save appearances, but truth. To some, this smacked of heresy.

The Protestant Reformation was the greatest threat faced by the Catholic Church in the Sixteenth Century. To combat that threat, the Church launched the Counter Reformation, epitomized by the Council of Trent and the installation of the Inquisition. The chief goal of this movement was to counter the kind of free thinking exemplified by proliferating Protestants. The edict of the Council asserted that only the Church Fathers were allowed to interpret Scripture. Within the Church, the issue of interpretation versus literal reading was fought between different camps. For instance, Dominicans tended towards literal reading of Scripture, while the Jesuits were willing to interpret Scripture relative to the modern world. However, after the Council of Trent, any reinterpretation needed Church approval. One of the consequences of the Council was the dogmatization of Aristotle according to the views of St. Thomas Aquinas, made several centuries before to reconcile Aristotelian thought with Scripture. It was not that Aristotle, a pagan, understood the word of God, but rather that Aristotle, interpreted by Aquinas, had defined the natural world in a manner that was considered true and consistent with Scripture. New ideas that worked against the existing worldview needed approval by the Church followed by careful integration into the current dogma—they could not be asserted out of hand by just anyone. That was heresy. The penalty for heresy was often torture and death, as in the case of Bruno Giordano, who had such bazaar ideas that the Inquisition got rid of him by burning him at the stake in 1600.

The danger of new ideas in the Catholic world was recognized and understood at all levels. For instance, the Grand Duchess, the mother of Cosimo II, was concerned that these great discoveries made by Galileo in the Medici name went against Scripture. During a state dinner the Grand Duchess interrogated Benedetto Castelli, close friend of Galileo's, concerning the consequences of a Sun-centered world. Castelli held his ground admirably, providing well-practiced arguments for a nontechnical audience, but he wrote to Galileo about the encounter. Galileo, concerned that such a close supporter as Christina might have qualms about accepting a heliocentric world, wrote a detailed letter to Castelli outlining his most compelling arguments as an aid for future encounters. The clarity of the letter impressed Castelli so much that he made copies to distribute to friends and supporters. As the copies multiplied, they eventually fell into the wrong hands. In 1614, a little-known priest in Florence denounced Galileo and his heliocentric worldview as heretical from the pulpit. Around the same time, Galileo's activities were reported to the Inquisition in Rome.

There were no immediate consequences from the denouncement, other than fueling Galileo's bile and his determination to bring Italian science and culture into the modern century with the Sun firmly at the center. Within a year, he had decided to go to Rome, against the advice of his friends, to campaign for acceptance of the heliocentric Universe. He was, and always would be, a devout Catholic, and he fully believed that heliocentrism and Catholic faith were compatible. After all, the Jesuits believed in an interpreted Bible, so he just needed to get the Church Fathers to revise their interpretation. Earlier in the year he had written a letter to the Grand Duchess stating his position: "I hold the sun to be situated motionless in the center of the revolution of the celestial orbs while the earth rotates on its axis and revolves about the sun. They know also that I support this position not only by refuting the arguments of Ptolemy and Aristotle . . . especially some pertaining to physical effects whose causes perhaps cannot be determined in any other way, and other astronomical discoveries; these discoveries clearly confute the Ptolemaic system, and they agree admirably with this other position and confirm it."[12]

Once in Rome, he pushed his case wherever he went and to whomever he met, especially the astronomers at the Jesuit College. Several of them were readily accepting of the orbits of Mercury and Venus around the Sun, as well as the spinning Sun, and all were using the Copernican system as an essential aid in their calculations. It seemed but a small step to accepting the Copernican system as reality rather than as a model. While in Rome, Galileo stayed at the Villa Medici, the Tuscan embassy, as the guest of the ambassador. His activities made his host nervous. In a letter to the Medici court in Florence, the ambassador reported, "He is passionately involved in this fight of his and he does not see or sense what it involves, with the result that he will be tripped up and will get himself into trouble, together with anyone who supports his views. For he is vehement and stubborn and very worked up in this matter and it is impossible, when he is around, to escape from his hands. And this business is not a joke, but may become of great consequence, and the man is here under our protection and responsibility."[13]

A recurring criticism that plagued Galileo's efforts to promote the Copernican system was the lack of *proof positive*. For instance, stellar parallax would have been proof positive for the revolving Earth, but so far even Tycho's most accurate instruments had detected none. On the other hand, Galileo believed that the daily tides were just such a proof positive, caused by the sloshing of the water as the Earth wobbled underneath it. He was mistaken, not recognizing the influence of the Moon, but he quickly wrote up a compelling booklet with his arguments for the moving Earth, and by extension, a Copernican system. He gave the booklet to a Tuscan Cardinal with instructions to pass it on to Pope Paul V. Paul did not read it, but by now Galileo had made such a pest of himself, and had caused such a noise, that everyone was talking about his heliocentric theory. Paul convened a panel to investigate whether the heliocentric system was formally heretical. Galileo suddenly realized that he had pushed too far. In his overzealousness to reason the merits of Copernicanism with learned scholars, convinced that all would flock to him once they understood how simple the Copernican model was, he had kicked the sleeping dragon.

To serve as head of the panel the pope appointed Robert Bellarmino, later to become St. Bellarmino, a Jesuit theologian of fearful reputation who had presided over the committee of the Inquisition that had burned Giordano at the stake. Bellarmino did not buy into Galileo's theory of the tides, nor his spinning Sun. He was convinced that the absence of stellar parallax supplied strong evidence against a moving Earth. To top it off, the Copernican epicycles were no more accurate than the Ptolemaic epicycles, removing even that possible advantage of the Copernican model for celestial calculations. (He was not aware of Kepler's improvements, partly because Galileo had not made noise about them.) Perhaps most conclusively, though he accepted the need to interpret Scripture, he saw no need to change the current interpretation based on a few intriguing observations by a Tuscan astronomer. To bolster his eventual decision, Bellarmino enlisted Francesco Ingoli, a Jesuit astronomer of the Roman College, to write an open letter to Galileo listing all the advantages of the Ptolemaic system and all the deficiencies of the Copernican.

Galileo received the letter and was prepared to reply, when he was stunned to learn on 23 February 1616 that the Inquisition had banned Copernicus. The wonderful *Revolutionibus* was to be placed on the index of prohibited books. On 26 February Galileo was called to Bellarmino's office where the cardinal told him that he could no longer hold or discuss, or teach in any way, the model of Copernicus. Two priests were present, implicitly as witnesses, although this did not occur to Galileo in his shock over this direct blow against him. The interdict on Copernicus was a blow to many more than only Galileo. Several of the astronomers at the Jesuit College had been eagerly open to the Copernican system, but now had to abandon any effort or thought in that direction. Kepler blamed Galileo directly for the disaster, needlessly stirring up the hornet's nest, driven more by his self-importance than concern for the benefit of the nascent field. Galileo was angry and bitter, blaming the duplicity of the Jesuit astronomers and their cowardice. He retreated to Florence, momentarily beaten, but not defeated. He looked for a new opportunity, which came in the form of light wisps in the night sky.

Cometary Prelude to a Dialog

The year 1618 was unusually blessed to witness three comets. Comets are rare and were believed to be portents, so all astronomer's eyes were drawn to the newcomers. The critical question about comets was where they fell within the celestial spheres. Were they between the Earth and the Moon, or above the Moon but below the Sun? Orazio Grassi at the Jesuit Roman College had good reason to believe the latter, and he gave a public lecture during which he alluded to Galileo's ideas, without naming him. He was open-minded about the Tychonic system, recognizing the necessity for Mercury and Venus to revolve around the Sun, within whose framework the motion of the comets made some sense. The lectures were highly praised, and an account of them was published. When Galileo received the written account, he immediately recognized the reference to himself, and although Grassi had not been critical, Galileo of course took offense. He was still bitter at the defection of the Roman College from the Copernican model and the role that the Jesuits Scheiner and Ingoli had played in the Interdict of 1616. He was also frustrated with his inability to reply openly after his warning from Bellarmino. Piqued and galled, Galileo responded through the name of his colleague Mario Guiducci, publishing a scathing and insulting diatribe against Grassi.

Grassi recognized the thinly veiled hand of Galileo behind the Guiducci pamphlet and responded in kind late in 1619. He weighed Guiducci (Galileo) on three accounts, using a metaphorical balance, and found them wanting. Grassi was no fool, and he understood the context of Galileo's reply, even though he could not understand why Galileo wrote in such an "exasperated and angry a spirit."[14] The first weighing exposed Guiducci as Galileo, and Grassi chided his manners, telling him that he "much preferred to lose a friend than an argument."[15] The second weighing pointed out that Galileo was trying to have it two ways, by rejecting Aristotle when it suited him, and by embracing Aristotle when it suited him too. This was referring to Galileo's obsession with Copernicus, while rejecting comets (as Aristotle did) as celestial objects. On this count, Grassi was far in advance of Galileo, recognizing that comets were luminous objects moving in the heavens. The third weighing related to the translucence of the comet's tail that allowed stars still to be seen. Galileo argued that stars cannot be seen through flame, and hence the tails could not be made of fiery material. This was the right conclusion, but the wrong premise, which Grassi knew and pointed out, because it *is* possible to see through flame. Grassi did the experiment himself, putting Galileo, the arch experimentalist, to shame. Grassi scored more than a few good points against Galileo and may have felt he had won the argument, but he did not count on the depth of Galileo's scorn for the Tychonic system nor his literary flair for lampooning an opponent.

It took Galileo two years to craft his response to Grassi, but when he did, it brought forth *The Assayer*, one of the most famous polemics in the history of science. Apart from clever quips and crafty caricatures of Grassi, *The Assayer* is a manifesto of the scientific method. Above all, Galileo said, "philosophy is written in this grand book—I mean the Universe—which stands continually open to our gaze, but it cannot be understood unless one first learns to comprehend the language and interpret the characters in which it is written. It is written in the language of mathematics, and its characters are triangles, circles, and other geometrical figures, without which it is humanly impossible to understand a single word of it; without these, one is wandering around in a dark labyrinth." Galileo stated his fundamental proposition that direct observation trumps the testimony of experts. This was in response to Grassi's use of ancient wisdom on which he built his arguments as the common practice of Scholastic science. In Galileo's turn to modern science, measurement became the foundation, and mathematics became the tools to wrestle the truth from the measurements. By the 1620s, Galileo had already made extensive experimental investigations of motion, and hydraulics and the strength of materials, all based on meticulous measurement. His experience in the laboratory he extended to the stars and more generally to the investigation of the book of nature.

Galileo's *Assayer* marked a decided change in the prosecution of science, and it followed on a period of decided change in his life and times. In early 1619, before he experienced the barbs of Grassi's balance, his longtime mistress Marina Gamba died, leaving his children motherless, followed by the death of his own mother in 1621 along with his patron, pupil and friend the Grand Duke Cosimo II. The same year took Bellarmino as well as Pope Paul V. Although Bellarmino and the Pope had been central players in the banning of Copernicus, each had absolved Galileo of wrongdoing at the time, and the Pope had personally promised to protect Galileo for the rest of his life. These little protections were now gone, but Galileo had reason to rest assured, because one of his long-time supporters and friends, the Cardinal Barberini, had just been elected Pope as Urban VIII. The *Assayer* was dedicated last minute in 1623 to the new Pope, and Barberini bees decorated the front piece of the book. The new Pope had the book read out loud at dinnertime and listened with interest and amusement at Galileo's barbs sent with adroit wit and cunning at the hapless Grassi.

Galileo believed that this favorable turn was a new opportunity to help bring Italian science into step with the rest of Europe by adopting the obvious advantages of the Copernican system of the world. He travelled to Rome in 1624 and was granted several audiences with Urban. He had first met Cardinal Barberini in 1611, and they had kept occasional contact since then, for instance when Galileo was in Rome in 1616 trying to forestall the edict against Copernicus. At that time, cardinal Barberini had even participated in keeping the word "heresy" out of the final edict. There is no record of what was discussed or concluded during their meetings, but several key points can be surmised from Galileo's letters. Urban made several positive statements about the status of Copernicus, saying that had he been Pope at that time that the edict would never have been issued. However, he also made it plain that he did not believe that earthbound astronomical observation would ever be able to prove definitively whether Copernicus was right or wrong. His model was interesting and useful, but it would remain always thus. Urban did agree with Galileo's concern that the outside world must not think Italian science and astronomy backwards and ignorant. It was important to let the world know that the ban on Copernicus had nothing to do with lack of understanding of the Copernican system, but was purely based on scriptural necessity. However, what was most important to Urban was the subtle but deep theological point that God was unlimited in what He could cause or create, and worldly manifestations were mere impressions of our limited understanding of God's designs. Urban may have encouraged Galileo to write a book on the issue, impressing on him the need to demonstrate the intelligence of Italian science and the unlimited abilities of God.

Galileo left Rome reinvigorated and re-inspired. He floated a trial balloon testing the new attitude in Rome with a long-delayed reply to Ingoli's letter of 1616 that had weighed so heavily against Copernicus. Ingoli had become ever more deeply entrenched in the defense of Ptolemy and had been put in charge of rewriting (defacing) Copernicus when *de Revolutionibus* was removed from the Index. Galileo's letter went through several drafts and was sent to Cesi of the Lyncean Academy for additional edits. Copies began circulating around Rome, and one copy even made its way into Urban's hands. Ironically, the letter somehow never made it to Ingoli, its intended mark, but its purpose was realized. No denouncement ensued from the Holy See, and Galileo began work in earnest on a manuscript that he tentatively titled *Dialog on the Tides*. This choice of title shows how convinced he was that the tides were proof positive of the double motion of the Earth—its rotation on its axis and its revolution around the Sun. Yet as the project progressed over the next five years, its scope steadily expanded to become a broad examination of the two chief world systems: Ptolemaic and Copernican. He completely ignored the Tychonic systems as idiotic, and Kepler's ellipses were just distractions.

Galileo fashioned the manuscript as a dialog among three friends. The dialog was a popular literary device of the times, providing an entertaining format for readers and giving the protagonists free license to discuss whatever their author fancied. Interestingly, his father had chosen a dialog structure for his polemic on musical tunings. The three friends—Salviati (Galileo's alter ego), Sagredo (the intelligent host) and Simplicio (a somewhat thick-skulled Aristotelian)—met over four days for entertaining discussions. The centerpiece of Galileo's book was the fourth day on the tides. The three days leading up to the fourth were spent softening up the audience. Day One established the inadequacies of Aristotelian science. Day Two removed arguments against the daily rotation of the Earth, and Day Three removed objections against the yearly revolution of the Earth around the Sun.

In his 500-page manuscript, Galileo brought out every conceivable evidence supporting the double motion of the Earth. He invoked the motion of the planets with their wide yearly variations in distance to the Earth and their retrograde motions. He invoked sunspots again, but with much greater insight into their complex motions that he explained as the effect of the axis of the Earth tilted by 23 degrees. He invoked the clearest presentation yet of Galilean relativity, anticipating Einstein by three hundred years by stating that in a closed room on a moving ship that no possible experiment could discern absolute motion. He invoked Copernicus' arguments for the immense distance to the stars, explaining why no yearly parallax could be observed, while predicting that some small residual parallax may one day be discovered (as it was two hundred years later by Friedrich Bessel). And finally, he invoked his theory of the tides, which, although he included effects of the Moon as an improvement over his theory first written in 1616, still missed the main effect of the Moon's gravitation. For us today, it is a little deflating reading Galileo's *Dialogue*, with all his excellent reasoning and inspired analogies, that the centerpiece of his argument has the wrong physics. Nonetheless, Galileo may be forgiven for missing what Newton fifty years later would not.

Galileo finished the manuscript of his *Dialogue Concerning the Two Chief World Systems* on Christmas Eve 1629. There were still important changes and additions to be made, and he sent sections to Cesi in Rome to get feedback. Most importantly, he needed to craft the denouement with respect for the intentions of Urban VIII. He had a choice of three possibilities as the mouthpiece for Urban's final word on the omnipotence of God: Salviati, Sagredo or Simplicio. Throughout the Dialogue, Salviati had argued fervently for the Copernican system, so it would have been out of character, and dramatically anticlimactic, for him to retrench at the end. Sagredo was too passive, and it would not have made dramatic sense to have him make an uncharacteristic impassioned plea at the end of four days. This left Simplicio as the most dramatically acceptable candidate. And so Simplicio (whose name in Italian is dangerously close to Simpleton) makes the final address—Urban's fine point on the power of God and man's inability to know his designs. It was a mistake.

Orbit Out of Control: Galileo's Tragedy

Galileo rarely miscalculated—he was a mathematician after all. But he miscalculated terribly the intense single-minded ambition of his supposed friend Matteo Barberini and his deep-laid plans for the ascension of the Barberini family. Barberini applied every advantage at his disposal to enhance the name of the Barberinis, their importance, their wealth and their power within Roman society. He sprinkled his relatives, close and far, into the leading ranks of papal bureaucracy. He engaged the sculptor Bernini to fill Barberini villas and loges with Bernini masterpieces at the expense of the Vatican. He increased the wealth of the Barberini family by over 100 million scudis (Galileo had earned 60 scudi per year as an entry-level professor at Pisa), plunging the Vatican into deep debt. He levied taxes upon the people of Rome to support his building projects, expanding St. Peter's to further its glory in the name of the Barberinis. He even went to war, sending Vatican storm troops, led by his nephew the Cardinal, against the small province of Castro because they opposed him. In the end, Castro was obliterated off the face of the Earth. In what possible scenario would the Pope allow Galileo to tarnish his name?

Galileo, after all, did not figure in his grand scheme. Galileo was a diversion, an interest, a hobby. Barberini enjoyed the arts and sciences, and Galileo had been an invigorating correspondent. But as soon as Galileo became a source of slander against the Barberini name, he became an enemy, and Barberini was as ruthless as he was without scruple. Many within the Italian states interceded for Galileo, and even Barberini's nephew the Cardinal petitioned the pope on Galileo's behalf, but it was clear that Galileo would be made an example of. His offense was not the teaching of Copernicus, nor the teaching of a Sun-centered Universe. His offense was making the pope look foolish. Copernicus was on the Index for only four years before he was rehabilitated by Ingoli and removed. But Galileo would be on the Index for more than 200 years, finally exiting the list in 1835, long after even the Church taught heliocentrism.

Galileo had no comprehension of this side of Barberini's character. He was blind to it, in his own single-minded way. Otherwise, he never would have risked making Simplicio the fool of the Dialogue. Galileo had been so secure in the conviction of his own greatness, a greatness that everyone around him lauded and proclaimed, that he could not imagine that he could be taken down. He made the situation worse because he failed to respect his enemies, treating them with scorn as fleas with inconsequential bites. Unfortunately for Galileo, Pope Urban VIII was one of the most consequential people in the world in that day. The Pope had supranational power—in a Catholic nation with a ruler who ruled by divine right, the right to rule came from God, and God was Catholic in half of Europe. As God's Catholic representative on Earth, the Pope had power over more than half of Europe and power over the very rulers who would seek to shield Galileo.

It must finally have dawned on Galileo that he was in deep peril only late in the game. It may have been when he was ordered to Rome though he lay sick in bed, but he had been investigated by the Inquisition before and had weathered it. Or it may have been during the first day of questioning before the Inquisition. On that day the questions, initially straightforward and seemingly unthreatening, took a scary turn. Galileo answered a question about his appearance in 1616 before the Inquisition and the injunction he had received at that time. Then he was asked,

"That said injunction, given to you then in the presence of witnesses, states that you cannot in any way whatever hold, defend, or teach the said opinion. Do you remember how and by whom you were so ordered?"

"I do not recall that this precept was intimated to me any other way than by the voice of Lord Cardinal Bellarmino," Galileo replied. "And I remember that the injunction was that I might not hold or defend; and there may have been also 'nor teach'. I do not remember that there was this phrase 'in any way'."[16]

Galileo then explained how he had obtained an affidavit from Cardinal Bellarmino on 26 May 1616, which he brought with him and presented to his questioners. Galileo obviously felt that this affidavit represented some form of protection, stating for the record that Bellarmino was aware of Galileo's activities, and conversely that Galileo was aware of the precept, and that as long as Galileo stayed within its bounds, all would be well. But later, when he was asked,

"When you sought permission from the Master of the Sacred Palace to print the said book, did you reveal to the same Most Reverend Father Master the injunction previously given to you concerning the above-mentioned directive of the Holy Congregation?"[17]

Galileo's heart must have sunk, for he had not mentioned the injunction when the book had been inspected for the imprimatur in Florence. Furthermore, it was clear from the way the questions were running what the real focus of the inquiry was about. In an off-the-cuff attempt to deflect this line of questioning, Galileo dissembled,

"I did not happen to discuss that command with the Master of the Sacred Palace when I asked for the imprimatur, for I did not think it necessary to say anything, because I had no doubts about it; for I have neither maintained nor defended in that book the opinion that the Earth moves and that the Sun is stationary but have rather demonstrated the opposite of the Copernican opinion and shown that the arguments of Copernicus are weak and inconclusive."

Was this an outright lie, in the face of the Inquisition, in the face of possible torture or death? Biographers of Galileo answer that he did not lie, because he could simultaneously hold, and not hold, two opposing ideas—that the Earth moves, according to a corporeal world, and that it does not, according to a theological world.[18] One of the main reasons why Galileo was so devastated by his eventual conviction of heresy was his sincere belief that he was, and always had been, a faithful Catholic.

He truly believed that he had adhered to the precept, expressing Copernicus' theory as a hypothesis. At the end of his second day of questioning, he eagerly suggested that he could publish a revised version of the Dialogue with new Days tackling new problems of Nature, while using that opportunity to clearly lay out the hypothetical nature of the Copernican model. The commissary must have responded with some compassion to Galileo's plea, for they allowed him to return to the Tuscan embassy where he was surrounded by friends. As he arrived at the embassy at the end of his second day of questioning, the Tuscan Ambassador saw a tired old man climb slowly out of the carriage, a man sick in his body and in his heart, worn by worry over what he had done, yet certain, at the same time, that what he had done was of the utmost importance.

Galileo was questioned four times in the offices of the Holy Inquisition over two and a half months, the final time officially being threatened with torture if he did not tell the truth. However, Galileo was steadfast that he had never believed in the Copernican theory, holding it only as a hypothesis. Despite the overwhelming evidence in the *Dialogue* that he, at the very least, not only accepted the logical consequences of the Copernican system, but embraced them with enthusiasm, Galileo was not tortured into admitting his duplicity.

There was no need for torture because on 22 June, 1633, in the presence of the assembled Cardinals and witnesses of the Inquisition, Galileo was condemned and ordered to kneel and read his abjuration, as his own words, although every word was written by the commission. Galileo at this point had no alternative, so he complied. He agreed (so said the words on the paper he held before his eyes) to abandon the false opinion that the Sun is the center of the world and unmovable, and that the Earth is not the center of the same and moves, because this opinion was against Holy Writ. He further promised to never again assert in speech nor writing any future opinions that might cast suspicion upon him (in effect agreeing never to publish again). He also agreed to abide with whatever punishment was ordered by the Holy Office, which had not yet been communicated to him. This must have filled him with uncertainty and apprehension. Giordano Bruno had been burned at the stake only 33 years before (when Galileo was just beginning his fruitful years at Padua) for his heresy, which was the worst possibility, or he might be imprisoned in the dungeons of the Holy Office for the remainder of his life. Fortunately, Cardinal Barberini, previously unable to sway his uncle towards leniency in the Galileo affair, was able finally to convince Urban to allow Galileo to be imprisoned in the house of the Archbishop of Siena, Ascanio Piccolomini, a longtime admirer of Galileo's.

The political purpose of Galileo's trial was to make a clear demonstration of the power of the Church over vainglorious thought. This was the time of the Counter Reformation and the middle of the Thirty Years War that was devastating central Europe, as centralized papal power was being tested and questioned on a continental scale. Urban used the Galileo verdict as a warning to all good Catholics who might let their thoughts wander down speculative avenues. As fast as couriers could be saddled, the word of Galileo's verdict was sent out across Catholic Europe. Also, as fast, copies of Galileo's *Dialogue*, which had enjoyed only tepid sales, came into high demand, with prices per book soaring. One copy was smuggled across the Alps to Strasbourg, where it was translated into Latin (and hence readable to audiences outside Italy) and dispersed widely. If Galileo previously had been known only to a few experts outside Italy, his conviction by the Inquisition launched him to worldwide fame.

All of this was lost on Galileo, who was in despair. Until the last minute, he had believed that he would be exonerated. But he had not grasped, and still could not grasp, the larger political situation, nor the bitter sense of betrayal that Urban felt. When Galileo arrived in Siena, back in his beloved Tuscany so close to home, but infinitely removed from his daughter Marie Celeste and from his normal pursuits, he may have been suicidal. Piccolomini did all he could to help him, distracting him from his inner anguish by inviting intellectuals and supporters to his house, and encouraging Galileo to reengage with his former interests. One of those former interests had been with him since his first days at Pisa. Across forty years, Galileo had slowly pieced together a body of work on motion, but the work was incomplete, spread out in various unfinished manuscripts. Now, encouraged by Piccolomini, he had the time and opportunity to bring it all together, to finish his greatest work—a new science of motion.

3

Galileo's Trajectory

> *. . . the horizontal motion remains uniform . . . the vertical motion continues to be accelerated downwards in proportion to the square of the time, and that such motions and velocities as these combine without altering, disturbing, or hindering each other . . .*
>
> Galileo Galileo, *Two New Sciences* (1638)

Motion was always on Galileo's mind. He saw motion in his father's stringed instruments, vibrating in rational resonances. He saw motion in the lantern high above in the Duomo di Pisa, swinging with fixed regularity. He saw motion in the heavens as his Galilean moons orbited Jupiter. When he received his first teaching post at the age of 25 at the University in Pisa in 1589, he spent more time thinking about and studying motion than was perhaps wise for a young instructor of mathematics. Mathematics, after all, had little to do with the physical world at that time. Aristotle had expressly rejected mathematical applicability to the messy and corruptible gyrations of the earthly domain. But Galileo saw mathematics as the best way to understand simple behavior, like the falling of two balls of equal material but unequal weight, where he deeply suspected that Aristotle had gotten it wrong.[1]

Form is the Function

Galileo was the recipient of a long tradition of scholasticism in Europe that extended back to Charlemagne's *Charter of Modern Thought* of 787 when he invited the Irish scholars, who had helped safeguard Greek and Roman learning for western culture, to establish centers of learning in every abbey in the realm of the Franks. The Irish arrived in France with their valuable texts none too soon, because the monasteries and abbeys of Ireland were soon plagued and plundered by two hundred years of Viking invasions. The scholastic tradition in medieval Europe began with an eye to the past, resurrecting lost Greek and Latin manuscripts,

Galileo Unbound. David D. Nolte, Oxford University Press (2018).
 DOI: 10.1093/oso/9780198805847.001.0001

driven by the practical need to supply educated clerics who could read the Latin bible to the masses. But eventually the time came for the schoolmen to turn their eyes the future.

In the early to mid fourteenth century, a succession of extraordinary mathematicians overlapped at Merton College at the University in Oxford. These men are called the Oxford Scholars or the Oxford Calculators. They represented the pinnacle of a scholastic tradition at Oxford that had begun nearly a hundred years before. These Oxford Scholars included Thomas Bradwardine, William Heytesbury, Richard Swineshead and John Dumbleton. They forged a new path beyond Neoplatonic science, with its old focus on time-invariant properties, to explore what was called *the latitude of forms*, or Greek κίνησις (kinesis, from which we get the word *kinetics*). The Oxford Scholars sought to describe things that change in time, things like temperature or density or intensity—forms that had a range, or latitude, of behavior. The values of these forms (*intensio*), and their time dependence (*remissio*), are easily recognized today as topics of physics, and the Oxford Scholars were among the first "modern" physicists to try to describe the physical world using the meager mathematical tools that they had. For instance, Heytesbury is credited with proving the mean speed theorem for uniformly accelerated motion, at that time called *uniformly difform* motion.

While the Oxford Calculators were pursuing kinetics (the description of *how* things moved), a parallel effort was under way at the University of Paris concerned with dynamics (the explanation of *why* things moved). The leader of this school of thought was Jean Buridan (1295–1363), a French priest teaching in the faculty of arts at the University. Buridan lived life on the edge. He was widely admired (or vilified) as a free thinker who was perhaps a little too free in his personal life, leading to numerous apocryphal stories of late-night antics and illicit romances. Buridan must have been aware of the work of the great Persian philosopher Avicenna (Ibn Sīnā) who had commented on the late Hellenistic philosopher Philoponus (c. 490–c. 570) of Alexandria. Philoponus was critical of Aristotle's rule that continued motion is only possible if there is a continued force driving that motion. In the case of a rock thrown by hand or some other means, Aristotle speculated that the air displaced by the rock exerts a force on the rock that allows its motion to continue, until it is finally dissipated through resistance. Philoponus suggested an alternative in which the rock acquires a quality, an inclination to move, imparted from the throwing agent. Avicenna expanded on this idea by suggesting that this impressed inclination in horizontal motion, known as violent motion, decreases until it is spent, and that natural vertical motion then resumes as Aristotle's objects seek their natural place (rocks fall to earth, and hot air rises).

Buridan rejected the automatic decrease in the impressed motion, giving the impressed inclination to move the name *impetus*, declaring[2]

> . . . after leaving the arm of the thrower, the projectile would be moved by an impetus given to it by the thrower and would continue to be moved as long as the impetus remained stronger than the resistance, and would be of infinite duration were it not diminished and corrupted by a contrary force resisting it or by something inclining it to a contrary motion . . .

Buridan associated the speed of an object with its impetus, as well as its weight. In these pre-algebra days, he could not take the obvious next step to suggest that impetus was mass multiplied by speed, but his intuition was correct and provided the first semiquantitative theory of dynamics. He took an even bolder step by applying his theory of impetus to accelerated vertical motion, which heretofore had been Aristotle's sacred natural motion. Buridan suggested that the increasing speed of falling objects was due to their acquiring successively more impetus. Buridan did not have the mathematical tools to enable him to identify fall as uniformly accelerating, but he did succeed in removing Aristotle's final cause, *telos*, from the study of motion.

One of Buridan's pupils at Paris was a young German scholar from the kingdom of Saxony, who became known as Albert of Saxony (1320–90). Albert received his master of arts and became a member of the faculty in Paris, where he remained for over ten years, making important advances in the application of Buridan's impetus theory by applying mathematical analysis to Buridan's theory of increasing impetus during fall. He was responsible for founding the University of Vienna, where he became its first rector in 1365, and his widely distributed publications went far to establish the legacy of Buridan and the Paris impetus theory.

Handicapped by lack of proper algebraic symbolism, Albert talked his way through the problem of free fall and arrived at a solution that had the correct qualitative behavior in which speed successively gets larger, but not the correct quantitative behavior that would have allowed him to recognize uniformly accelerated motion in his law of fall. His incorrect numerical results argued against uniformly

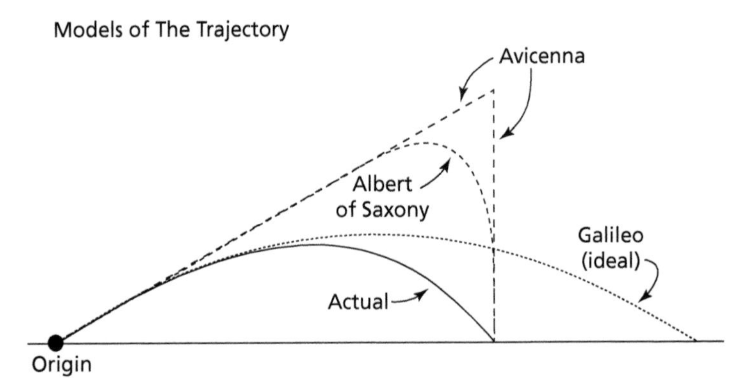

Fig. 3.1 Models of the parabolic trajectory, from Avicenna, Albert of Saxony and Galileo, with the actual trajectory that includes air resistance

accelerated motion, which may explain why it took another 200 years before Galileo demonstrated uniform acceleration as the correct law of fall. Nonetheless, Albert's work was a significant advance, applying mathematical analysis to a dynamical physics problem. Increasing impetus was the cause of fall (dynamics), and increasing speed was the kinematic consequence. An additional contribution of Albert to kinematics was his rejection of the sudden change from violent motion to natural motion that had been assumed by Avicenna. Albert proposed that between violent and natural motion there was an intermediate case as the object gradually lost its impetus and began to curve downward. He only proposed this as a qualitative argument, suggesting that the path of a projectile had some continuous smooth form, shown in Fig. 3.1, although the form was too complicated to define without coordinate geometry. Fortunately, the first visual symbols of motion were being developed at just that moment.

Contemporary with Albert at Paris was the Norman mathematician Nicole Oresme (1320–82), born in Normandy and trained at Paris under Buridan. Like Bradwardine, who was an advisor to Edward III and archbishop of Canterbury, Oresme was both a courtier and a cleric, becoming a close confidant of Charles V of France, a benefactor who rewarded his services with a series of increasingly influential appointments in the church. He was eventually made bishop of the cathedral in Lisieux, a small town in Normandy between Caen and Rouen, where Henry II of England is thought to have wed Eleanor of Acquitaine in

1152. Oresme divided his time between the church in Lisieux and the college in Paris where he advanced his academic studies of mathematics, geometry and impetus theory, publishing prodigiously. During this time, he made the clearest statement of a famous thought experiment (known as an *imaginatio* in Oresme's time) whose origins date back to Plutarch around the first century. The thought experiment involves an object that is dropped down a long channel that passes through the center of the Earth, shown in Fig. 3.2. Oresme correctly proposed, using impetus theory, that the object would gain increasing amounts of impetus as it fell, until it passed the center. Then its impetus would be dispelled by amounts equal to those gained in the fall, rising to the surface of the far side of the Earth, and falling again towards the center, repeating the motion, oscillating back and forth. Oresme was able to reason out the correct qualitative behavior of the falling object using Buridan's impetus theory, providing one of the earliest examples of the utility of thought experiments—a technique that Einstein would use to significant effect 600 years later.

Oresme was an expert on the Elements of Euclid as well as many other aspects of Greek geometry. As Oresme studied Buridan's impetus theory and the latitude of forms, he struggled to find a geometric

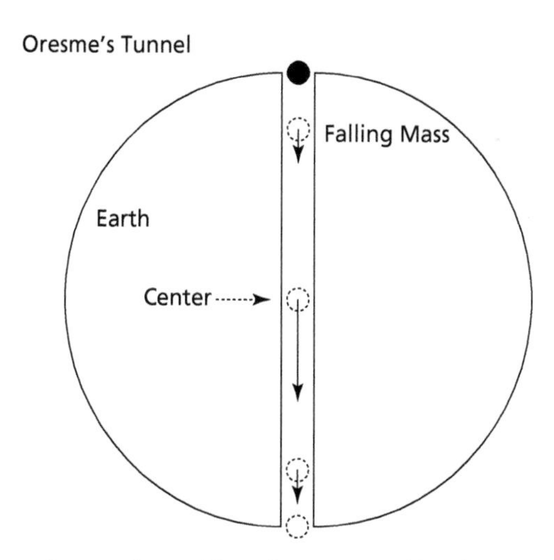

Fig. 3.2 Oresme's tunnel, a problem first suggested by Plutarch concerning a mass falling through a tunnel through the center of the Earth, passing beyond the center to rise to the far surface and return

approach that captured the essence of change, but first he had to overcome a persistent mindset of mathematicians towards the objects of geometry, because Greek geometry was static. The geometric forms of lines and planes and Platonic solids were solid and hence immovable. The problem facing Oresme was how to use static geometry to describe things that change in time. The conceptual breakthrough—the stroke of genius—was to lay down a horizontal line whose length denoted duration of time. Oresme called this line *longitudines*, and with this simple device, a time-varying quantity became a geometric plane figure. The quantitative value of a chosen form, called its intensity or *intensio*, was drawn as a vertical line that Oresme called the *latitudines* perpendicular to the *longitudines*. For example, a form that was constant in time was drawn as a rectangle, shown in Fig. 3.3, while a form that decreased uniformly to zero was a right triangle. When the uniformly changing form was speed, i. e., uniformly difform motion, Oresme understood that the distance traveled was simply proportional to the area of the geometric figure. Proving the mean speed theorem of uniformly accelerated motion became as simple as using the area of the triangle to find the distance traveled. Oresme had turned dynamic motion into static geometry! Geometric figures could stand in for the dynamic

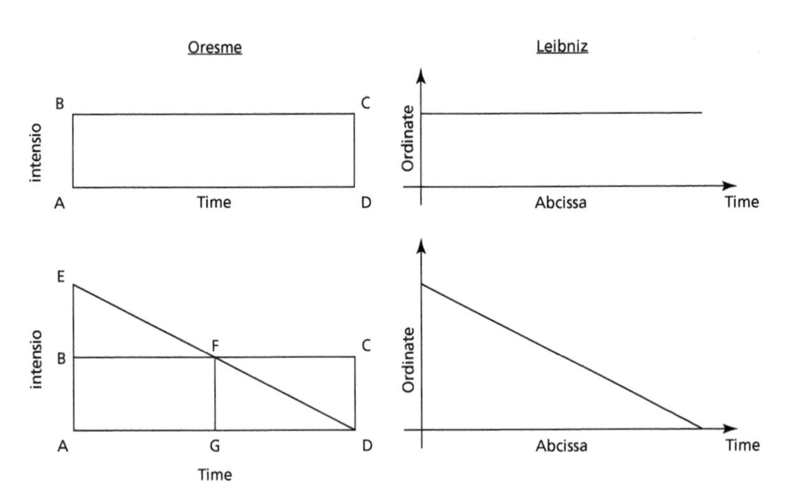

Fig. 3.3 Oresme represented the latitude of forms as geometric figures: a constant intensio was a rectangle (top); a steadily decreasing intensio (constant *remissio*) was a right triangle (bottom). Three hundred years later, Leibniz graphed functions according to abscissa and ordinate

motion, subject to all the applicable laws of Euclid for the development of proofs. By his stroke of genius, Oresme had freed time to expand into space.

The half-century between 1320, when Bradwardine arrived at Oxford, and 1370, when Oresme retired to Lisieux, was a time of brilliant intellectual ferment. The Oxford and Paris schools were in close synchrony, advances made at one place being watched and amplified at the other, back and forth. The intellectual climate of England under Edward III and France under Charles V were conducive to open exploration of new ideas. The church, always policing for new heresies, had not yet adopted Aristotle as a sacred source, so those tilting at Aristotle could do so with impunity. Unfortunately, stability was fleeting in medieval politics, and Charles V the Wise was succeeded by Charles VI the Mad, and Edward III was succeeded by the ten-year-old Richard II who also may have gone mad late in life. Whether for the failing political environment, or for social and economic pressures, the flower of impetus theory wilted. It lived on in the manuscripts of Heytesbury, Swineshead, Albert and Oresme, which were widely disseminated through Europe, stored in monastic libraries, but not improved upon for 200 years, until flowering again in Renaissance Italy.

The Arc of War

The French cannon shells arced mercilessly through the deluge of rain above the heads of the advancing troops as the Venetian defenders fell back on the outskirts of Brescia, the jewel of Lombardy. Nothing now could stop the French forces from sweeping into the city, barefoot, their boots and socks long lost to the sucking mud. The city leaders had rebuffed Gaston de Foix, the French commander, inflicting surprisingly high casualties on the French after refusing to surrender, but the tide had changed, and Foix decided to make an example of this northern Italian town, revolting against French rule in 1512, lest other towns make similar costly stands. As the Venetian defenders retreated, the French took out their wrath on the citizens.

Twelve-year old Niccolo Fontana's hand was likely gripped in panic by his mother as they fled through the drenched chaotic streets to the basilica hoping for sanctuary, struggling to find a safe haven in the nave already packed with women and children. The great doors of the church could not save them, crashing in as Gascon troops and German mercenaries fell upon the masses, cutting and slashing with swords and sabers, heedless of cries for mercy. We can imagine Niccolo torn from his mother's grasp as a cutting sweep of a sabre sliced open his face, and he fell unconscious covered in blood, left for dead as the soldiers moved on in their evil harvest. Niccolo's mother, saved somehow from the carnage, would have searched among the corpses for her child, finding him motionless with a mutilated face, unrecognizable under the black gore, but somehow still alive. She swept him up and fled to a dark corner, nursing him back to life. After hiding for five days of unceasing slaughter and pillage, the remnants of the people of Brescia emerged to a ruined town that never again recovered its former greatness, even after the end of the wasteful Italian wars that cursed the first half of the sixteenth century.

Niccolo Fontana had lived, but he never recovered from the wounds that caused him to stammer for the remainder of his life. Because of his stutter, and to honor his mother who had saved him, he proudly adopted the moniker Tartaglia (stammerer), by which he is known across the world today. He also grew a thick beard to cover the deep wound on his face, giving him the appearance of a wise sage. Tartaglia's family was poor, his father murdered when he was still a small child, and there were no means to send the precocious child to school, so he taught himself. By the time he struck out on his own, settling in Venice, he had mastered the secrets of the Latin mathematics texts then beginning to circulate outside the closed walls of academic universities, and he began making a living as a mathematics tutor. It was an inauspicious beginning for a quirky mathematician who would, although he never lived to see it, become a catalyst for the development of the parabolic trajectory.

Tartaglia was adept at thinking on his feet, and he supplemented his meager income by competing now and then in public mathematics competitions with sizable monetary prizes for the winners. Some of the questions posed to the contestants required finding the unknown solution to cubic equations. Many cubic equations at that time could be solved only by trial and error, by guessing an answer and plugging it in to test it, and contestants with good intuition could often converge on the right answer, winning by being the fastest. Tartaglia, ever searching for a competitive edge to improve his winnings, discovered a new set of cubic equations that could be solved directly without guessing. This made him a minor celebrity among the informal mathematics community in northern Italy, helping him gain students to tutor and providing him with a better income.

With his newfound financial security, Tartaglia was able to expand his tutorial activities to include the publication of textbooks. Professionals in northern Italy during the 1500s—engineers and surveyors and practitioners of military arts—were unusually well educated and literate in Italian, but not Latin, learning from a tutelage system that operated outside the walls of academia. Therefore, there was a growing need for technical and mathematics books written in Italian, and Tartaglia saw a business opportunity when he was approached by an artillery engineer who asked him whether there was a mathematical solution to the problem of artillery ranging.

The art and science of ranging projectiles in hunting and in war is as old as our tool-making species. Among the earliest tools are spearheads dating from half a million years ago, used by Homo erectus before our own species, Homo sapiens, emerged. Arrowheads have been dated as old as 64,000 years, although the first firm evidence of the propulsion of arrows by a bow dates only to 10,000 years ago. The Greeks and Romans were the first to learn how to launch large masses long distances using catapults and ballista. By the time of Tartaglia, cannons had been in use in Europe for two hundred years, becoming standard weapons in warfare by the time Constantinople fell in 1453 to the Ottoman Turks. The importance of cannon and artillery had grown into a specialty for a professional class of military engineers. The central problem for an artillery officer was how to set the range of a cannon, which would be facilitated by understanding something of the shape of the trajectory. Tartaglia was in a unique position, at just the right time in the rise of professional engineering, to use his mathematical abilities to aid them and make some money in the process.

In 1537, Tartaglia published his first book at his own expense, titled *Nova Scientia* (New Science) on the mathematical (mostly geometrical) aspects of trajectories with explicit application to artillery ranging. There is an interesting dichotomy in *Nova Scientia* between Tartaglia's intuition and his mathematical models. The standard Aristotelian treatment of trajectories for over a thousand years considered motion to be composed of two types that could not be mixed: natural motion and violent motion. Natural motion is free fall, the vertical motion as objects seek their proper place. Violent motion is anything that is not natural, such as horizontal motion or upwards motion. Natural motion is pure and eternal, while violent motion dissipates and ultimately ceases. Therefore, in the Aristotelian view, a thrown object will travel in a straight line executing violent motion until the motion is spent, and then will fall straight down in natural motion. Albert of Saxony had already modified that view by introducing a curved part that connected the two straight motions. Tartaglia's intuition guided him to state in the introductory pages of his book that a trajectory is not composed of straight lines, but is everywhere curved downward as soon as the projectile leaves the hand or cannon. Yet, in the body of the book he performs his mathematics of projectile motion on two straight lines joined by a segment of a circular arc. In his defense, he does state that he adopts this line-and-arc approach as a simplification, what we call a *model* today, to arrive at important results easily, with acceptable accuracy. By adopting a model, the ease of use trumps rigorous accuracy, which is as true today as it was then.

Nova Scientia was pragmatic and useful, correctly demonstrating that maximum range is achieved by elevating the cannon muzzle to an angle of 45° above the horizon, while shorter ranges can be struck by two complementary angles, one greater and one lesser than 45°. He fit the curved motion of a true projectile with his line-and-arc approach, condensing years of artillery experience and rules-of-thumb to mathematical detail. However, Tartaglia had no way of knowing that the curved arc of the true projectile was a parabola, nor that the vertical motion was uniformly difform. The impetus theory of Buridan and the Oxford Scholars had lost sway by the early Renaissance, and there is no indication that Tartaglia was aware of impetus or of the mean velocity theorem. Therefore, Tartaglia's role in the development of the parabolic trajectory would have been minor, if based on *Nova Scientia* alone, but he had caught the publishing bug. *Nova Scientia* was a modest financial success, so Tartaglia looked for more opportunities.

The books of Euclid that had been circulating around Medieval Europe for over a century were still written in Latin, which was not accessible to the non-Latin-reading professions that were bourgeoning in Renaissance Italy. Tartaglia decided there was money to be made, so he painstakingly began to translate the Latin books of Euclid into Italian, offering his own commentaries on the side. The translations of books I–IV were straightforward, and Tartaglia's commentaries helped to make the subject easy to understand by anyone willing to apply themselves. But as he was working on a medieval Latin version of Book V, Tartaglia hit a snag—a couple of definitions made no sense, which was odd, because Euclid was usually the epitome of clear and lucid logic. Tartaglia then found a different Latin version of Book V that had been printed in 1505 in Venice, based on original Greek texts that had retained the original definitions by Euclid. This printing had been overlooked by the academics, so Tartaglia translated the correct definitions into his Italian version, including his commentaries.[3]

The definitions in question pertained to ratios, which were of profound importance to the foundations of mathematics as well as to mathematical sciences like the science of motion. Euclid had shown how to handle the ratio of continuous quantities, quantities that did not need to be whole numbers, having any magnitude (except zero or infinity). Whether through bad translation, or lack of understanding, the medieval texts had stripped away these key innovations by Euclid and had replaced them with inconsequential definitions, throwing back the theory of ratios to the rudimentary level of the Pythagoreans. Tartaglia restored the definitions to their correct form, in readable Italian, accessible to architects and engineers, with helpful commentaries showing the utility of ratios of continuous quantities, publishing his improved version of Euclid in 1543. The restored definitions of continuous ratios opened the door to deeper applications of geometry to the physical world, and fifty years later they provided Galileo with a useful tool as he searched for fundamental relationships in the motion of projectiles along their trajectories.

Around the time that Tartaglia was restoring Euclid, Renaissance scholars rediscovered Archimedes. Some translations of Archimedes had been known to medieval scholars, but their focus on Aristotle had caused the more modern and useful advances made by Archimedes to be largely ignored. Only in the sixteenth century did the importance of Archimedes become apparent, and he began to emerge from the shadows. Always looking for financial payoff, Tartaglia published a practical book in 1551 on the raising of sunken vessels, which was a central concern for Venice with its shallow waterways, and he included a translation of Archimedes' *On bodies in water*. In his signature commentaries, Tartaglia proposed using Archimedean principles that objects of equal specific gravities would fall (or rise) with equal speeds through water. This simple statement, made without mathematical proof, had profound consequences for Galileo's discovery of the Law of Fall, but only after percolating through a half century of mathematicians, plagiarists and military engineers.

Sometime around 1546 Tartaglia took on a sixteen-year-old Venetian aristocrat as a pupil, a young Giovani Battista Benedetti (1530–90) who had never attended school beyond the age of seven, but who had learned to appreciate the deep mysteries of the physical world through his father. Benedetti was curious and eager to master mathematics from such an esteemed teacher as Tartaglia, but he was soon disappointed. Tartaglia was impatient and distracted by the tepid reception of his latest treatise *Quesiti et Inventioni diverse de Nicolo Tartalea*, and he was indifferent and dismissive of the young man. Benedetti succeeded in learning the first four books of Euclid from Tartaglia, but then they parted ways with no lost feelings. Benedetti continued to mature and grow in his personal explorations of mathematics. By the age of twenty-two, he had achieved a level of proficiency with geometric constructions that surpassed even Tartaglia. Benedetti realized that Tartaglia's 1551 proposal of equal rate of fall of objects of equal specific gravity through water should apply just as well for objects falling through air, and he decided to stake his claim for this physical insight by putting it into print in 1554, providing the mathematical derivation of equal rates of fall through air based on principles of Archimedean buoyancy.

Benedetti's theory of fall, in clear contradiction of Aristotle, caused a stir, not least in Rome where Aristotle had somehow become entrenched as revealed truth. He was invited to Rome to lecture and defend his ideas. Many were not convinced, supposing that Benedetti had made a mistake, though none were sufficiently proficient in mathematics to show where. In the audience was a shrewd though unscrupulous Belgian, Jean Taisner, who recognized the seed of greatness in Benedetti's work and decided to make it his own. He acquired a copy of Benedetti's book (an early edition that had a mathematical error later corrected by Benedetti) and returned home to Antwerp where he republished the entire book, with the error, under his own name. Taisner's plagiarized book became widely read outside Italy, with copies distributed to the lowland countries as well as in English translation. By the 1570s, the ideas of constant rate of fall, as well as the mathematical derivation, were widely known across Europe.[4] One of Taisner's many readers in the Netherlands was a mathematical and engineering genius who had a knack for clear thinking and a productive energy that could not be contained to one field of endeavor—the remarkable Simon Stevin.

Singular intellects are singularly rare. Where they come from and what shapes their talent cannot be known. Biographers seek to uncover the roots of genius, to find formative events, but in the end, the emergence of brilliance remains a mystery. The Flemish mathematician, physicist, and engineer Simon Stevin (1548–1620) was singular, although his beginnings were not. He began a bastard child in Bruges, but after moving in 1581 to the city of Leiden in the United Provinces (the block of low-country lands that resisted Spanish rule), he embraced and participated in a cultural and scientific renaissance that took hold in the free states. He entered Leiden University in 1583 at the late age of 35, where he became friends with Prince Maurits, the son of William of Orange, who was the leader of the Dutch revolt. When William was assassinated in Delft in 1584, Prince Maurits was elected staatsholder of the United Provinces, and Stevin was installed as military engineer, participating in the planning and execution of several key victories over the Spanish. Stevin remained in the service of the Prince for the rest of his life, while also pursuing mercantile success, providing him the kind of leisure that frees up one's time and one's mind for more creative activities.

Stevin began to write books. He wrote books on diverse topics with clear and original insight—works on mathematics, mechanics, astronomy, navigation, military science, engineering, music theory, civics, dialectics, bookkeeping, geography, and house building. Among his most influential books was a pamphlet on the utility and simplicity of decimal fractions, which is credited with the widespread introduction of the decimal system to Europe. A book on hydraulics is cited as the first significant work on the topic that went beyond Archimedes, and his book on algebra introduced key symbolic innovations of "+," "–" and "√." These were the beginnings of the symbolic language that would transform mathematics in the next century.

In 1586, while working in the town of Delft, he came upon Taisner's plagiarized version of Benedetti's law of equal fall. He studied the mathematical proof and recognized that Taisner's exposition had an arithmetical error (the error that Benedetti had corrected in his later edition). Uncovering the error must have caused Stevin to pause, piquing his curiosity in the rate of fall, as he began thinking how to test the hypothesis. Stevin was an engineer, a practical profession, and it must have seemed to him that physical demonstration was superior to abstract reasoning. In Delft a famous bell tower stands attached to the Oude Kerk at the side of the old Delft Canal. The land by the banks of the canal was soft, and the weight of the bell tower made the land subside underneath, causing the tower to lean slightly.[5] The leaning tower of Delft provided the perfect spot to test the rates of fall. With a friend standing below, Stevin simultaneously dropped two lead balls, said to have been a factor of ten different in weight, from a height of 30 feet up the tower. While no apparatus existed in that day to measure the time of fall, it was easy to listen for the respective thuds of the weights as they hit the ground. The result of the experiment was two nearly simultaneous thuds, at least to within the uncertainty of the times he let the weights go. Here was a clear and undeniable demonstration of rates of equal fall, demonstrating just as clearly that Aristotle was wrong. Stevin's experiment from the leaning tower of Delft predated by three years Galileo's own demonstration (if at all) of the rate of equal fall from the more famous leaning tower of Pisa.

Inclined to Move

When Galileo began his lectureship in Pisa in 1588, his former professor, Buonamici, who had a keen interest in the science of motion, was then one of Galileo's colleagues, and Galileo began his own studies on motion, writing down his thoughts and findings in a manuscript he titled *De Motu* (On Motion). This was the start of a project that would take nearly 40 years to complete. Although he already was displaying anti-Aristotelian tendencies, his approach to motion while at Pisa remained solidly within the tradition of Aristotle and his former teachers, with one notable exception—his adoption of Archimedean principles that superseded Aristotle. Galileo's introduction to Archimedes was likely through Tartaglia's Italian translation.[6] He also probably was introduced to Euclid's fifth book, corrected and translated into Italian by Tartaglia, through his tutor Ostilio Ricci who had been one of Tartaglia's pupils.[7]

Armed only with principles of buoyancy and the ratios of continuous variables, Galileo's first steps to understand projectile motion were initially far from the mark, but already contained novel seeds that would bear fruit years later. Galileo focused on the role of acceleration in fall, an aspect often neglected by his predecessors. At first, he believed that acceleration occurred only at the start of the downward motion, and that after a short time the motion became uniform. This view was based on his experience and understanding of falling bodies through water, where terminal velocity is reached rapidly. Because he was treating a fall through air using the same principles, he arrived at the same conclusions. Nonetheless, he recognized that a crucial element in the problem of fall was the role played by forces just prior to, and just after, the object is let go. He recognized that a set of forces must be in balance just prior to the moment of release, and he described in *de Motu* how a weight resting on a plane must have an upward force that cancels the downward tendency of the weight.[8] This recognition of action-reaction pairs anticipates Newton by nearly a hundred years. Galileo's sense of this balance was only a brief glimpse of a principle that Newton later made one of his founding axioms, but it showed Galileo's sensitivity to nuance.

In these early stages, Galileo was careful not to assume too much. For instance, he knew that the density of a material was a crucial property when considering buoyancy. Therefore, as he collected his arguments to support the equal rate of fall of different masses, he only made this claim for different masses of the same material. If the dropping of weights from the Tower of Pisa ever took place, he certainly did this with objects composed of the same material, probably using lead like Stevin in Delft, not comparing lead to wood. The uncertainty about whether Galileo ever dropped weights from the Tower of Pisa comes from the lack of direct evidence for the event. Although drafts of *de Motu* from this time exist, they say nothing of the experiment. The tale is told by one of Galileo's last pupils, Vincenzo Viviani, after Galileo had died. Viviani relates how Galileo had told him of the experiment sometime during their last years together, probably in connection with the publication of *Two New Sciences*, although Galileo makes no mention of it in that book. Viviani was a respected scientist in his own right, having also been a pupil of Torricelli, and he is famous for making one of the first accurate measurements of the speed of sound. Viviani likely would not have fabricated the tale, although it is not as clear whether Galileo, in his final years, might have been telling a yarn to a young admirer. What is clear is that performing such an experiment was consistent with how Galileo thought and worked at the time he was in Pisa.

When Galileo moved from Pisa to Padua in 1592, he hoped to continue his work on *de Motu*, but his new surroundings distracted, and his mind turned to new schemes, partially driven by his perpetual need for money. He invented a military and geometric compass, selling the instrument to augment his meager salary. This activity honed his craftsman's skills building fine instruments that would attract Sarpi several years later with his proposal to have Galileo improve on the weak Dutch telescopes. Yet Galileo remained active in his general scientific interests as well. During the early years at Padua he constructed his theory of the tides in support of Copernicus' theory for the double motion of the Earth (rotation on its axis and revolution about the Sun). He also wrote a short treatise on mechanics *Le Meccaniche* in 1596 in which he explored the principle of horizontal inertia as a type of motion independent of vertical motion. In Aristotelian dynamics violent motion (horizontal motion) and natural motion (fall under gravity) were linked as natural motion was suspended while violent motion was taking place. This is why the Aristotelian trajectory was composed of two straight lines—violent motion until it was expended, followed by natural motion downwards. Galileo's treatment of horizontal inertia would be a crucial element in his final theory of the parabolic trajectory.

As a young mathematician and academician, Galileo developed strong ties to a broad group of intellectual friends including Paolo Sarpi (the almost-assassinated anti-Jesuit theologian) in Venice and Guidobaldo, Marchese del Monte, from Urbino. These three all had an intense interest in the study of motion and exchanged letters and ideas freely. For instance, before leaving Pisa, Galileo had been working on the problem that the rate of fall was too fast to make accurate measurements, and he realized that it could be slowed down considerably by performing experiments on inclined planes. If the planes were inclined at low angles, the rate of fall could be made as small as one wished. He communicated this idea in a letter to Guidobaldo who began his own experiments on inclined planes. When confronted with the problem of tracing out the path of a ball on the plane, Guidobaldo hit on the idea of using a ball coated in ink or carbon dust. As it rolled across paper placed on the surface of the plane, it left behind a fine trail that marked out the ball's trajectory. Guidobaldo communicated this breakthrough in a letter to Galileo around 1600, noting that when throwing the ball obliquely up the plane, the trace was symmetric and looked like an inverted hyperbola. Galileo immediately reproduced this fundamental finding for himself. The symmetry of the trajectory was a consequence of the conservation of transverse motion. All the vertical speeds when going up the plane were reversed when going down the plane. It was the first accurate glimpse of projectile motion.

The three correspondents, Sarpi, Guidobaldo and Galileo, continued working on the science of motion into the early years of the new century, exchanging their ideas in frequent letters (some of which have survived, but the existence of others can be inferred). In 1602, Galileo returned to a problem that was distantly related to the law of fall—the isochrony of the pendulum. Viviani in his biography recounted how Galileo first became aware of the motion of equal times in the Duomo in Pisa while watching the large hanging lantern. He measured the time of oscillation by comparing it with his pulse and noticed that the period did not vary even if the amplitude did. This early observation of isochrony is anecdotal, but by 1602 Galileo was working on the motion of the pendulum theoretically and experimentally. He worked with large pendulums many feet long with heavy weights and nearly frictionless pivots, easily verifying by demonstration that the period of oscillation was independent of amplitude, at least for small enough motions. He tried to work out the theory using motion along cords of circles, but here he was stumped, and he communicated his findings to Guidobaldo.

Guidobaldo was not convinced of Galileo's experimental results and certainly not in his partial attempt at mathematical proof, because he did not believe that larger distances could be traversed in equal time. Despite Guidobaldo's inability to accept Galileo's theory of the pendulum, Galileo was not one to let other's opinions stand in his way, and he proceeded to construct accurate timepieces using long pendulums. He used these pendulums to calibrate the rate of water clocks he built in his workshop. These water clocks were key to his subsequent highly accurate investigations on motion through an ingenious experimental approach in which Galileo literally weighed time.

By 1604, Galileo was engrossed in experimental studies of the rate of fall. The rolling of inked balls along inclined planes had given way to shallow groves recessed into the plane along the direction of steepest descent, down which he carefully rolled brass balls about an inch in diameter. The planes were inclined at angles around 20 degrees, considerably slowing down the rate of descent compared to the vertical, making the possibility of time measurements more accurate. From Galileo's notebooks, one can see that he was struggling to make sense of the physics of fall, even when slowed down on the shallow inclined planes. He understood that descent required the velocity to increase from rest to a final velocity at the bottom of the plane. He was aware of Oresme's latitude of forms and the mean speed theorem, but as a geometer, Galileo was locked into thinking of the path length as the important variable rather than time. In his early attempts at understanding fall, he was seeking to establish the hypothesis that the speed of fall was proportional to the distance of fall. This is, of course, incorrect, and it caused Galileo serious problems because in his experiments the distances traversed in specified times were not agreeing with his theory.

It is a testament to Galileo's originality that he expected experiment to agree with theory. Prior to Galileo, the predominant sentiment followed Aristotle who believed that the inaccuracies of experiment made experimental art far inferior to the pure truth of logical thought. Yet Galileo had broken free of this mindset, partially because of his understanding that the complexities of experiments could be controlled and minimized, allowing the experiment to approach the truth and not just the expectation. Therefore, the disagreement of his measured times with his theory of speed proportional to the distance of fall must have spurred him on to make ever more accurate measurements of the time.

This is where the calibrated water clocks became so important. At the moment he released a brass ball, he opened the flow into a small vessel, and as soon as the ball reached a specified position down the track, he closed off the flow. Then he weighed the vessel and subtracted off the weight of the empty vessel. Even more impressive, since he repeated the experiments many times, he accounted for the small remnant of water that wet the insides of the vessel after the bulk had been emptied. By using highly accurate balances to weigh the water, he was able to measure time in his experiments to within 3 percent error. This level of accuracy would have been impossible with any time-keeping instrument of his day, but between the pendulum clock's isochronicity, and the subsequent calibration of the water clock, and the use of the accurate balance, Galileo was performing the most accurate measurements of kinematics ever achieved up to that time.

And the results were conclusive—his theory of fall was wrong! He had no choice but to abandon the proportionality of distance and speed. The problem was what to replace it with. Here we see that Galileo was not entirely modern. With our current familiarity with calculus, we do not hesitate to think of rate of change as a principal property of a physical system. Yet Galileo did not have that advantage. He was working with distances traversed in certain times and then trying to back out the speeds without knowing anything about derivatives.

At this juncture in the story of Galileo's discovery of the law of fall, we come to a fork in the road. The historical experts who have pored over Galileo's notebooks with many fine-toothed combs disagree on what happened next. The problem is that Galileo never dated his notebooks (take note all you graduate students!) between 1604 and 1609, and they were not compiled in chronological order. Furthermore, most of what Galileo himself said about the discovery was from his *Two New Sciences* published thirty years later and ordered for dramatic literary effect, not for historical accuracy. One theory is that Galileo, ever the geometer, constructed a geometric argument that showed that the speed of fall increased as the square root of distance.[9] However, another theory is that he was guided by Guidobaldo's demonstration of the inverted conic section, which Galileo could have improved upon sufficiently to recognize it as a parabola.[10] By combining the parabolic arc with conservation of transverse motion, one comes immediately to vertical distance increasing as the square of time $s \propto t^2$, and after some simple geometry to $v \propto \sqrt{s}$. Therefore, in one time-ordering, the rate of fall was discovered first, followed by the parabolic trajectory. In the other time-ordering, the parabolic trajectory was discovered first, followed by the law of fall.

For those of us who are experimentalists, we would say that the two probably went hand in hand, partial advances on one hand offering confidence and consistency to the other, back and forth. The progression of thought, as you struggle to understand experimental results, is rarely linear, and parallel tracks are usually the best way to make progress. Diagrams written in a lab notebook can just as easily be derivations before the fact as justifications after the fact. So, in our modern journey tracing the legacy of Galileo's trajectory, we lose nothing by taking a holistic view of his discovery.

What is clear is that Galileo arrived definitively at the law of fall—distance proportional to the square of time—supported by highly accurate measurements of distance and time. He discovered the law of successive distances in equal times which yields the famous sequence of odd numbers (1, 3, 5, 7, 9, ...) as the differences of squares ($2^2 - 1^2 = 3$, $3^2 - 2^2 - 5$, $4^2 - 3^2 = 7$, $5^2 - 4^2 = 9$, ...). The simplicity and the beauty of these results must have pleased him, while establishing the squared-time law absolutely. Just as clear is that Galileo arrived at an equally definitive demonstration of the parabolic trajectory of projectiles. Because of the relevance of free flight to the practicality of ranging of artillery, Galileo rightly considered his discovery of the parabolic trajectory as his greatest achievement in mechanics.

The rapid descent during free flight might at first seem to render these experiments much more difficult to perform than the work on the inclined planes, but even as an experimenter Galileo thought like a geometer. His notebooks record how he projected the brass balls horizontally off the end of the table after having rolled down the inclined plane from accurately measured heights. Rather than attempting to trace the parabolic arc directly, he placed a horizontal board at selected distances below the top of the table, and either used wax on the board to record the impact location of the ball, or used colored balls that left a trace where they hit the board. When he changed the height of the recording board by equal amounts, the distances of impact on the board increased as the square root of the change in height. To a geometer, this was clear and irrefutable evidence that the trajectory was parabolic with the apex at the precise location where the ball left the horizontal table. He performed additional experiments when the ball was launched from the tabletop at oblique angles and found that the path of the trajectory continued to have parabolic form. Galileo drew an accurate diagram of these parabolic trajectories in his notebook, shown in Fig. 3.4. These experiments, when combined with the law of fall, were irrefutable proof for the conservation of transverse motion and the independence of violent (horizontal) and natural (vertical) motion.

By 1609, Galileo had arrived at his definitive theory for the law of fall, with speed increasing proportionally to time, and had established the parabolic trajectory as well as the independence of orthogonal motions. These are fundamental milestones in the history of physics, laying the foundation for the next generation of scientists, if only Galileo would publish them. Up to this point Galileo had published nothing during his career, despite being in his early forties. Galileo did write some of these things down in his letters. Unfortunately, Guidobaldo had died in 1607, so Galileo was not able to share his crowning success with his long-time friend and supporter. Sarpi, on the other hand was as active and as interested as ever, but had just learned about the Dutch telescopes, and he was intent on seeing what Galileo, the craftsman, could do to improve them, which distracted Galileo for the next thirty years. As we

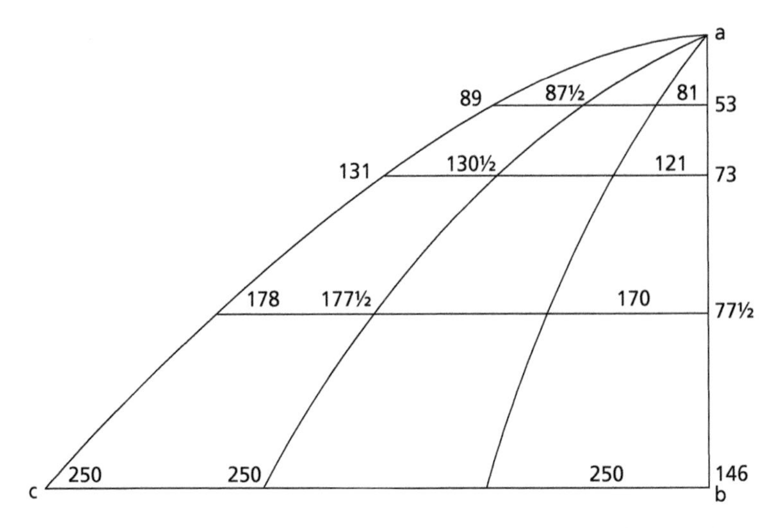

Fig. 3.4 Diagram from folio f81 of Galileo's notebooks showing the parabolic trajectory of balls projected with differing speeds from the edge of a table[11]

saw in the last chapter, it took the Pope in Rome to give Galileo the time and leisure to finally record his discoveries for posterity.

Saved from the Ashes

As house arrests go, the appointments of the residence of Archbishop Piccolomini in Siena were more than comfortable and the life there was intellectually invigorating. The Archbishop was a steadfast friend and supporter of Galileo and spared neither expense nor effort to make his confinement agreeable. He invited the illuminati to dine and to participate in engaging discussions with Galileo. An invitation to spend time at the Archbishop's residence in the presence of such an esteemed guest was a coveted honor. Galileo must have felt some comfort in the Archbishop's support and encouragement to offset the despondency he sometimes felt for being cast so abruptly aside, feeling as if his life and reputation had been erased from the book of the world. Recognizing the need for Galileo to have a constructive outlet, the Archbishop encouraged him to write.

Galileo had always wished to return to his work on motion. Periodically through the years since Pisa, he had taken up the task, only to be distracted by one thing or another. Now there was nothing barring his way, and he routed out the old manuscripts he had carried with him, some dating back over 30 years. In addition to his interest in motion, Galileo had also become interested in the properties and strength of materials, expanding the topic of mechanics to include statics as well as kinematics. As he took up his pen, he created a new dialogue among his old friends Sagredo, Salviati and Simplicio to span a time of five days. This discussion began in the open air, before the famous ship works of the Aresnal in Venice, as the friends discussed and argued their points and counterpoints. Galileo had matured as a writer, perhaps sobered by his trials. Simplicio is no longer a simpleton or comic foil, but represents Galileo's thinking during his days in Pisa. Sagredo represents Galileo during his Paduan years, while Salviati is the sage of the current day, older and wiser, sadder and less flamboyant, although glimpses of the old combatant still showed through.

Because of the blanket interdict against his publications, Galileo must have known that what he wrote could never be published openly. But it is a mark of writers that they write, in spite of themselves and in spite of the chances for seeing their work in print. Whether Galileo wrote because it was therapeutic, or because of a deep sense of destiny, or as an act of rebellion, we cannot guess. Nonetheless, he wrote, completing drafts of the first two days of the dialogue while enjoying the Archbishop's stimulating house in Siena. Unfortunately, word of his good treatment made its way to Rome via the hands of his detractors, and it became clear that Galileo could no longer enjoy Piccolomini's hospitality. Rather than leaving the location of his next incarceration to unfriendly forces, a preemptive move was made to bring him back to Pisa where he could be close to his eldest daughter and her Abbey that relied on his support. His move back to his house at Arcetri, in the hills outside Pisa, was approved in late 1633, but with renewed admonitions of austere treatment that forbade him the steady stream of visitors that he had enjoyed while in Siena.

Despite hopes raised by his return, Galileo's life took a turn for the worse on several points almost immediately. His beloved daughter Marie Celeste died suddenly shortly after his return. On top of that tragedy, his eyesight began failing. He asked permission to visit a doctor in Florence, but the vindictiveness of Pope Urban VIII quelled even that request. The Pope also made clear that all of Galileo's works, past, present and future, were forbidden. Galileo despaired of love and life, yet somehow continued to work on his two new sciences. The dialogues were completed on topics of the isochronicity of the pendulum, uniform motion, uniformly accelerated motion, and the parabolic trajectory. These were all discoveries that he had established in Padua by 1609 just before Sarpi brought him the first Dutch telescope, but he laid them down in print for the first time in his *Two New Sciences*.

Galileo began to explore the possibility of publication in the Republic of Venice. Sarpi was already gone, but there were others in Venice who would willingly defy the Pope. Unfortunately, even that rebellious republic could not afford to antagonize Urban with impunity, so Galileo was turned away. Next was the possibility of publication in Protestant Germany, but around this time Galileo's old foe Scheiner returned to his homelands, and although he did not have the power to block publication, he could still cause trouble, so again Galileo turned away. He was beginning to falter in his hope that his work would ever see the light of day, when an old pupil of his, Francois de Noailles, who had risen to the highest levels of the French government as French ambassador to the Papal States, requested to meet with Galileo.

Galileo had never been allowed to travel during his house arrest, not even to Florence when his eyesight began to fail, but suddenly he was given permission to travel 30 miles to the small village of Poggibonsi where he would meet his old pupil. The unexpected leave to travel must have mystified Galileo, but he relished the chance. He could not have known that at the end of the Thirty Years War, with France declaring war on Spain, the Pope could not afford to lose the allegiance of France even with an imagined slight, and so had no choice but to allow the meeting.[12] What Noailles hoped to accomplish, and what Galileo chose to bring with him to the meeting, are open to historical interpretation. By one account, Galileo brought a handwritten draft of his *Two New Sciences*, which he loaned to Noailles, forgetting to retrieve it when Noailles left for Paris. By another account, this transfer never took place, although the meeting provided Galileo with an alibi when his book appeared suddenly in print in the Netherlands with the Elzevier imprint, supposedly without his knowledge.[13]

In fact, Louis Elzevier, the publisher from Leiden, had secretly visited Galileo at his home in Arcetri, and it is most likely that he spirited away the majority of the manuscript, with Galileo furnishing additional parts through clandestine channels in Venice. Galileo was working on a Day 5 of the dialogue concerning the exchange of forces during collisions, but could not bring it to a close. After several long delays, Elzevier lost patience and informed Galileo that he could wait no longer and would proceed with the publication of four Days, plus an appendix that Galileo had already provided. All he needed was a dedication from Galileo. Even this risked Galileo's safety, and hence he used the ruse of the meeting with Noailles to dedicate the book to the very same Noailles, claiming that he had no knowledge how Noailles had managed to have the thing published, but as the deed was already done, he graciously dedicated the book to his former pupil. Publication of the book in Italian in 1638 was a great success. Fifty copies were sent to Rome and sold out in a day. Even the Pope's nephew, the Cardinal Barbarini, managed to procure a copy. Across Europe, Galileo's famous incarceration made him a celebrated figure and a target of curiosity, and made the book an instant best seller.

Galileo's *Two New Sciences* was his last work. By the time he received a copy of his own book he could not read it—he had gone completely blind. Yet, he left behind an enduring legacy, as generations of future scientists would confess their debt to his deft insights and foundational work. The wide-ranging discussions in *Two New Sciences* raised more questions than they answered, but this is part of the importance of the work, providing fruitful directions for younger minds to search. This was Galileo's stated purpose, which he set forth in the preface of the Elzevier publication. With the book he hoped

> . . . to set forth a very new science dealing with a very ancient subject. There is, in nature, perhaps nothing older than motion, concerning which the books written by philosophers are neither few nor small; nevertheless I have discovered by experiment some properties of it which are worth knowing and which have not hitherto been either observed or demonstrated. Some superficial observations have been made, as, for instance, that the free motion of a heavy falling body is continuously accelerated; but to just what extent this acceleration occurs has not yet been announced; for so far as I know, no one has yet pointed out that the distances traversed, during equal intervals of time, by a body falling from rest, stand to one another in the same ratio as the odd numbers beginning with unity.

Galileo continues with his favorite conclusion, concerning projectile motion, noting that

> . . .[I]t has been observed that missiles and projectiles describe a curved path of some sort; however no one has pointed out the fact that this

> path is a parabola. But this and other facts, not few in number or less worth knowing, I have succeeded in proving; and what I consider more important, there have been opened up to this vast and most excellent science, of which my work is merely the beginning, ways and means by which other minds more acute than mine will explore its remote corners.

Here is Galileo's legacy as well as his challenge to future generations. Galileo had framed the context of the new mathematical science, even if he did not solve all the problems. The key change from Aristotelian philosophy was the active participation of the scientist in the exploration of natural phenomena using an array of tools—mathematical, geometrical and experimental. By taming time, Galileo was able to perform the most accurate measurements of motion ever attempted, freeing him of his longstanding misconception of the dependence of speed on distance of fall. This was a paradigm shift for Galileo. He was foremost a geometer and had clung to geometric dependences until his water clocks forced him to recognize the central role of time in his law of fall. Yet Galileo also had demonstrated that the parabolic trajectory was an entire family of geometric curves—his diagrams in his notebooks displayed this view directly. The geometric character of trajectories, as a group, would become one of the central concepts of modern dynamics after Riemann, Poincaré, and Einstein. Galileo's geometric trajectories would also become the central structure of phase space supporting chaotic dynamics, including extensions to evolutionary dynamics and the wealth of nations and beyond.

Galileo may have thought that other minds more acute than his might be hard to find, but other minds as acute as his did indeed take up the thread of his thoughts. Descartes, Leibniz, Huygens and Newton were his intellectual descendants, owing him, in part, their methods as well as where they looked to uncover secrets of nature. When Newton wrote that the reason he was able to see farther than those before him was because he stood on the "shoulders of giants," it is clear that he was referring by oblique omission to the old master.

4

On the Shoulders of Giants

What Des-Cartes did was a good step. You have added much several ways, & especially in taking ye colours of thin plates into philosophical consideration. If I have seen further it is by standing on ye shoulders of Giants.

From a letter written by Isaac Newton to Robert Hooke (5 February 1676).

Along the arc of scientific progress, there are moments when the pieces of a puzzle have been gathered but not yet assembled. Those who helped gather the pieces saw each one as distinct and important, but could not see how to connect them. At such a moment, a figure appears and, as if by magic, swiftly assembles the pieces into a clear picture. The magic—the genius—is the ability to see the picture in its whole as the panorama takes a form easy to recognize, and the detailed intricacies of the individual pieces simply provide color and texture—like the structured pixels of a Chuck Close portrait or a contemporary photomosaic. Sir Isaac Newton (1643–1727) was one of these pivotal figures.

Conversely, there are times in the history of science when progress is made steadily by many contributors building ideas upon ideas, taking wrong turns, but eventually finding guiding principles to bring them back on track. As the pieces of the puzzle are assembled, a bigger picture emerges, and just as in a family jigsaw puzzle, there is the last person who fits in the last piece to finish the puzzle. Lagrange was one of these figures, synthesizing a hundred years of progress after Newton to give physics a theoretical foundation that persists to this day.

The character of physical law at the two ends of the eighteenth century could not have been more different, nor the personal characters of these two scientists—Newton and Lagrange. Newton, on the one hand, was a giant in his time, an acclaimed genius who, as he grew comfortable in his public role, slowly relinquished his science. Lagrange, on the other hand, was a respected mathematician as well as statesman who

Galileo Unbound. David D. Nolte, Oxford University Press (2018).
 DOI: 10.1093/oso/9780198805847.001.0001

shrewdly navigated the French Revolution without losing his head—a feat that was beyond the powers of his friend Antoinne Lavoisier. Newton's Laws were based on forces acting on individual masses, a *local* description of physics captured in the "free-body diagram" of introductory physics classes today. Local forces may be transmitted to a larger system through sequences of collisions, or through extended bodies as stress and strain, but the central idea is infinitesimal—as in Newton's infinitesimal calculus. Lagrange's laws, on the other hand, were a *global* description of physics where a global property of the entire system (smallest possible time or minimum mechanical action along a path) is unaffected by small variations in the executed trajectory. These are the two bookends of the hundred years from Newton to Lagrange: local forces and infinitesimal calculus on the one side, contrasted with nonlocal descriptions and variational calculus on the other. We begin with Newton.[1]

The Plague on You

Scientists often suffer from mixtures of insecurity, ego, jealousy and delusions of grandeur—the greatest scientists suffering these failings to a degree greater than most. Hence, what may be the most significant book in the history of physics, Newton's *Principia*, owes its existence to a jealous rivalry between two difficult geniuses: Isaac Newton and Robert Hooke (1635–1703). Newton and Hooke were contemporaries, although Hooke was senior to Newton by seven years. Both were members of the Royal Society of London that had been founded in 1660 by a royal charter from Charles II when both men were in their 20s. Hooke was an experimentalist and a keen observationalist, while Newton was the consummate mathematician. Hooke relied on his intuition to gain insight, while Newton relied on axioms. This difference in approach matched their difference in character, and both men recoiled from the presence of the other. They did share one character trait—they were both convinced of their own greatness, and Hooke, at least, was keen to prove it over the other.

In January of 1684, a meeting took place between Hooke, Edmund Halley and Sir Christopher Wren. Halley was famous at the time (before his prediction of the return of "Halley's" comet) as "the southern Tycho" for his astronomical expedition to St. Helena where he mapped the stars of the southern sky. Christopher Wren was famous as the architect who was rebuilding London after the Great Fire of 1666 (and building his masterpiece, St. Paul's Cathedral, completed in 1710). This may seem an odd meeting among an experimentalist, an astronomer and an architect, but it is a measure of the times that all three were deeply interested in, and active in, the problem of gravitation and the motion of the planets. The topic of discussion that day turned to the likelihood that gravity followed an inverse square law. Wren, in good humor, offered as reward a book of 40 shillings to the one who could prove it. Hooke boasted that he had already accomplished the proof, and when the three departed, it was settled that Hooke would send his proof along later. Months passed, and Hooke never produced the proof.

By August of 1684, Halley was working on problems of cometary motions, and he went to visit Newton at Cambridge, where he was Lucasian Professor of Mathematics. Halley mentioned to Newton about the meeting with Hooke and Wren earlier that year, and that Hooke had claimed to have a proof of the inverse square law, which had never materialized. Newton, in a manner similar to that of Hooke, claimed that he had already proved it, but then confessed that he had "lost" the derivation. When Halley left, he strongly encouraged Newton to "find" it. One can only conjecture that neither Newton nor Hooke had sufficiently proven the inverse square law at the time they claimed. It is possible that each went to work on the problem in earnest only after the fact. While Hooke never produced his proof, Newton delivered to Halley, in November of 1684, a nine-page manuscript titled *De motu corporum in gyrum* (On the Motion of Bodies in an Orbit) that Halley recorded into the transactions of the Royal Society. This short manuscript derived Kepler's 2nd Law, and proved that Kepler's other two laws were consistent with the inverse square law. It has never been reported whether Newton received the award of the book of 40 shillings from Wren, but it is clear that Newton succeeded where Hooke did not, much to Hooke's chagrin.

A large part of the mythology surrounding Isaac Newton is the story that Newton experienced the *Annus Mirabilis* (miraculous year) from 1665 to 1666 while on leave from Cambridge during an outbreak of the bubonic plague. He later laid claim, during that year, to having solved the binomial theorem, invented the calculus, uncovered the spectrum of white light and discovered universal gravitation. While all of these claims have been shown to be partly true, only the claims on the binomial theorem and colors of light are known to be entirely true. Newton made these claims many years afterwards, when he was embroiled in the priority dispute over the invention of the calculus with Leibniz, and still chafing under the accusation, by Hooke, that Newton had somehow stolen Hooke's priority on the inverse square law.

The nearly twenty-year path to Newton's *De motu* from 1666 to 1684 was more gradual than a sudden miraculous year, and it drew from the progress others were making during the same time. Newton undoubtedly was very active in 1666, and he did derive the centripetal force for uniform circular motion (force proportional to the square of the speed and inversely to the radius of the circle) in that year. By 1669, he had uncovered that Kepler's 3rd Law (that the ratio of the square of the period to the cube of the radius of a planetary orbit are constant for all planets) was consistent with an inverse square law.[2] During the succeeding years, Newton was involved in many activities, including optical studies and his theory of colors, but he continued to delay publication of his ideas. For instance, he lost priority for the derivation of the acceleration of circular motion when Christiaan Huygens (1629–1695) published his 1673 treatise on clocks and the physics of the pendulum. However, Huygens was thinking in terms of centrifugal (center-fleeing) force that was restrained by the string of the pendulum. The following year the irascible Robert Hooke (1635–1703) published *An Attempt to Prove the Motion of the Earth* in which he rejected Huygens centrifugal force and explained the circular motion of the planets in terms of a force of attraction (Newton's centripetal force from 1666 that Hooke was unaware of) as well as a law of inertia in which bodies continue along straight lines in the absence of forces. The law of inertia (uniform motion unless acted upon) had also been proposed by Galileo and Descartes, and Descartes had suggested that quantity of motion was equal to the quantity of matter times its speed. However, Hooke did not specify whether the attractive force was an inverse square, and it was not universal because it was generally believed that comets moved in straight lines past the Sun and were not affected by the same laws that bound the planets in their orbits. Newton obviously had different ideas.

Because Newton was an obsessive note taker and keeper, it is possible today to follow his mathematical scribblings as well as his modes of thought. He was strongly influenced by Descartes in his thinking, and his early ideas on planetary motion were dominated by Descartes' vortex theory of the aether. However, this changed around 1680, as Newton rejected the aether vortex theory and began to think in terms of central forces that acted at a distance. By the time of Halley's visit in 1684, Newton's thinking on the planetary orbits had matured, and he was primed to show that the solutions to the orbital equations, subject to an inverse square force, were Kepler's ellipses. As soon as Halley saw Newton's *De motu*, he pressed Newton to expand on it and publish it properly, which he offered to help edit and even to pay for out of his own pocket. Newton was ready for the challenge, and he launched himself completely into his work. His ability to focus is legendary, and he was consumed by the task, barely eating and barely sleeping across several years, driven forward with zeal to document a new science for posterity.

Newton's original goal had been to fill out *De motu*, by giving it more background and rounding out the derivations, but the task kept expanding as Newton, in some sense working backwards, found ever deeper underpinnings of motion. What had been *De motu* eventually became Book III of the *Principia*, but Newton's greatest contribution to science was Book I, where he states his definitions and posits his axioms of motion. By this time, he had stripped motion down to its essentials, with eight definitions and three axioms—known today as Newton's three Laws of Physics. The first law was the law of inertia. The second law stated that changes in quantity of motion are caused by impressed forces (See Fig. 4.1 for a facsimile of Newton's first two axioms.). The third law treated action and reaction pairs of forces.

The First Law was already known to Descartes and to Hooke. When a ball on a string is executing circular motion, the string supplies the force to turn the ball's path into an arc. If the string is cut suddenly, the ball moves off in a straight line tangential to its circular path. In contrast to the First Law, the Second Law used more ambiguous wording. Newton was probably thinking of instantaneous collisions and impulses rather than continuous changes in the quantity of motion. In other words, he was stating $\Delta(mv) \propto F$, while today we say that Newton's second law is $F = dp/dt$. It is clear that even if the second law was related to collisions, Newton applied the continuous form in the solution of many problems, including the elliptical planetary orbits. Of the three laws, the Third Law is the most subtle, and continued to challenge scientists into the early nineteenth century. The third law is the law of action-reaction pairs—of the back-action of bodies to the impulse. It would become a central feature of modern dynamic theory when D'Alembert, in the eighteenth century, used Newton's third law to place impressed forces

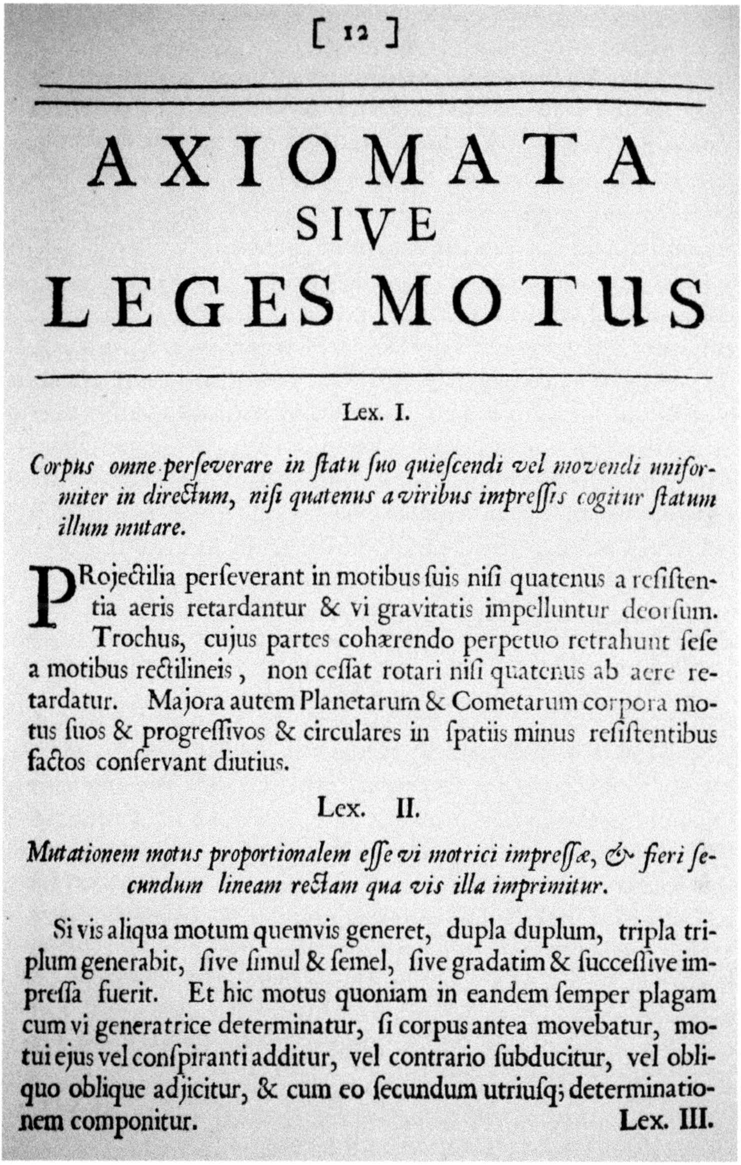

[12]

AXIOMATA
SIVE
LEGES MOTUS

Lex. I.

Corpus omne perſeverare in ſtatu ſuo quieſcendi vel movendi uniformiter in directum, niſi quatenus a viribus impreſſis cogitur ſtatum illum mutare.

PRojectilia perſeverant in motibus ſuis niſi quatenus a reſiſtentia aeris retardantur & vi gravitatis impelluntur deorſum. Trochus, cujus partes cohærendo perpetuo retrahunt ſeſe a motibus rectilineis, non ceſſat rotari niſi quatenus ab aere retardatur. Majora autem Planetarum & Cometarum corpora motus ſuos & progreſſivos & circulares in ſpatiis minus reſiſtentibus factos conſervant diutius.

Lex. II.

Mutationem motus proportionalem eſſe vi motrici impreſſæ, & fieri ſecundum lineam rectam qua vis illa imprimitur.

Si vis aliqua motum quemvis generet, dupla duplum, tripla triplum generabit, ſive ſimul & ſemel, ſive gradatim & ſucceſſive impreſſa fuerit. Et hic motus quoniam in eandem ſemper plagam cum vi generatrice determinatur, ſi corpus antea movebatur, motui ejus vel conſpiranti additur, vel contrario ſubducitur, vel obliquo oblique adjicitur, & cum eo ſecundum utriuſq; determinationem componitur.

Lex. III.

Fig. 4.1 Newton's First and Second Laws, presented through exposition rather than equations

and inertial forces on equal footing, and subsequently Einstein in the twentieth century enlisted D'Alembert's principle to help explain the deflection of light by gravity even though gravity can exert no force on light.

Whereas Book I established the principles, and Book III applied them to orbital mechanics, Book II of the Principia was a sideline devoted to the motion of bodies through resistive media. This seems a strange volume to add between the axioms of motion in Volume I and the applications to gravitation in Book III, theoretical on the one hand and mathematical on the other. But in Book II Newton was taking a jab at Cartesian physics, especially Descartes' vortex theory of the aether. He knew that he needed to give clear evidence that the motions of the planets and comets could not be caused by vortices, and that motion through aether would behave differently than the motion observed astronomically. Newton put forward important new ideas in Book II on hydrodynamics (some types of fluids are called "Newtonian" today), but it does not have the depth or impact of the first and third books. Nonetheless, it stands as a conscious effort to try to fend off what Newton knew would be criticism from the Continent.

Pride and Prejudice

One of the most bitter priority battles was the dispute between Newton and Leibniz over the invention of the calculus. The infinitesimal calculus was the greatest invention of its time—more important even than Newton's *Principia*. The calculus brought the solution to problems that had stood unsolved for over a thousand years. Newton and Leibniz had arrived at the calculus independently, with Newton first, although Leibniz tended to publish faster. This nearly simultaneous discovery was not an accident, because each was partially aware of the other, and each knew some of the results of the other, if not the methods. The time was ripe for the development of the calculus, because many of the preliminary elements had already been put in place, and the chief unsolved mathematical problems of the day tended to focus on slopes or areas of curves.

The difficulty between Newton and Leibniz arose many years after their discovery, but while they were still alive, when Leibniz pushed too hard to claim priority for his own efforts over Newton. This caused a backlash from English scientists, and Leibniz' position turned out to be weakened because of an ill-timed visit he had made to the Royal Society at just the time that he and Newton were at the cusp of their respective discoveries. Some of Newton's results had been deposited with Oldenburg, the secretary of the Society, and an accusation was made, these many years later, that Leibniz had seen those materials. Inquiries were launched, committees were convened, nationalistic prejudices flared—and nothing was resolved, even though the English committee convened by the Royal Society ruled in favor of Newton. Nationalistic pride, at the time, pitted English science against Continental science. The French, Dutch and German mathematicians and scientists tended to be more cosmopolitan with closer ties to one another than to the English. This division was to continue for a hundred years after Newton, with Newtonian views in England competing against Cartesian (from Descartes) science on the continent. England, during this period, was isolated geographically as well as intellectually, and by the late eighteenth century English mathematics and science had been largely surpassed by Continental science even though Newtonian physics was adopted early on by influential French scientists and promoted over Cartesianism. The French were better at adopting and using Newton than the English. One of the chief early Newtonians was the French physicist and philosopher Pierre Louis Maupertuis.

Maupertuis (1698–1759) was born in Brittany, in the timeworn walled coastal city of Saint-Malo, the son of a government official from a merchant family that made their fortune through privateering as French corsairs during France's many wars. He was short in stature, with a round boyish face that he never shed even into adulthood, but his eyes were sharp and eager, signaling a quick wit and a strong assurance of his importance in the world, though he was plagued to the end by the fear that the world would fail to recognize his importance. He had refined social skills, despite his middle-class background, that made him at home in the burgeoning salon culture of Enlightenment Paris. As a young man, he entertained a short military career as a Musketeer, but that life did not live up to his swashbuckling imagination, so he retired to a social life of cafés and salons where he excelled at barbed repartee and weighty intellectualism. At this time, he was inexplicably infected by an enthusiasm for mathematics, although he had shown little promise in his early schooling, and his leisure time became devoted to learning the latest advances of mathematics under the informal tutelage of several famous Parisian mathematicians.

Deciding that an academic life was what he wanted, he applied for, and in 1723 he was appointed to, a position in the Paris *Académie Royale des Sciences* as adjoint mécanicien. His early mathematical work and his impact on intellectual life were not remarkable, but this changed in 1728 when he visited London for six months. His time there was an epiphany. He found the intellectual life and philosophy in England to be surprisingly at odds with the views of the French. He experienced what Voltaire later alluded to in one of his essays:

> A Frenchman who arrives in London will find things to have changed greatly, in philosophy as well as in other areas. He left the world full, and now finds it empty. In Paris the Universe appears to be composed of vortexes of subtle matter, in London these are nowhere to be found. With us, the pressure of the moon causes the tides, but for the English it is the sea that is drawn towards the moon ... for you Cartesians everything comes about through an impulse that one can scarcely understand, for Monsieur Newton it is through an attraction, whose cause is equally unknown; in Paris you imagine the earth shaped like a melon, in London it is flattened on both ends. For a Cartesian, light consists of air, for a Newtonian it travels from the sun in six and a half minutes.... You can see how strident the contradictions are.[3]

Voltaire was referring to Newtonianism versus Cartesianism, especially with respect to the nature of vacuum and light, and the prevalence of Descartes' theory of vortices in French physics. In 1644, Descartes had rejected the concept of a vacuum, and instead filled space with some sort of matter that he called the aether. Because he thought that the aether was partially or wholly incompressible, the only free motion would be vortex motion, just as in the case of swirls in water. As the aether in the vortex spun, it accumulated in the outer reaches of the vortex. Just as a helium balloon experiences buoyancy that moves it from higher atmospheric density to lower, the aether exerted a buoyant force on celestial objects. Descartes proposed that the Earth was at the center of a vortex, and the Moon experienced a buoyant force that pushed it towards the Earth. This buoyancy of bodies within the aether was the cause of Cartesian gravitation. Because Descartes was concerned with causes, his theory provided an explanation of gravitation, while Newton's theory merely posited the existence of an attraction without an explanation of what caused the attraction. In this sense, Descartes provided an explanation for gravity, while Newton did not, and the French favored Descartes. However, Maupertuis was smitten by the

Newtonian view of the world, and he returned to France determined to champion Newtonian physics. To accomplish this end, Maupertuis entered the University of Basel in Switzerland in 1729 for formal training in mathematics under Johann Bernoulli (1667–1748), and he lived in Bernoulli's home. Bernoulli was a staunch supporter of Descartes and Leibniz, so Maupertuis kept his newfound Newtonianism mostly to himself. While completing his degree, Maupertuis befriended one of Bernoulli's younger pupils, Johann König, an association that later caused him great personal and professional distress. In 1732, armed with his new degree, as well as the delusions of grandeur that grip so many young scientists, Maupertuis set out to make his mark in science and mathematics. The topics that caught his interest were truly peripatetic, ranging across biology, philosophy, metaphysics, music, physics, and theology, although it was in geodesy that Maupertuis achieved the fame he so yearned for.

In 1736, the Paris Royal Academy, in the name of Louis XV, charged Maupertuis to lead an expedition to Lapland far to the north in Sweden (current Finland) to determine whether the Earth was flattened at the poles. This was a point of great interest at the time that pitted the physics of Descartes against the physics of Newton, and nationalistic feelings exerted considerable force on the beliefs of the principal players in the drama. According to Newton's theory, the Earth should be a slightly flattened sphere, known as an oblate spheroid, because its spin causes a centrifugal bulge, which is the correct situation. However, the vortex theory of Descartes predicted that the Earth would be an elongated (prolate) spheroid like a melon because the pressure exerted by the Moon on the aether, and hence on the Earth, squashed the sphere into the shape of a melon. The French astronomer Jacques Cassini (1677–1756) had made astronomical measurements that supported the Cartesian results, but the Newtonians were not convinced. The resolution of the shape of the Earth was not just an idle curiosity, because entire worldviews hung in the balance.

Maupertuis organized and executed the expedition to the cold climate of Lapland. The expedition members suffered considerable discomfort from insects in summer and cold in winter, but through great physical effort, they acquired the needed data. On the trip home the ship carrying them was shipwrecked, but they managed to save their scientific materials. Once safely back in Paris, Maupertuis's friend and fellow physicist on the journey, Alexis Clairaut (1713–1765), analyzed the data, and it became clear that the Earth was slightly flat—Cassini and the Cartesians were wrong. Maupertuis wrote up and published the findings in a booklet that doubled as technical report and adventure travelog. This made Maupertuis a celebrity even outside scientific circles, not unlike the celebrity that Einstein would enjoy two hundred years later when a similar expedition was launched by Eddington to test the predictions of General Relativity. Unfortunately, Maupertuis was not gracious in his success, and he took the opportunity to launch vicious attacks against Cassini, in private as well as in public communications. The ferocity of the attacks shocked even Maupertuis' friends, and it earned him the undying enmity of Cassini who held the grudge hard and would later take his revenge.

The Principle of Least Action: Opening Act

Today, the controversy over the shape of the Earth has faded into obscurity, and Maupertuis might not even be remembered. However, Maupertuis holds an eternal place in the history of physics because he conceived of the Principle of Least Action that, in modern terms, describes how physical processes adopt configurations and follow motions for which a physical quantity called Action has an extreme value—either a minimum or a maximum. Maupertuis, in his day, used a different language to describe his principle, claiming that Nature (meaning God) sought to be most efficient in all efforts. When expressed this way, the Principle is teleological, meaning that it operates through final causes—it seeks a desired or optimal result, as in "an object takes a path so that it can minimize its action." Teleological explanations in physics are rejected today, but in Maupertuis' time, God was very much a part of daily life, and the Universe was considered to act with a purpose. In this way the teleological view becomes also a theological view, a view which Maupertuis took to extremes in later years, although it dimmed his legacy in modern eyes.

Minimum principles in physics had a long history even by the time of Maupertuis. In surveying, principles of shortest distance date back to ancient Egypt, and Hero of Alexandria (AD 10–70) used the principle of shortest distance to explain the law of reflection of light by a mirror (for which the incident and reflected angles are equal but opposite). On the other hand, the related phenomenon of refraction, where light is deflected when it passes from one medium to another more or less dense medium, could not be solved as a path of least distance, which clearly would be a straight line. Some other principle, or set of principles, was required to solve this problem.

The law of refraction, known today generally as Snell's Law, had multiple origins. The deflection of light when it entered water was known in antiquity, and was a topic of Arabic science, especially in Ibn Sahl's *On Burning Mirrors and Lenses* (984) and Ibn al-Haytham's (Alhazan) *Book of Optics* (1021). Thomas Harriot in 1602, and the Dutch physicist Willebrord Snel van Royen in 1621, each provided an accurate mathematical description in terms of sine functions, but neither published. The first openly to publish the law of sines was Descartes in his 1637 *Discourse on Method*, where he used it as an example to demonstrate the utility of his new analytic geometry. Controversy, as usual, erupted in later years on this point, as some (especially the Dutch) argued that Descartes had been aware of Snel's results. This accusation has never accumulated any evidence and may have been motivated by national pride on the part of the Dutch engaged at this time in their bitter struggle for independence from overbearing European neighbors. Although the French today still call the law of refraction the Law of Descartes, Huygens later made a clear case in his 1678 *Traité de la Lumiére* for its prior (but unknown) discovery by Willebrord Snel. Huygens also derived the law using what is known today as Huygens' Principle based on the wave nature of light. However, the predominant theory of light at the time was corpuscular, and hence it was subject to the same laws of mechanics as matter.

In February of 1662, Pierre de Fermat (1601–1665) wrote a paper *Synthesis ad refractiones* that sought to explain Descartes-Snell's Law by considering the time it took for light to travel, as particles, from a point in the first medium to a point in the second (see Fig. 4.2). He discovered that if he inverted the Cartesian assumption (that light traveled *faster* in the denser medium), and instead assumed that light travels *slower* in the denser medium, then the path taken by a ray of light expended the least amount of time. All other possible paths took longer time, even paths very close to the actual path. He was understandably excited about his discovery and considered it important and beautiful. He said that "the reward of my work has been most extraordinary, most unexpected, and the most fortunate that I have ever obtained.... I was so surprised by a happening that was so little expected that I only recovered from my astonishment with difficulty. I repeated my algebraic operations several

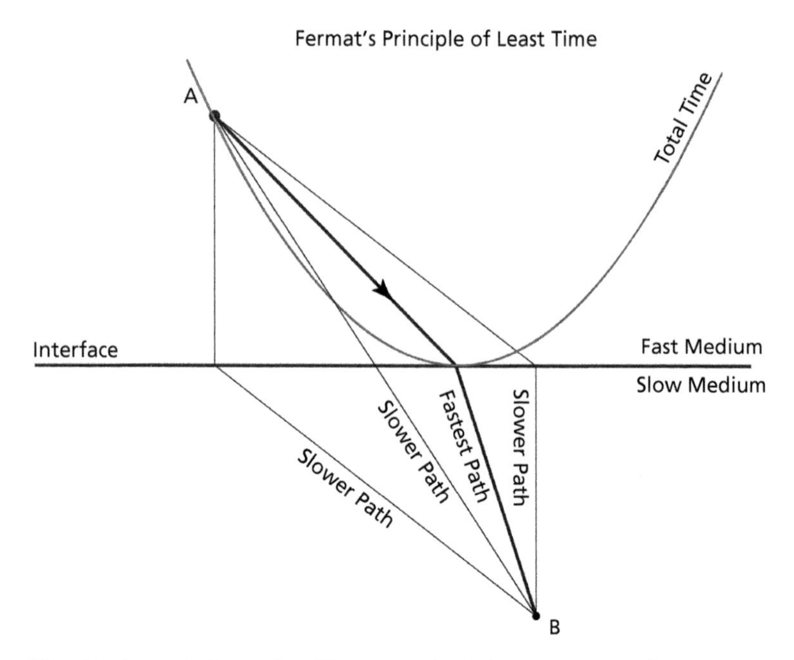

Fig. 4.2 Fermat's Principle of least time for light to propagate from A to B. The parabola shows the total time as a function of the intersection of the path with the interface. The minimum time occurs on the actual path taken by a light ray

times and the result was always the same, though my demonstration supposes that the passage of light through dense bodies is more difficult than through rare ones—something I believe to be very true and necessary, and something which M. Descartes believes to be the contrary."[4]

His satisfaction was short lived as numerous Cartesians attacked his assumptions on the speed of light in dense media, viewing his successful derivation of Snell's Law as fortuitous, i.e., wrong. Nonetheless, Fermat had unleashed a new idea onto the world, the idea of a minimum principle. Minimum principles would play a central role throughout the history of physics, eventually providing the basis for understanding trajectories in Einstein's general theory of warped space-time, and for understanding quantum trajectories, but first it needed Maupertuis to draw it forth.

In the early 1700s, the problem on most physicists' minds was the problem of static equilibrium. From a Newtonian point of view, equilibrium was a consequence of balanced forces, but it was sometimes difficult to identify all the forces, especially for systems of many particles and many constraints, and every problem required its own particular solution. There was a search for more general principles that could be applied without the need to identify all the component forces. This set the stage for Maupertuis.

In 1740, Maupertuis read a paper before the Paris Academy, *Loi du repos des corps* (Law of repose of bodies), on a new approach to the problem of static equilibrium that identified it as a minimization problem. Maupertuis defined a quantity that was the sum of the product of the masses of numerous particles, subject to a power-law central force, multiplied by their distances from the force center. When this quantity was minimized, the static equilibrium configuration of the system was uncovered. Today, we recognize this as the minimum in the potential energy of the system, but at that time the concept of potential energy was only beginning to be defined, and had yet to be given a name. Here was a general principle, a minimization of a quantity that described the correct solution to a physical problem. It was the first step along a road in which minimization principles not only allowed one to *find* correct solutions, but also might provide an *explanation of* those solutions.

Maupertuis was just as earnest as a philosopher as he was as a mathematician. At that time, Maupertuis was considering philosophical origins of pain and pleasure. He published a theory on the topic where he proposed that two quantities had to be considered collectively—these were the intensity of the feeling and the duration of the feeling. It was the product that mattered, not either individually. If pain was experienced at a low level but over a long time, this was just as "bad" as an intense pain experienced over a short time. To minimize pain, one sought to minimize the joint experience, the product of the intensity and time. These philosophical ideas were on his mind as he began to consider the mathematical problem of the law of refraction. He believed that the correct path between the two points in the different media was neither the shortest path, nor Fermat's shortest time (which was still considered to be wrong at that time), but was perhaps the minimum of some joint property.

To translate this minimum principle of joint properties to the physics of refraction, the problem was to find one quantity that represented the intensity, and another quantity that represented the duration, and to define their product that was to be minimized. Maupertuis chose the speed of particles of light to be the intensity, and the distance traveled by the particles to be the duration. The quantity to be minimized was *speed multiplied by distance*. Maupertuis called this quantity the *Action* of the light path, and in 1744 he presented a paper before the Paris Academy *Accord des différentes loix de la nature qui avoient jusq'ici paru incompatibles*, (The harmony of various natural laws which have thus far appeared to be incompatible) that showed that the law of refraction was equivalent to the path of least action. The concept of "Action" had been used without well-defined meaning in the writings of Newton, for instance in his third law of action and reaction. Leibniz and his student Wolff had introduced a more specific definition of Action during their work on dynamics, but they had not defined it sufficiently nor succeeded in utilizing it in a predictive manner, at least not in their known publications. Maupertuis took the name and provided it with a clear mathematical expression that could be analyzed and minimized. By using the word "Action," with its historical context from dynamics, Maupertuis was clearly considering the propagation of light to be a mechanical phenomenon.

Maupertuis expanded the concept of action and presented a paper in 1746 that extended the principle to include particles with mass. With this principle, he was able to derive the laws for elastic and inelastic collisions as well as the principle of the lever. The universality of the principle appealed to his nature, leading him to remark,

> the laws of movement thus deduced [from the Principle of Least Action], being found to be precisely the same as those observed in nature, we can admire the application of it to all phenomena, in the movement of animals, in the vegetation of plants, in the revolution of the heavenly bodies: and the spectacle of the Universe becomes so much the grander, so much the more beautiful, so much more worthy of its Author.[5]

He began to believe that he had discovered a unifying principle of physics that governed all else, proudly stating that:

> after so many great men have worked on the matter, I hardly dare say that I have discovered the principle on which all the laws of motion are founded; a principle which applies equally to hard bodies and elastic bodies; from which the motions of all corporeal substances follow.[6]

The simplicity of the role of action in mechanics was astounding—the perfect path—and Maupertuis was duly astounded and impressed with himself. He considered that his Principle of Least Action would be the underlying and universal principle from which all the specialized rules and partial laws could be derived and explained. Even beyond physical theory, Maupertuis sought to establish the Principle of Least Action as a metaphysical (philosophical) law—as the final cause of all motion and equilibrium. This discovery surely was his life's greatest work and the path to his immortality. However, fate had other plans for him.

Of Friends and Forgers

Around the time that Maupertuis returned from Lapland with Clairaut and was being hailed as a hero in France, the new King of Prussia, Frederick II, sought to establish a scientific academy in Berlin to rival the Academies in London and Paris. Frederick the Great had an imposing charismatic character, wielding wide influence across Europe and counted Voltaire as one of his supporters. Voltaire recommended Maupertuis for the position as first president of the Berlin Academy. Frederick approached Maupertuis who was understandably honored but reluctant to take such a bold move. Nationalistic pride was balanced against the growing cosmopolitan character of European intellectualism. There were many examples of top scientists across Europe being courted by kings and queens to add to their constellation of courtiers: the Frenchman Descartes attracted to Sweden by the queen Christina, the Swiss Euler attracted to St. Petersburg in Russia by Catherine I the wife of Peter the Great, the Dutch Huygens attracted to the French Academy in Paris, and now Maupertuis attracted to Berlin by Frederick. In 1746, Maupertuis became the first president of the Prussian Academy. Maupertuis was correct to hesitate leaving Paris, even if the lure and prestige of his new position was too hard to resist. As soon as Maupertuis announced that he would take the position, his old enemy Cassini succeeded to have Maupertuis ejected from the French Academie des Sciences. Only after Cassini's death in 1756 was Maupertuis reinstated to the Academy. While the prestige and power that came from being the president of the Berlin Academy suited Maupertuis' ambitions, one of the greatest benefits of moving to Berlin was the close association it promised with Leonhard Euler, the greatest mathematician of his time.

Euler (1707–1783) was a Swiss-born mathematician who achieved such fame that he was attracted by powerful kings and queens to the seats of power in Europe. His father was a Protestant minister who could not afford a formal education for his son, but through home schooling and tutors, Euler learned so quickly that he was able to enter the local university in Basel before he was the age of fourteen. In his autobiography, Euler described how he found an opportunity to be introduced to a famous professor, Johann Bernoulli, who gave him "valuable advice to start reading more difficult mathematical books on my own and to study them as diligently as I could: if I came across some obstacle of difficulty, I was given permission to visit him freely every Saturday afternoon and he kindly explained to me everything I could not understand."[7]

Euler excelled in mathematics as a natural talent, ultimately becoming the towering mathematician of his era, participating in or originating many new fields that included the development of functions, variational calculus, complex analysis, graph theory, analytic number theory as well as transcendental functions. Euler was sensitive to notation, and he introduced the concept of a function and the expression f(x). He introduced the letter i as the symbol to stand for the square root of -1, and he established the Greek letter π as the ratio of the diameter of a circle to its diameter. He was also a master of applied mathematics and produced major works in physics and astronomy. The first book of many he published in his career was his *Mechanica sive motus scientia analytice exposita* (1736). This work was a synthesis of Newton and Leibniz and is one of the most influential works in the physics of dynamics for the simple reason that it put a modern face on dynamics. Anyone who has read Newton's *Principia*, even translated into English, realizes that it is unrecognizable as the physics we know today, entombed as it is in geometric approaches and expository prose. Euler re-derived and re-expressed Newton using the fruitful calculus notations of Leibniz. The high place Newton holds today, as the father of Newtonian physics, was made possible by Euler. We see Newton through the lens of Euler.

At the time when Frederick offered the head of the Berlin Acadamy to Maupertuis, Euler was in the process of inventing a new mathematical tool, the variational calculus, that he used to calculate, among other things, paths of least action, working these out a short time before Maupertuis' contribution. Maupertuis was foremost a philosopher, and while he had some mathematical talent and skill, he did not have the depth of someone like Euler. Conversely, Euler was a pure mathematician with little philosophical inclination. The Principle of Least Action, as Maupertuis conceived of it, was primarily a teleological principle, but it required deep mathematics to supply the proof. Maupertuis provided the philosophical motivation while Euler provided the mathematics. Their intellectual compatibility extended to personal compatibility, and Euler would be a staunch supporter of Maupertuis throughout the trials to come. The worst trials came at the hands of an old acquaintance.

Former friends can become the worst of enemies because the sense of betrayal cuts deep. Causes of betrayal vary, but often arise from jealousy, envy or ego. When groups of friends are involved, the social dynamics can get very complicated, especially where love and denial are concerned. In this context, a pretty and brilliant French Marquise became the unlikely nexus that entangled the life and fate of Maupertuis with a group of rivals. Emilie Marquise du Chatelet (1706–1949) was the wife of a wealthy Marquis in the east of France. Her father recognized her mathematical talents at an early age and helped her receive a good education in math and science, unusual for a young woman of that age, with the help of colleagues associated with the French Academie des Sciences. While still a teenager, she became fluent in Latin, Italian, Greek and German and devised gambling strategies that helped provide money to buy her books. She also had a flair for performance, played the harpsichord, sang opera, danced and was an amateur actress. In short, she was an exceptional young woman whom exceptional men found intoxicating.

In 1733, seven years into her marriage to the Marquis and after giving birth to three children, two of whom survived childhood, du Chatelet sought to resume her mathematical studies. She attracted the thirty-five-year-old Maupertuis, at that time a rapidly rising member of the French Academy who was still a few years away from his debut on the world stage with the expedition to Lapland to measure the flatness of the Earth. Du Chatelet and Maupertuis developed a complex relationship of tutor and student, advisor and friend, lover and mistress. The affair was brief but ended amicably. Afterwards, du Chatelet continued her studies with Maupertuis' friend Clairaut with whom he had studied in Basel under Johann Bernoulli and with whom three years later he would travel to Lapland. The notorious French satirist Voltaire (who was forced to flee Paris more than once during these years for fear of imprisonment) circulated through the same social scene, frequenting the same café's and salons as Maupertuis, Clairaut and du Chatelet. Voltaire and Maupertuis became close friends and mutual supporters, sharing a strong affinity to Newtonianism. Maupertuis helped guide Voltaire through the mathematical intricacies of Newton's *Principia* and the problems of gravitation as Voltaire wrote a pointed defense of Newtonian philosophy against Cartesianism. Around this time, Voltaire became du Chatelet's steadfast paramour and partner for the remainder of her short life. They set up house together, apparently with her husband's tacit agreement, in Cirey, which became a hub for the French intellegencia.

When du Chatelet needed of a new tutor, Maupertuis introduced her to his acquaintance Johann Samuel König around 1740, as she was about to publish her first work of physics, her *Institutions de Physique*. The *Institutions* was a remarkable work aimed at bringing the difficult parts of Newton's physics to an accessible level. It ostensibly was intended to instruct her 13-year old son, but it contained original contributions with important observations on early ideas of kinetic energy. One of the most important features of the book was the use of Leibniz' calculus to replace Newton's geometric approach. Its publication created a stir in Parisian society and launched her reputation as the top (perhaps only) female physicist in France. König, ever struggling in someone else's shadow, could not resist taking credit for his association with du Chatelet, and spread rumors in Paris that some of the original parts of the book were his. Because du Chatelet was a woman, these assertions by a man carried more weight than was their due. But she had her own powerful supporters, none less than her steady companion Voltaire, and König's rumors were forgotten as du Chatelet became famous and respected within scientific as well as literary circles. However, ten years later, König would strike again, with more damaging and lasting effects.

After Maupertuis supported König's election to the Berlin Academy of Sciences, König submitted a paper for publication in the Berlin Academy, giving Maupertuis a copy for his preapproval. Whether because he considered König a friend, or because he thought König was harmless, Maupertuis approved its publication without bothering to read it. Only after the paper was in print did Maupertuis become aware that König was asserting that Maupertuis had made a critical error in his development of the Principle of Least Action, casting doubt on the validity of the principle. What is more, König claimed to have seen a letter written by Leibniz in 1707 to the Swiss mathematician Hermann in which Leibniz described an original principal of least action. In a single stroke, König not only accused Maupertuis of making a serious professional mistake, but he stripped him even of his priority for the discovery of the Principle of Least Action, the pinnacle of Maupertuis' career and reputation. Maupertuis was stunned and horrified. While he had had his share of professional battles during his days in Paris, provoking many of them himself and even relishing in them, nothing he had experienced came close to this attack on his very core. His greatest achievement was cast into doubt by a former friend and ally.

As the president of the Berlin Academy, Maupertuis had nearly absolute power and a group of powerful allies, chief among them Euler. He brought this power to bear on this upstart who dared challenge him, placing Euler in charge of an official enquiry into the validity of König's assertions. Euler prosecuted this responsibility with sincerity and integrity. The mathematical error that König claimed Maupertuis had made was investigated by Euler himself, with his deep understanding of variational principles, and found to be baseless. Furthermore, the purported letter from Leibniz was demanded, but König could only provide copies, not the original. König had seen the original in the possession of a Swiss collector of manuscripts, Samuel Henzi. In an ironic twist, Henzi had since been put to death over trumped up charges of treason, and his political enemies had burned all of his papers. The absence of an original letter, as well as the absence of any mention of least action, or any principle even closely aligned with it, in Leibniz' published works, of which there were an amazingly complete and large number, Euler found no supporting evidence for König's claim for Leibniz' priority. One possible conclusion of the König affair was that the letter was a forgery.

Up to this moment, Maupertuis had remained aloof of the affair, appropriately letting his friend Euler prosecute the matter. However, Maupertuis could not restrain his natural tendencies towards lashing out at those who criticized him, and he used his power as president of the Academy to pressure Euler into declaring, as the official finding of the Academy, that König had forged the letter. Of course, there was no proof of forgery, and over a century later the matter remained unresolved. It did become clear, through historical research around the turn of the twentieth century, that all the letters, even those in Henzi's possession, were copies, and several other copies surfaced around that time. However, no original was ever found. It may have been one of the plethora of ideas on mechanics that Leibniz studied, but it never surfaced in his other work, and hence was entirely unknowable to Maupertuis. In the end, Euler issued an official report exonerating Maupertuis as the originator of the Principle of Least Action. The prestige and respect that Euler commanded gave considerable weight to his report. This might have been the end of the matter, if Voltaire had not inserted himself into the fray as the perpetual gadfly that he was and relished being.

Emilie du Chatelet, Voltaire's closest companion whom he loved dearly, died in childbirth at the age of 43 in 1749. The child was neither Voltaire's nor her husband's the Marquis'. Voltaire was still recovering from this loss when Frederick attracted him to come to Berlin and stay at his palace in San Souci. Maupertuis and Voltaire had been friends for more than ten years, and Maupertuis often visited Voltaire and du Chatelet at their retreat in Cirey. However, shortly after Voltaire arrived in Berlin, they began to fall out. Others who knew them remarked on their obvious incompatibility, and Maupertuis' growing pomposity and egotism were precisely the attributes that Voltaire could lampoon. Voltaire was first and foremost a satirist, and the squabble with König provided excellent fodder for his art. Furthermore, Maupertuis had used the considerable power at his disposal to quell the underdog König, and Voltaire's career had been founded by championing the downtrodden.

Voltaire put pen to paper and produced an anonymous pamphlet that harshly ridiculed Maupertuis as well as the actions of the Berlin Academy. Frederick was furious at seeing his Royal Academy ridiculed, and he published his own pamphlet defending it and its president. Voltaire was not to be outdone by a King masquerading as an author, and he published a short satire called the *Diatribe du docteur Akakia* that lampooned the president of the Academy, depicting him as a buffoon. Voltaire did this against Frederick's explicit orders, and used a forged royal permission to have the book printed under Frederick's very nose. Frederick had every copy impounded and burned them all, personally, in his private rooms. He sent the ashes as a "refreshing little powder" to Maupertuis who was then sick in his rooms.[8] However, Voltaire was able to spirit away one surviving copy to publishers he knew in Leiden, and he returned with a stack to Berlin while he sent another shipment to France. The anonymous pamphlets by Voltaire and Frederick, whose identities could not remain anonymous for long, fueled public gossip. The affair soon spread across Europe, fascinating the public at multiple levels: King squabbling with his courtier, tyrants quelling underlings, Prussians versus the French. The feud touched on central tensions of the Enlightenment that pitted absolutist rule and control of information against the free flow of ideas—not unlike issues at play today fueled by the revolution of the Internet.

The aftermath of this battle was a wake of destruction with collateral damage on all sides. Frederick's Academy came away looking biased and second best to the Paris Academy. Frederick appeared closed-minded and dictatorial. König was viewed broadly as an ungrateful and second-rate scientist. The usually apolitical Euler, who wanted only to be left alone with his mathematics, was bloodied. Maupertuis' reputation never fully recovered, and he never again enjoyed the level of personal respect that he so dearly yearned for. The Principle of Least Action, that was to be his greatest contribution to science and philosophy, that was to establish his name for all time in the pantheon of the likes of Galileo and Descartes and Newton, had been irretrievably sullied over a petty squabble. The reputation of Voltaire, alone of the principals, seems to have emerged unscathed.

The arc of Maupertuis' life was a wild ride. He managed his early career with alacrity and with a keen eye on self-promotion. He reveled in political battles, which he usually won. Even before espousing the Principle of Least Action, he was among the elite in the Paris Academy with a rare dual membership in the Academie Francaise. But his ambitions cried out for more. The move to Berlin should have been the apex of his career, placing him at the head of a major scientific institution with almost total control and the favor of a King, but he had over reached. His time in Berlin was marked by many stumbles, not the least of which was his handling of the König affair. Perhaps his greatest error, in terms of how posterity would view him, or rather dismiss him, was what he considered to be his most profound work—he used the Principle of Least Action in his *Cosmologie* to prove the existence of God. This was audacious even in his day, when such theological endeavors were not uncommon and often were linked to science. But today, his proof of God's existence is viewed with scorn, and his status as a serious scientist is seriously questioned. His name is virtually unknown among physics students today, unlike the names of Euler and Lagrange that headline many a chapter in physics textbooks.

Lagrange's View

Sprawling at the foot of the Alps in modern-day northern Italy lays the region of Piedmont, surrounded on three sides by the Italian Alps. On the fourth it gives issue to the river Po that drains to the east in the Adriatic Sea. Piedmont was part of the Roman province of Cisalpine Gaul, where Hannibal first engaged the Roman armies of Scipio in 218 BC after crossing the Alps. By the early eighteenth century, Piedmont-Savoy was an independent kingdom ruled by the Dukes of Savoy. The official language and influence in Piedmont was French, although the dukes considered themselves Italian. Giuseppe Lodovico Lagrangia (later Joesph-Louis Lagrange (1736–1813)) was born into an Italian family in Piedmont the same year that Euler published his *Mechanica*. He was the first of eleven children, only two of whom survived to adulthood. Lagrange was serious and reserved, with a mild almost timid disposition, yet with an intense ability to focus on difficult mental tasks. Although of minor nobility, the family fortune was lost due to speculation by his father, and Lagrange was steered by his family towards a lucrative career in law, although his attachment to the law never stuck. During his studies, two especially able teachers introduced him to mathematics, and, as he showed a striking affinity for the subject, he never turned back. At the early age of nineteen in 1755 he was appointed to a temporary teaching post at the artillery school in Turin where he was the first to teach calculus at an engineering school. As so many teachers have done before and after, he could find no appropriate textbook for his students, so he launched a project to write one. During this project, he thought deeply about the foundations of the calculus, and took the opportunity to add to the subject, giving lectures to his overwhelmed students following his development of variational calculus.

The object of the variational calculus is to maximize or minimize the integral of a function. For instance, the distance along a path between two points on a surface is an integral of the line element ds over the path. A fundamental question is what path minimizes the integral to yield the path of least distance, also known as the geodesic curve between the two points. On a flat plane, the path of least distance is a straight line. On a sphere, the path of least distance is a "great circle" route. Many mathematical problems go beyond geodesic curves that need to find the function that gives extremal (maximum or minimum) values to an integral, endowing variational calculus with a fundamental generality. The master of variational calculus at that time was Leonhard Euler. Euler had written his *Methodus inveniendi* in 1744, which Lagrange read thoroughly. In 1755, Lagrange wrote from Turin to Euler in Berlin to tell him of the progress he was making in the variational calculus, describing a new approach to variational calculus that went beyond Euler's efforts.[9] Euler responded enthusiastically to Lagrange's letter, sending off his reply with high praises for the young (only nineteen years of age) mathematician. Euler stated that Lagrange had taken the theory of maxima and minima nearly "to the peak of perfection."[10]

Between 1756 and 1759 communications were cut off between Turin and Berlin because of the Seven Years War (part of the French and Indian War in Canada) with hostilities between Frederick's Prussia and Austria. When communications were established again in 1759, Lagrange sought to have his papers published in Berlin in the Academy Proceedings, but Euler wrote back telling him of Maupertuis' death, and of the inability of the Academy to undertake publication at that time. Lagrange and several colleagues had recently formed their own Academy in Turin, and so Lagrange published his papers on the variational calculus and the principle of stationary action in the first proceedings of the *Miscellanea Taurinensea* 1760–1761. Given his previous support of Maupertuis' Principle of Least Action in his letters to Euler, his published paper on the principle of extreme action is surprisingly mute on Maupertuis or any teleological minimization principle. He also put off an invitation from Euler to leave Turin and to join the Academy in Berlin, expressing hesitation to be in Berlin while Euler was still there. Some historians have speculated that Lagrange felt badly used by Euler when publication in the Berlin Academy was denied. It is also possible that Lagrange had by this time rejected Maupertuis' teleological least action principle and would have felt awkward interacting with Euler, who had been one of Maupertuis' staunchest supporters. This obstacle was removed when Euler returned to St. Petersburg in 1766, and Lagrange promptly moved to Berlin to take over Euler's position as the director of Mathematics in the Berlin Academy.

Lagrange spent twenty years in Berlin, never to return home to Turin. These were the most fertile years of his life, which saw a large number of papers published in the proceedings of the Berlin and Turin academies. Most notably, it was during this time that he wrote his monumental *Mechanique Analytique*. He worked with great effectiveness and increasing maturity. While he was still in Turin, he had considered how to construct analytic mechanics based upon the Principle of Least Action, and one of his early papers was on the principle. However, in Berlin he realized that the Principle of Least Action, even if expanded to include maxima as well as minima, was not sufficiently general. It had limitations, and his goal was to begin with first principles that encompassed the greatest generality to tackle the widest possible range of physical problems.

As he dug deeper into the origins of the Principle of Least Action, Lagrange uncovered an old principle that Johann Bernoulli had called the principle of virtual velocities in 1717 in a letter written to the French mathematician Pierre Varignon. Varignon was the author of *Projet d'une nouvelle méchanique* (1687) (Project for a New Mechanics) that was published the same year that Newton published his *Principia*. Varignon's *Project* was a treatise on static equilibrium in which he used the composition of forces. It was modern for its day and paralleled some of Newton's work on forces. In the principle of virtual velocities, Lagrange found a fundamental starting point—it was a first principle. However, the principle of virtual velocities only applied to problems of static equilibrium, while Lagrange sought to construct a dynamic theory. To accomplish this, he invoked an additional principle first expressed by D'Alembert in his *Traité de dynamique* (1743) (Treatise on Dynamics).

Jean le Rond D'Alembert (1717–1783) was a central player in the French Enlightenment. He is best known today for his collaboration with Denis Diderot on the great Encyclopédia that sought to record for posterity the current state of the sciences, arts, and crafts. D'Alembert's introduction to the Encyclopédia is considered one of the finest examples of French prose as well as the clearest proposition of the ideals of the Enlightenment. His name is mentioned in the same breath as Voltaire and Rousseau by humanists, and in the same breath as Euler and Lagrange by physicists. But, unlike Maupertuis, ambition and fame did not seem to have been his major motivation—he was driven more by an underlying love of the life of letters and ideas.

D'Alembert began life as a bastard child abandoned by his mother on the steps of a church in Paris when only a few days old, but later was retrieved by his father who set him up in a foster home. The foster home turned out to be a loving and nurturing environment for the precocious lad, and he flourished. In school, he was gripped by a love of mathematics that he first discovered in a book of lectures by Varignon. He drifted through studies of law and passed the bar, but this did not engage him, so he returned to his love of mathematics and classics, and he taught himself all that he could. D'Alembert was a natural autodidact, and he submitted two papers to the Académie des Sciences at the age of twenty-two. One was on mathematics, and the other was on the physics of fluids. They were modest papers, but caught the attention of important French scientists such as Clairaut, and he was soon elected to the Académie in 1741.

By the late 1740s, D'Alembert had become known among the salons of Paris as a man of wide-ranging talent. In addition to his interests in mathematics and physics, he was a master of prose and a Latin scholar, and he held his own as a philosophe. Around this time, Denis Diderot was involved in constructing a great encyclopedia that would capture all that was known, and categorize all of knowledge into a series of volumes. He enlisted D'Alembert to edit the sections on science and math, and D'Alembert contributed his own articles numbering over a thousand. The Encyclopédia was the epitome of the Enlightenment, but the Age of Reason was not entirely a hospitable home for it.

The French Enlightenment was the revolution before the Revolution. The royal excess and bureaucracy of France were collapsing under their own weight, while the grip of the Church on behavior and thought was strangling the progress of ideas. Part of the mission of the Encyclopédia was to break free from the mythology and superstitions that imprisoned the greater part of the population, and the Church was an implicit target. Diderot and D'Alembert were not openly hostile to the Church, but the underlying philosophical currents in many of the articles went against religion. When the first volume of the Encyclopédia was published in 1751, it needed to pass the censors, and luckily, the head censor at that time was supportive of the philosophes. However, as additional volumes appeared, and several took more anticlerical tones, the Church used its influence with Louis XV to suppress the publication and to declare the whole enterprise seditious. The charge of sedition made the Encyclopédia more popular than ever, and sales soared, but sedition in that age was punishable by death, and D'Alembert anticipated his arrest at any time, fearing for his life. Although he was never arrested, the prospects for it hung over him like a blade. He was ultimately forced to resign from the Encyclopédia after he wrote an article for the seventh volume in 1757 that criticized the Calvinists in Geneva. Although the Church in France had no love of the Calvinists, the article questioned belief in Christ, which stepped over the line. Even D'Alembert's friends, fearing for his safety, advised him to recant, which he did before he resigned. When he later died in 1783, the Church had its revenge—as a nonbeliever he was buried in an unmarked common grave.

D'Alembert made two monumental contributions to physics. One was his study of the physics of music and vibrating strings in which he introduced the first wave equation into physics. The wave equation is a partial differential equation that goes beyond the simple ordinary differential equations that were the topic of physics at that time. The second-order differential operator is known today as the D'Alembertian operator. The more important contribution, published in 1743 at the age of twenty-six, was his *Traité de dynamique* (Treatise on Dynamics) in which he developed his own equations of motion. Central to D'Alembert's approach to physics was a distrust of the concept of absolute forces. Even though forces were central to Newtonian physics, and the force of gravity was well defined, most other forces were difficult or impossible to study. For instance, rigid bodies hold their shape through internal forces that are inaccessible to experiment, and even forces between bodies are hard to quantify. Therefore, D'Alembert sought to define a framework within which any force of any kind in a system of bodies could be defined in terms of the effects on the system. This led him to propose what is known today as D'Alembert's Principle.

D'Alembert's Principle is, at first sight, a tautology, but it is subtle, and its importance to physics runs deep. We are all familiar with Newton's second law $F = ma$. This law states that if a force F acts on a mass m, then the mass will accelerate with acceleration a. The problem with this, in D'Alembert's view, was that a system of many masses shares a plethora of forces, many of which cannot be known directly. How then does one treat the system? His answer was to balance force with the inertial response of the system. D'Alembert's Principle is expressed symbolically as $F - ma = 0$. The subtlety in D'Alembert's Principle is that he places forces and their inertial effects on an equal footing. It is no longer necessary to speak of external forces and inertial forces as different quantities. This principle treats fictitious forces, like centrifugal force in circular motion, as real forces. In this view, it is impossible to tell the difference between an external force and a fictitious force, and D'Alembert's Principle anticipates Einstein's Equivalence Principle, in which it is impossible to tell the difference between an accelerating reference frame and a gravitational field. D'Alembert's Principle was not particularly useful for solving problems, but it removed the need to track every action-reaction pair (from Newton's third law) in a complex system. Action-reaction pairs simply cancel because of the constraints imposed by the rigid shape of an extended body, providing conceptual simplification of complicated problems.

Lagrange was well acquainted with the established works of physics of his day, and as he treated the Principle of Least Action, beginning with Bernoulli's static equilibrium approach of virtual velocities, he arrived at an equation that he recognized as D'Alembert's Principle. In Lagrange's derivation, D'Alembert's Principle had balanced forces by their inertial response to generate a type of dynamic equilibrium—in other words, statics! Therefore, dynamic problems could be converted into an equivalent static equilibrium and thus be addressed by the principle of virtual velocities. This was an epiphany for Lagrange, because it meant that all of known physics could be derived by combining two fundamental principles: the principle of virtual velocities and D'Alembert's Principle. This was the turning point, and while in Berlin, Lagrange began work on his great treatise of analytic dynamics, building the entire edifice of physics upon these two principles.

In the midst of his studies, in 1786, Frederick II died, and the erstwhile vibrant intellectual life in Berlin collapsed under the less enlightened rule of Frederick's successor Frederick Wilhelm II. Lagrange moved from Berlin to Paris, carrying his finished manuscript with him. The *Mechanique Analytique* was published in Paris in 1788 around 100 years after the publication of Newton's *Principia* (see Figs. 4.3 and 4.4). It was the culmination of a century of work by many brilliant scientists and mathematicians, among them Newton, Leibniz, three Bernoullis (Jakob, Johann and Daniel), Euler, Maupertuis, D'Alembert, Clairaut, and du Chatelet, each supplying their piece to the puzzle that Lagrange finally assembled. The ideas early in the century were vague, especially concerning the nature of material forces, while concepts of energy and work were even more tenuous. As the century progressed, understanding deepened and profound results were discovered, like Least Action. Finally, in the hands of Lagrange, these all were brought together under

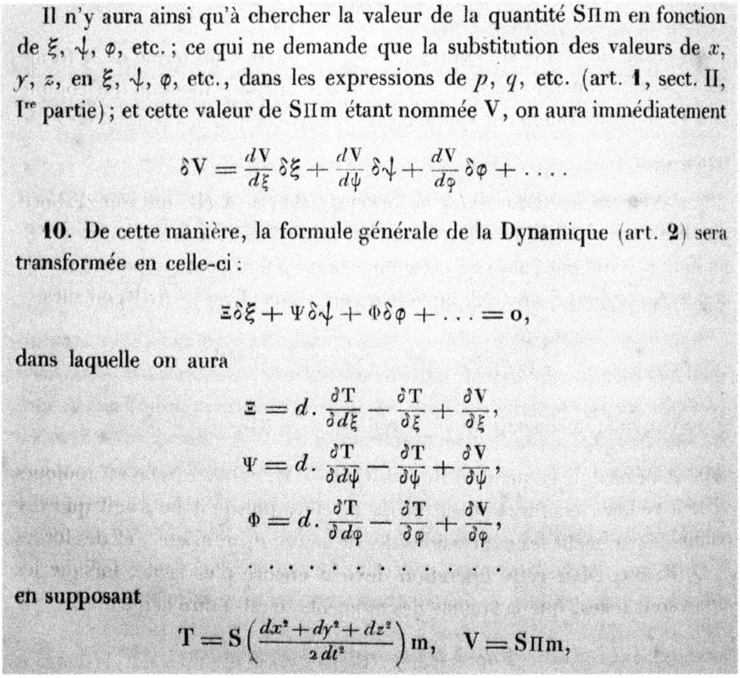

Il n'y aura ainsi qu'à chercher la valeur de la quantité SΠm en fonction de ξ, ψ, φ, etc. ; ce qui ne demande que la substitution des valeurs de x, y, z, en ξ, ψ, φ, etc., dans les expressions de p, q, etc. (art. **1**, sect. II, Ire partie) ; et cette valeur de SΠm étant nommée V, on aura immédiatement

$$\delta V = \frac{dV}{d\xi}\delta\xi + \frac{dV}{d\psi}\delta\psi + \frac{dV}{d\varphi}\delta\varphi + \ldots .$$

10. De cette manière, la formule générale de la Dynamique (art. **2**) sera transformée en celle-ci :

$$\Xi\delta\xi + \Psi\delta\psi + \Phi\delta\varphi + \ldots = 0,$$

dans laquelle on aura

$$\Xi = d.\frac{\partial T}{\partial d\xi} - \frac{\partial T}{\partial \xi} + \frac{\partial V}{\partial \xi},$$
$$\Psi = d.\frac{\partial T}{\partial d\psi} - \frac{\partial T}{\partial \psi} + \frac{\partial V}{\partial \psi},$$
$$\Phi = d.\frac{\partial T}{\partial d\varphi} - \frac{\partial T}{\partial \varphi} + \frac{\partial V}{\partial \varphi},$$
$$\ldots\ldots\ldots\ldots\ldots$$

en supposant

$$T = S\left(\frac{dx^2 + dy^2 + dz^2}{2dt^2}\right)m, \quad V = S\Pi m,$$

Fig. 4.3 Lagrange's derivation of his famous equations, and the definition of potential.[11] In Lagrange's notation, *T* and *V* stand for kinetic energy and potential energy, respectively. The symbol *d* stands for the time derivative *d*/*dt*, and *S* stands for summation. The Greek capital Π stands for the potential function. He uses the phrase *etc.* explicitly to show that the number of generalized coordinates is not restricted to three. Although he eschewed geometry, he was taking the first steps into higher dimensional dynamical spaces

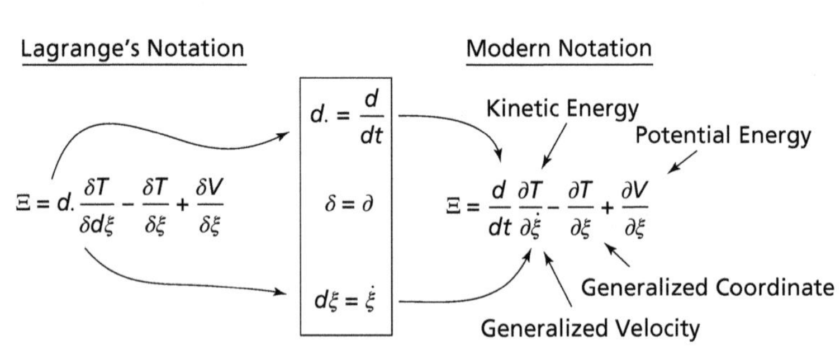

Fig. 4.4 Correspondence between Lagrange's notation in 1788, and modern notation used in today's textbooks. The "dot" on the generalized velocity is carried over from Newton's fluxions

a single theory that allowed the known results of the day to be derived from first principles.

The Fourth Dimension and Beyond

Despite the demotion of the Principle of Least Action to a secondary principle by Lagrange, it continued to intrigue scientists and mathematicians. Lagrange himself devoted a section of the *Analytique* to the derivation of the Principle of Least Action from his analytical theory. The principle ultimately came to be understood as a consequence of dynamics rather than a cause. It could be invoked, under allowable circumstances, to provide a fresh viewpoint for understanding the behavior of a physical system. Furthermore, the action integral only needed to be an extremum rather than a strict minimum. Therefore, the efficiency of the principle disappeared, and it became the principle of stationary action, where the phrase *stationary* means that slight deviations in the dynamical path keep nearly the same values for the action integral.

A young Richard Feynman, as a graduate student working on his PhD thesis under the direction of John Wheeler at Princeton University, resurrected the principle of stationary action for quantum systems. What Feynman discovered was that classical trajectories were the limiting behavior of quantum systems whose many paths interfered constructively and destructively, depending on the value of a quantum action integral. For instance, around the stationary configuration of the action integrals, the phases of nearby quantum trajectories add together constructively. Therefore, classical trajectories are the paths of maximum constructive quantum interference along the paths of stationary action. This astounding view has led to a wide range of philosophical discussions on the nature of trajectories in quantum systems, which will be explored in Chapter 8.

A singularly important invention in the *Analytique* was Lagrange's use of generalized coordinates. In previous work, Cartesian coordinates were used to describe position (x, y, z) as each body was located in a physical three-dimensional space. Likewise, the same bodies were acted upon by forces that were broken down into three Cartesian components. Calculating the resulting motion meant keeping track of all components for each separate body, and in the end, each body executed its own separate trajectory through 3D space. Lagrange dispensed with this approach and instead conceived of generalized coordinates, and with them, he invented configuration space. Generalized coordinates could have any number and could be chosen in the best way to simplify the description of the system. Though they formed the coordinates of a new multidimensional configuration space, generalized coordinates were an important step away from geometry and into the heart of algebra. Lagrange was proud of the fact that he did not use a single diagram in his treatise. For him, generalized coordinates were algebraic quantities rather than geometric, so there was no need to try to visualize the resulting trajectory in a high-dimensional space. At that time, there was no conception of any space other than ordinary 3D space, so a high number of coordinates made no geometric sense.

However, this demotion of geometry did not last long. Within a few years the principle of stationary action attracted the attention of Karl Jacobi, one of the towering figures in the history of nineteenth-century physics, who recast stationary action from a dynamical principle back into a geometric principle, generating the pure geometrical trajectory of a dynamical system through space with no need for a time variable. Jacobi's Principle of Stationary Action was the first decisive step towards the geometrization of physics, which ultimately led to Einstein's vision of the trajectories of suns and planets as geodesic curves through warped space-time geometries as gravity faded into the background as just another of D'Alembert's fictitious forces. To understand what Einstein saw, our minds must slip the surly bonds of 3-space and expand into the fourth dimension and beyond.

5

Geometry on my Mind

Everyone knows what a curve is, until he has studied enough mathematics to become confused . . .

FELIX KLEIN[1]

Motion takes place in space—trajectories are the tracks of mathematical points through it. But what is it? Since the time of Euclid, the properties of space have been defined in relation to geometric objects such as points, lines, planes and solid figures. This progression of objects steps up in dimension by integers starting with a point in zero dimensions. A point translating through space creates a line in one dimension. A line translating transversely to its length creates a plane in two dimensions. And a plane translating transversely to its surface creates a volume in three dimensions. No fourth dimension could be envisioned by Euclid's geometry. Geometry was prevented from expanding into higher dimensions by the lack of a language. This changed in 1637 when the French Philosopher René Descartes adapted algebra—the language of numbers—to treat geometry in his essay entitled *La Geometrie.*[2]

The combination of algebra with geometry gave birth to the new mathematical field of analytic geometry (also known as coordinate geometry), opening the door to mathematical developments such as the infinitesimal calculus by Leibniz and Newton, and the variational calculus by Euler and Lagrange. The high number of generalized coordinates required to capture even simple systems led Lagrange to deny geometry any pertinent role to play in rational mechanics, but within a few decades Joseph Liouville (1809–1882) and Karl Gustav Jacob Jacobi (1804–1851) developed a new multidimensional language to transform among many coordinates, though they did not realize it. In England, one of Britain's most prolific mathematicians of that age, Arthur Cayley (1821–1895), took up the multidimensional techniques of Liouville and

Galileo Unbound. David D. Nolte, Oxford University Press (2018).
 DOI: 10.1093/oso/9780198805847.001.0001

Jacobi and made a decisive step in 1843 by calling them for what they were—geometric coordinates of four dimensions and beyond.

This progression to increasing dimension merely scratched the surface of what would become possible in the future of geometry. Flat spaces would become curved, tangible spaces would dissolve into abstract, and integer dimensions would shift into fractional. Curved, fractionally abstract spaces enable new types of dynamical trajectories to be defined, from orbits around black holes (Chapter 7) to stock prices in world markets and the slow drift of evolutionary change (Chapters 10 and 11), from quantum trajectories (Chapter 8) to strange attractors in deterministic chaos (Chapter 9). Armed with the new geometry, generations of pioneers possessing a universal language, invented new sciences as the concept of trajectory expanded far beyond Galileo's simple parabola.[3]

Manifolds from Heaven

Sometimes the most divine glories of human thought are inspired by the basest earthbound needs. For instance, the birth of the mathematical theory of surfaces, by the French military engineer Gaspard Monge (1747–1818), was born of the art of war through the need to understand fortifications. Similarly, the birth of the mathematical theory of intrinsic curvature, known as differential geometry, grew from the geodetic needs of Carl Friedrich Gauss (1777–1855) to map the Kingdom of Hanover. Gauss was born into extremely poor circumstances in Braunschweig (Brunswick) Germany. His mother was illiterate, and his father was a crude and uneducated man who was severe and openly antagonistic towards his son's education, even after, or especially because of the recognition of Gauss' talents. By the time he was three years old he could already do complex mathematical calculations in his head, without being taught. Gauss used to tell the story, later in life, that he learned to figure before he learned to talk. His saviors in this numbing environment were his mother, who doted upon him, and her brother, who may have been a genius himself. He recognized Gauss' abilities at an early age and engaged with him, setting him puzzles and mind benders in which Gauss delighted. With great hope, Gauss was sent to a dreary school with a surly schoolmaster who loathed his hapless students—until the day when Gauss arrived. Even surly schoolmasters can sometime recognize genius, and he became a mentor, encouraging Gauss in his mathematical interests.

The reputation of Gauss spread so widely that Ferdinand Duke of Brunswick heard of him, and he provided Gauss with a stipend to attend the Collegium Carolinum and then the University of Göttingen that had been founded in 1734 by King George II of Great Britain who was also the Elector of Hanover. Gauss received his degree in 1799 with his famous thesis *Disquisitiones Arithmeticae* that established the foundations of modern number theory. After graduation, Gauss remained under the patronage of Duke Ferdinand, freeing him to study and compute as his will desired. His facility with mental calculations were bolstered by new numerical techniques that he invented, such as the principle of least squares, allowing him to calculate the trajectory of a new planet, Ceres, discovered by Giuseppe Piazzi on 1 January1801. Piazzi only had time for a few dozen nights of observations before Ceres became too dim as it passed behind the Sun. There was great uncertainty about when and where it would reappear. Gauss performed extensive numerical calculations—for him a delight—and predicted where to look for the new planet. On 31 December 1801, Ceres was found again precisely where Gauss had predicted.

Life, for once, was good to Gauss. He was making a name for himself. He had financial freedom. He married Johanna Ostoff in 1805 with whom he had his first child Joseph. Then, in 1806, Napoleon's armies swept like a wave of chaos across Europe. His friend and benefactor Ferdinand, Duke of Brunswick, at the age of seventy, was placed at the head of the Prussian army in a last-ditch effort to fend off Napoleon. The attempt failed at the battle of Jena, where Ferdinand was mortally wounded. Gauss was looking out the window of his little house on the main highway in Braunschweig when the ambulance carrying his dying friend passed by. His days in Braunschweig were over, but with his fame still ascendant, in 1807 he accepted the position as director of the observatory at Göttingen.

The town of Göttingen is on the southern border of what was then the Kingdom of Hanover. In 1811, Gauss was charged with making the first geodetic survey of the Kingdom, an activity that occupied him for more than ten years. He was struck by the mathematical difficulties of creating a two-dimensional map of a three-dimensional topography. Monge had earlier developed an approach for describing surfaces that relied on a coordinate system that was attached to an underlying flat plane. Gauss realized that the curvature of the Earth made this approach only approximate, and that a coordinate system could be attached to the surface to better capture the geometric description of the hills and valleys of the Kingdom of Hanover. This insight launched Gauss on a program to find a formal approach to the description of surfaces that was freed from the Euclidean plane, and even from Euclidean geometry.

In 1827, Gauss published his *Disquisitiones generales circa superficies curvas* (General Investigations of Curved Surfaces).[4] It contained several theorems that proved that one could determine the curvature of a surface even if all measurements of angles and distances are confined to the surface without recourse to, or even knowledge of, the three-dimensional space in which the surface was embedded. For instance, a geodesic triangle on a spherical surface is constructed by starting at a point and proceeding along a "straight" line that, on a sphere, is a great circle route. At some point along this line, turn through an angle and proceed again along a new great circle route. Then at some point along this second segment, turn and proceed along a great circle route that intersects the starting point. The result is a triangle composed of three segments of great circle routes. One of Gauss' theorems states that, if you measure and sum the angles of each of the three vertexes of the triangle, there is an excess above the number π that is equal to the total curvature or the sphere times the area of the triangle.[5] Aspects of this result had been known previously, but Gauss defined the total curvature (known today as Gaussian curvature) and generalized the result to find the curvature of surfaces that were non-spherical.

Of greater importance, inaugurating modern differential geometry in his work of 1827, was a theorem that Gauss called the *Theorema Egregium* (Remarkable Theorem). This theorem introduced the idea that the properties of a surface could be defined solely using the local properties of the surface without needing to extend a significant distance along a line segment. Gauss defined an infinitesimal line element on the surface that he called the first fundamental form[6] where the surface has two coordinates that define locations of points on the surface. If the surface is flat, then the Gaussian coordinates are the same as Cartesian, but for a general non-flat surface, the Gaussian coordinates can take any form that captures the surface. By calculating the first fundamental form, the curvature of the surface at that point can be determined uniquely.

The idea that local measurements can capture intrinsic properties of a surface, without regard to the embedding space, was a fundamentally new concept. Curvature had previously been viewed in terms of the radius of a virtual large sphere whose surface was tangent to the surface of interest at a specified point. This is geometry in the large—the global properties, or large-scale properties, of a geometric object. What Gauss had introduced was geometry in the small—geometry depending only on local properties without needing to invoke an extended part of the geometric object. For instance, geodesics had previously been seen as the line that had the "shortest" distance between two separated points—geometry in the large. With Gauss' new viewpoint, a geodesic could be seen as a "straightest" trajectory that moves every point in a steady direction without needing to aim at some far target—geometry in the small. This new viewpoint would become an essential part of Einstein's theory of General Relativity, seeing complex trajectories as simple geodesics through complex geometry. But first, the very notion of geometry needed to be defined and generalized and expanded.

Georg Friedrich Bernhard Riemann (1826–1866) was born as poor as Gauss and possibly as brilliant. It may be hard to think of one of the towering figures of the field to be a shy and reserved boy, but everyone has to grow into their calling. Riemann was helped by many sympathetic teachers. The story of his schooling is the lucky story of a succession of mentors taking a keen interest in a keen mind. He was able to master mathematics at any level, devouring a 900-page treatise on the theory of numbers by Legendre in only a week when he was still in high school. By the time he appeared at the University of Göttingen, he was ready to commit his life in service to mathematics. Despite Gauss' presence at Göttingen, the level of mathematics education there was not high, so he transferred to the University of Berlin where he was able to take lectures from some of the towering figures of geometry—Steiner, Jacobi, Dirichlet and Eisenstein. These are names in the history of mathematics that stand out even today, and it is near miraculous to find them all at the same place at the same time. Riemann appears to have overcome some of his shyness in Berlin, for when revolution swept across Europe in 1848, he was briefly a member of a loyalist student group, spending a nervous night at the royal palace guarding the nervous person of the king, Friedrich Wilhelm IV, lest marauding mobs stormed the palace. Luckily for the king, and for the future of mathematics, they didn't.

To obtain his doctorate in mathematics, Riemann returned to Göttingen in 1849 to study under Gauss and Wilhelm Weber (1804–1891). He had a surprisingly good talent for experimental science, spending a good deal of his time on experimental physics with Weber. His mathematical activities focused on mathematical physics because he was an intuitive thinker and was grounded by a pragmatic approach to phenomena. This allowed him to dispense with the distractions of details and see a phenomenon in its naked form. This was part of his genius, as it had been Galileo's. The other part of his genius was an obsessive compulsion for perfection that allowed him to produce works that even the always-critical Gauss praised as perfect.

In November of 1851, Riemann finished his doctoral dissertation on the *Foundations for a General Theory of Functions of a Complex Variable* and presented it to Gauss for his consideration. Gauss' report to the faculty glowed with unusual praise for a student. He stated that Riemann had a "truly mathematical mind, and of a gloriously fertile originality. The presentation is perspicuous and concise and, in places, beautiful."[7] Riemann successfully defended his thesis at a public event (never an easy prospect for the ever-shy Riemann), and he hoped to be able to begin teaching at Göttingen.

To become an instructor in the position of *Privatdozent* (without pay, relying on tips and fees from students), it was necessary to prepare a *Habilitationsschrift* (probationary essay) followed by a probationary lecture presented to the faculty. Riemann began working on his essay, but was drawn more and more into problems of theoretical physics rather than mathematics. He spent so much time on his physics pursuits that three years passed before he was finally able to complete his probationary essay on the use of complex trigonometry for the representation of functions. Following custom, a candidate would suggest three topics on which to prepare the public lecture. Riemann had thoroughly prepared for his first two topics on complex trig and on aspects of the mathematical theory of electromagnetic phenomena that he had been working on with Weber, but on an apparent whim he suggested a third topic on the foundations of geometry. He had toyed with this idea, but had never worked it through. Following custom, the faculty advisor would pick the first or possibly the second topic, but almost never the third. Gauss picked the third, and Riemann threw himself so thoroughly into the task that he became exhausted, withdrawing temporarily from his university activities. Finally pulling himself together, Riemann worked furiously for seven weeks following Easter of that year and was ready to ask Gauss for a time to schedule the lecture. Gauss originally suggested waiting until the fall, but at the last minute scheduled the lecture for the next day.

Riemann's lecture on the 10 June 1854 *Über die Hypothesen, welche der Geometrie zu Grunde liegen* (Over the Hypotheses which underlie Geometry[8]) was epic. The lecture contained few equations, but the concepts were groundbreaking, changing the way that we look at space. Geometry, Riemann said, is not restricted to our physical space, but consists of manifolds of ordered numbers! The word manifold comes from *Manigfältigkeiten*, a word coined by Gauss in 1851,[9] but Riemann expanded the concept, endowing it with a minimal set of properties from which all else could be derived. His theory came in three parts. First, manifolds were just ordered collections of *things*. For example, consider a set of 3-tuples (x_1, x_2, x_3), where each entry is able to vary in a regular sequential manner. This might be Cartesian 3-space, where each entry is the (x, y, z) coordinates of a point. However, it is also possible that these entries could be measurements, such as temperature or pressure. Or they could be colors, or enumerated elements of a class—it did not matter; they are all manifolds subject to geometric study. Second, Riemann defined a distance operation on this manifold. Because the entries vary in a regular manner, Pythagoras' theorem would apply for sufficiently small excursions, providing a squared distance known as a quadratic form. In 3-space, the squared infinitesimal line element is given by the familiar Pythagorean theorem, as in Fig. 5.1. Third, using the coefficients in the quadratic form, Riemann showed how to derive the intrinsic curvature of the manifold. For the properties of distance and curvature, the earlier work of Gauss was a special case limited to surfaces embedded in three dimensions. However, in Riemann's new system, much broader fields were opened up to geometric study, able to have arbitrarily large numbers of dimensions with coordinates of any kind.

The impact that Riemann's work had on number theory and function theory was considerably greater in his lifetime than his work on geometry and topology. Riemann never published his lecture on the foundations of geometry in his lifetime. It finally appeared in 1868, two years after he passed away of tuberculosis on a trip to Italy, through the efforts

Riemann's Metric

$$ds^2 = \sum_a \sum_b g_{ab} dx^a dx^b$$

In 3-space, this is ...

$$ds^2 = dx^2 + dy^2 + dz^2$$

In Minkowski Space-Time, this is ...

$$ds^2 = dx^2 + dy^2 + dz^2 - c^2dt^2$$

Fig. 5.1 Riemann's metric describing the length of a line element in a metric space (using a notation that came only later with the work of Ricci-Curbastro, but this captures the impact of Riemann's proposal). The metric in 3-space is the familiar Pythagorean theorem. The metric in Minkowski's 4-dimensional space-time can take on negative values

of his friend Richard Dedekind (1831–1916). The question naturally arises to ask how Riemann came to make such an impact on geometry if none of his works on the topic were published in his lifetime? The answer is that his ideas on geometry propagated "in the background" through contacts with colleagues. His new definition of geometry and his tools for studying new spaces spread quickly and were especially applicable for non-Euclidean geometries invented by Gauss, Bolyai and Lobachevsky. For instance, the unusual properties of hyperbolic geometry, caused by its intrinsic curvature in contrast to flat Euclidean spaces, were easily and naturally captured by Riemann's formalism.

Expansion into multiple dimensions was a natural outcome of Riemannian geometry. For instance, Riemann was visited in 1858 by the Italian geometer Enrico Betti (1823–1892) with whom had struck up a friendship and whom he visited in Italy in 1863 as his health began to fail and he sought respite in warmer climates. Their correspondence and friendship established a common ground between them. Modern terminology for n-dimensional spaces first appeared in "On spaces of an arbitrary number of dimensions (*Sopra gli spazi di un numero qualunque di dimensioni* (1871))[10], where Betti refers to the n-ply of numbers as a "space." Camille Jordan (1838–1922) adopted this terminology a year later in "Essay on the geometry of n-dimensions" (1872). In Germany, at the University of Erlangen in 1872, a young Felix Klein (1849–1925) broadened Riemann's concepts of space into a search for classes of different types of spaces, enlisting group theory (also known as Galois theory) to define how the different types of geometries should be classified. Klein's approach became known as the Erlangen Programm and created the scaffold for much of subsequent geometric research. Another important step in the "resurrection" of Riemannian geometry occurred when William Kingdon Clifford (1845–1879) in England translated Riemann's *Grundlage* in 1873, contributing an introduction that helped popularize Riemann's ideas to a broad English-speaking audience. Work in n-dimensional geometry became commonplace by the mid-1870s with work by Klein and Jordan and others. This was also the time when critics emerged, asserting that n-dimensional spaces were meaningless, as enthusiasts performed "experiments" on the fourth dimension. Concepts of higher dimension had become part of cultural literacy by the middle of the next decade.

A measure of how far these ideas had propagated into the commonplace is the publication in 1884, by Edwin Abbott, of *Flatland*; the popular account of how intelligent beings might sense higher dimensions. At the beginning of the twentieth century, Poincaré wrote:

> The geometry of n dimensions studies reality; no one doubts that. Bodies in hyperspace are subject to precise definition, just like bodies in ordinary space; and while we cannot draw pictures of them, we can imagine and study them.[11]

Poincaré, as usual, was prescient, endowing the multiple dimensions of hyperspace with a reality that had relevance for the real world. Within a few years, the trajectories of particles moving with speeds approaching the speed of light would require a fourth dimension for their description, and Riemann's manifolds provided the new fabric of space-time. Then, as Einstein grappled with the problem of generalizing relativity theory to non-inertial frames, Riemann's manifolds again were the foundation for a radical new viewpoint of warped geometries and the geodesic curves that thread through them, as we will explore in Chapter 7.

Yet, Riemann's manifolds were not the end of the abstract expansion of concepts about space and geometry. The next step occurred when Guiseppe Peano and David Hilbert conceived of abstract algebraic spaces where the coordinate axes of a manifold were replaced by mathematical functions spanning what, today, we call a Hilbert space. These abstract spaces are at the heart of quantum mechanics and quantum trajectories.

The Cat that Walks Through Walls

Hermann Günter Grassmann (1809–1877), a self-taught mathematician born in Stettin, Prussia, was stuck in a rut teaching second-rate students at a second-rate secondary school in the same position his father had held before him. Grassmann yearned for greater things, aspiring to create a grand new edifice of mathematics that would become like the universal language that Leibniz' himself had sought to define but had failed to achieve. Such delusions of grandeur are seldom realized. Oddly enough, Grassmann succeeded. Unfortunately, no one realized it in his lifetime.

Several years after taking up the teaching post in 1840 at the gymnasium (the German college preparatory high school that students attend for eight or nine years before entering the university), Grassmann was still teaching the entry-level courses to the lower-division students. To be allowed to teach the senior students required him to pass exams as well as submit a short essay on a technical topic. He chose to write a thesis on the physics of tides. Despite having no formal training in mathematics, he had been dabbling with strange new notations of his own invention, concepts that he called *forms* with the novel property that they denoted spatial orientations. To explain the tides, he began with the theories of Lagrange and Laplace and rewrote them using his oriented forms, depositing a dense 200-page thesis *Theorie der Ebbe und Flut* that was as original as it was long. Of course, the examination committee had no clue what they had received, but they passed him anyway, and allowed him to teach the senior courses at his high school.

Despite the lackluster reception of his ideas by middling examiners, Grassmann knew that he had created something extraordinary. He used his thesis on the tides to launch a major mathematical work on his theory of forms that could be extended to arbitrary degree. He published *Die Lineale Ausdehnungslehre, ein neuer Zweig der Mathematik* (Linear Extension Theory, a New Branch of Mathematics) in 1844. What Grassmann had invented were vectors! His *Ausdehnungslehre* introduced vector analysis, including vector addition and subtraction, vector differentiation, and vector function theory. His conception of vectors was entirely abstract, but he used concrete examples from geometry to illustrate his ideas. These examples took points into lines, and lines into areas, and areas into volumes, and volumes into the next higher form and beyond. Because the system was general and abstract, the higher geometric forms did not require visualization, but the rules for their construction were unambiguous. In this way, it was natural to imagine a multidimensional space of arbitrarily large dimension. Grassmann noted "One cannot here go further than up to three independent directions (rules of change), while in the pure theory of extension their quantity can increase up to infinity."

Geometry at that time was still rooted in Euclid's axioms, while Grassmann's program was creating an entirely new set of axioms upon which to build a logical structure. In 1844, there was nothing else like this. Consequently, Grassmann's *Ausdehnungslehre* was a flop. Few books were sold, and the remainder was torn up for scrap paper. Where Grassmann had hoped that his achievement would open doors to a professorship at a university, he only was able to advance to the position of "professor" at his secondary school in Stettin, where he remained for the rest of his career. In a final attempt to resurrect his theories, he completely rewrote the *Ausdehnungslehre* and published a second edition in 1862 at his own expense. He gave it a new title: *Die Ausdehnungslehre: Vollständig und in strenger Form bearbeitet* (The Theory of Extension, Thoroughly and Rigorously Treated), but this work fared no better.

Receiving little recognition in mathematics, Grassmann turned his attention to the study of languages. Ironically, he made a name for himself in the academic arena of historical linguistics when he demonstrated that the German language had phonetic constructions that were older than in Sanskrit. Up to this time, it had been thought that European languages were descended from Sanskrit. Grassmann showed that Sanskrit and German each derived from an older language—some older form that came to be known as Indo-European. This was a major achievement, and he received moderately wide acclaim, being elected as a member of the American Orientalists' Society and receiving an honorary doctorate from the University of Tübingen. By the time Grassmann died in 1877 he was no longer involved in mathematics or geometry, having turned away from the many disappointing years he had spent on his new branch of mathematics. This may have been the end of Grassmann's dreams, but it was not the end of Grassmann's work, which was destined to extend far beyond the ordinary three-vectors of three-dimensional geometry.

When his oldest son Justus Grassmann went to study mathematics at Göttingen in 1869, he took a copy of his father's *Ausdehnungslehre* to show to his professors, two of whom, Stern and Clepsch, took an interest, sharing it with Felix Klein as he was then gathering his thoughts on his Erlanger Programm of 1872. The year after Grassmann's death there was sufficient interest in his theory of extension that a second edition of the first *Ausdehnungslehre* was published, even though stacks of the second *Ausdehnungslehre* had never sold. Gibbs bought a copy around this time, and although he had already invented his own form of vector calculus that, with work by Heaviside, would be adopted universally as the modern tool it is today, Gibbs did not see or appreciate the broader context of Grassmann's theory of forms, within which vector analysis was just one small part. It took a pure mathematician, the Italian mathematician Giuseppe Peano, to recognize what Grassmann had truly achieved.

Giuseppe Peano (1858–1932) shared much in common with his compatriot Lagrange from a hundred years earlier. He was born in Piedmont, attended the University at Turin, began his teaching career at the Military Academy and founded an academic journal—all of which Lagrange had done before him. However, unlike Lagrange, who left Turin for Berlin and then Paris, never to return home, Peano never left home, becoming a full professor at the University of Turin and remaining there until he died. Peano was one of the founding fathers of mathematical logic and had a keen eye for the axiomatic foundations of logical systems. In the late 1880s, motivated by the ideas of Leibniz and Möbius, he became interested in Grassmann's theory of extension as an axiomatic system. Although Grassmann had not developed his theory of forms through an axiomatic approach, Peano supplied the rigorous background and definitions—the axioms—that allowed Grassmann's algebra to be seen as operating on elements—his forms—of a generalized space. In 1888 Peano published a small book titled *Calcolo geometrico secundo l'Ausdehnungslehre di H. Grassmann e precedutto dalle operazioni della logica deduttiva* (Geometric Calculus of the Second Ausdehnungslehre of Grassmann and results of Operations of Deductive Logic) that contained the first axiomatic description of a linear vector space.[12]

A linear vector space consists of elements that can be added to or subtracted from each other, for which the sum or difference is another element of the space, and can be scaled by multiplication with a scalar as the product is also an element of the space. In this way, elements of a vector space are able to "interact" with one another, producing resultants that are still part of the same space. A classic example of a linear vector space is the space of three-dimensional position vectors. Position vectors of objects contained in 3D can be added and subtracted, and their lengths can be changed, yielding other position vectors in 3D.

Linear vector spaces are characterized by their dimension, which is the number of independent "directions" in the space. Each direction has a basis vector associated with it, and a general vector can be represented uniquely as a sum, or combination, of scaled basis vectors.

It is no surprise that physicists use linear vector spaces routinely, because the addition or subtraction of vectors is a mainstay of any physics derivation, from the earliest introductory courses studying Galileo's parabolic trajectory to advanced graduate-level classes in theoretical physics. The principle of linear superposition allows the combined response of multiple influences to be calculated simply as the sum over the individual responses, weighted by the strengths of the influences. Superposition is one of the most powerful analytic techniques that physicists apply, from problems in free-body diagrams and electrostatics to electromagnetism and orbital dynamics, and it has fundamental importance for aspects of causality. A subtle but important contribution of Peano in his 1888 book was the inclusion of functions as elements of a linear space. Functions are mappings—linear vector functions take vectors from a space and map them onto other vectors in the same space. Thinking more abstractly, the functions are elements of the space, subject to the same properties of addition and scalar multiplication, yielding other functions of the same space. In this way, a linear space can be a space of functions—something that later would be called a Hilbert space.

David Hilbert (1862–1943) was one of the towering figures in mathematics around the turn of the twentieth century. Hilbert was the son of a judge and a merchant's daughter from the city of Königsberg in East Prussia on the Baltic Sea. Hilbert's early education was unremarkable, possibly because his parents and school had him concentrating on Latin and Greek rather than mathematics. He was not like William Rowan Hamilton (1805–1865) or Grassmann with their great facility for languages, and he eventually found his true path at which he excelled, graduating from the University of Königsberg with a degree in mathematics. He remained at the University to receive his doctorate, meeting a fellow graduate student, Hermann Minkowski, and they became fast friends. This friendship was to last for the remainder of Minkowski's short life, and had a profound influence on Hilbert's ideas. After receiving his doctorate, Hilbert traveled to Leipzig to visit Felix Klein and then to Paris, where he met Henri Poincaré, Camille Jordan, Gaston Darboux and Charles Hermite. On his return to Königsberg, he prepared his habilitation on the topic of invariant theory, a major interest at that time and a topic of special interest to Minkowski. Hilbert habilitated in 1886 at the University of Königsberg where he became a professor in 1892.

Felix Klein had by then taken the Chair of the Mathematics department at the University of Göttingen and was searching for talent. He had been impressed with Hilbert from their first meeting when Hilbert was no more than a postdoc, and he sought to attract Hilbert to Göttingen, which he finally succeeded in doing in 1895. Together, Klein and Hilbert created a mathematical powerhouse at Göttingen, restoring the fame of the department to the levels it had enjoyed during the days of Gauss and Riemann. Göttingen was the right place at the right time on the eve of two revolutions: the axiomatization of geometry in mathematics and the advent of relativity theory in physics. Hilbert published *Grundlagen der Geometrie* in 1899, which placed geometry on a solid foundation of abstract axioms. Such an axiomatization of mathematics became one of the central threads in the development of twentieth-century mathematics. Then, in 1902, Hilbert convinced the faculty to extend an offer to his friend Hermann Minkowski, who arrived in Göttingen only a few years before Einstein changed the world with his theory of special relativity to which Minkowski would later contribute the fundamental aspects of invariant theory, changing forever the perception of space and time.

Part of Hilbert's enduring fame today stems from an address he presented before the International Congress of Mathematicians in Paris in 1900. The title of the talk was *Problems of Mathematics*, and it outlined what he perceived to be the greatest outstanding mathematics problems at the beginning of the new century. He began:

> Who of us would not be glad to lift the veil behind which the future lies hidden; to cast a glance at the next advances of our science and at the secrets of its development during future centuries? What particular goals will there be towards which the leading mathematical spirits of coming generations will strive? What new methods and new facts in the wide and rich field of mathematical thought will the new centuries disclose?

In his oral address, he outlined ten major unsolved problems of mathematics, which he expanded to twenty-three problems in the printed conference proceedings.[13] It took nearly the entire twentieth century to solve most of Hilbert's problems, at least partially, and today only three remain completely unsolved, one of which is the Riemann Hypothesis concerning the zeros of the zeta function.

The second part of Hilbert's continuing fame today is the presence of his name on the geometric construction called "Hilbert Space." Hilbert space is a central concept in quantum theory that enables quantum wavefunctions to be decomposed into complete sets of orthonormal basis functions called eigenfunctions. Hilbert space is a linear vector space with infinite dimensions, and each basis vector (or direction) is represented by a function $f_n(x)$, of which there is an infinite number for $n = 1,2, \ldots,$ inf. Hence, Hilbert space is also known as a function space. The idea of Hilbert space has become so pervasive that it is routinely bandied about in hallways and in coffee houses around universities—wherever physics students and faculty congregate to discuss quantum physics. However, Hilbert personally had little to do with the geometric construction that bears his name today. Others constructed Hilbert space after Hilbert finished his work, and they named it in honor of him because of his influential contributions to its early stages, long before he recognized it as even a part of geometry.

This deserving honor stems from his work on the properties of solutions to integral equations. These equations contain an unknown function inside an integral—one famous example being the Fredholm integral equation that solves boundary-value problems like the modes of motion of a drumhead. Beginning in 1904 and continuing through 1910, Hilbert published five papers that systematically explored solutions that consisted of sets of real square-summable infinite sequences.[14] The theory and applications of quadratic forms runs like a thread through much of modern physics, such as the squared line element in Riemannian geometry and energy conservation in harmonic systems. It became especially useful in the hands of Hermann Minkowski who applied invariant quadratic forms in 1908 to the theory of space-time in relativity theory (see Chapter 7). The properties of Hilbert's infinite sequences shared much in common with Peano's theory of linear vector spaces, but Hilbert was unaware of Peano's work on the topic, and he failed to make the connection between his sequences and the geometric properties of vector space. Hilbert never referred to vector spaces or function spaces, nor used language that showed he was even thinking in those terms. Vectors at that time were still geometric, and Hilbert's work on integral equations, despite his being a major proponent of the "new" geometry, ironically was not "geometrical."

An effort was begun to axiomatize Hilbert space the way that Peano had axiomatized general linear vector spaces. The first steps in this direction were made by Hermann Weyl in his book *Space, Time, Matter*

(1918), which was limited primarily to aspects of space-time in relativity theory. Substantial and successful efforts were made a few years later by the American Norbert Wiener (1895–1964) and by the Polish Stefan Banach (1892–1945) in 1920 and 1922. The full axiomatization of Hilbert space was accomplished by the Hungarian-American John von Neumann (1903–1957) in 1927 in his work on the foundations of Schrödinger's and Heisenberg's quantum mechanics, which today are mentioned in the same breath, but at the beginning were opposite poles of bitter arguments, as we will see in Chapter 7.

The consequence of nearly a half-century of development, from 1888 to 1927, from Peano to Hilbert to von Neumann, was a fundamental transformation of the concept of space. The *directions* of space were taken over by abstract mathematical functions, and the familiar vectors of freshman physics, with their bold arrows pointing in the directions of forces or velocities, morphed into infinite series of *tuples*. This period brought the expansion in dimension and abstraction of dynamical spaces nearly to its modern limits, and the new spaces are almost unrecognizable when viewed from the original three-dimensional space of our daily existence. However, the description of trajectories within these new infinite-dimensional abstract spaces has stayed surprisingly the same since the time of Newton. A position vector tracing a particle trajectory in three-dimensional space is a set of numbers changing in time in response to impressed forces. After von Neumann, vector components in infinite-dimensional Hilbert space are still just numbers changing in time—the time-dependent coefficients of eigenvectors of the quantum-mechanical Hamiltonian. After all the years, and after all the abstractions, trajectories were still like the flight of a swallow—an unbroken thread linking the bird of now to the continuous set of all its past images in time. However, even this view of the trajectory was about to change, to increase in abstraction as integer dimensions dissolved into fractional spaces.

Fractional Worlds

When René Descartes constructed his first analytic curve in the 1630s using coordinate geometry for the first time, he visualized lines moving smoothly through space as their point of intersection traced out a continuous trajectory. It was literally a kinematic construction. Later, as Leibniz, Newton, and the Bernoulli's developed the infinitesimal calculus, they envisioned smooth curves of points whose tangents were always defined. However, this smooth world of continuous curves and ubiquitous tangents began to break apart as smooth functions were replaced with more complicated curves that could have sudden changes in slope at points where the tangent jumped discontinuously. For many years, it was believed that such functions always had a disconnected set of such points. Then, in 1872, a "monster" function appeared that was everywhere continuous, but nowhere had a derivative. This monster defied all the previous rules and expectations about the role and behavior of functions. This monster was the creation of the German mathematician Karl Weierstrass.

Karl Theodor Wilhelm Weierstrass (1815–1897) was born in Westphalia, on the western border of the Kingdom of Prussia, in the same year that Westphalia was born as a new province at the Congress of Vienna of 1815. The purpose of the Congress, chaired by the Austrian diplomat von Metternich, was an attempt to balance power in Europe after the chaos of the Napoleonic wars. Among of the consequences of the Congress was a consolidation of Prussian power and influence, which had the fortunate side effect that it established a stable environment across much of Germany that lasted over a century, fostering the intellectual rise of German science and mathematics in the nineteenth century.

Weierstrass showed an early interest in mathematics, but when he enrolled at the University of Bonn, his domineering father forbid him from pursing a career in mathematics, forcing him instead to take courses in business and finance to prepare the young Weierstrass to enter the ranks of the Prussian bureaucracy. Not unlike other rebellious youths, Weierstrass reacted by skipping classes and spending his time drinking and fencing, until he flunked out of the university. By 1839, with poor prospects for his future, Weierstrass was resigned to prepare to teach at a Gymnasium. As luck would have it, one of his teachers during this time was Christoph Gudermann, a leading mathematician interested in elliptic functions. Under Gudermann's guidance, Weierstrass became fascinated with functional analysis and resolved to become a serious mathematician.

Over the next decade, while teaching at a dreary Gymnasium in Braunsberg, Weierstrass filled what free time he had with studies of mathematics. In 1854, he published a paper titled *Zur Theorie der Abelschen Functionen* (On the Theory of Abelian Functions) in Crelle's Journal that outlined a method for solving problems with infinite series that converged uniformly to a function. This single paper created a firestorm of interest in mathematical circles, and offers for university positions began to arrive. Within four years, Weierstrass occupied a coveted faculty chair in the mathematics department at the University of Berlin, the top German university of its day. This was precisely the position that Hermann Grassmann was longing for at this same time as he was stuck teaching at his own Gymnasium in Stettin. Whereas both Weiertrass and Grassmann had begun as secondary school teachers, Weierstrass was launched to stardom with his single published paper, while Grassmann was overlooked after publishing his epic book. The mathematics world was ready for Weierstrass, but Grassmann was too far ahead of his time.

Weierstrass began lecturing at Berlin on his new approach to functional analysis, and his lectures became famous, attracting top students from around the world including the young Josiah Willard Gibbs in 1867, who was on his three-year European tour before returning to the US and to a faculty position waiting for him at Yale. Over the years, Weierstrass did not publish extensively, but introduced many new ideas in his lectures that were attended by a host of future major mathematicians, many of whom took official class notes that were later circulated and eventually published. Weierstrass influenced an entire generation of European mathematicians, and his teaching methods and style persist down to this day in the teaching of functional analysis. For instance, the "delta-epsilon" proofs that all mathematics students learn today were largely established by Weierstrass.

The main topics of interest to Weierstrass were questions of continuity of functions and problems of convergence, especially in the case of infinite series of points or functions. In this context, he was interested in special points of a function where derivatives do not exist. Many questions and arguments at that time concerned the nature of these points, whether they must be finite or could be infinite in number, and whether they could be dense or else must be sparse and isolated. Weierstrass, in 1872, was studying convergence properties of infinite power series and conceived of an infinite sum that was continuous everywhere, and yet when he calculated left- and right-limits of derivatives at any point, the derivatives failed to converge to the same value, no matter where he took his point.[15] In short, he had discovered a function that was continuous everywhere, but had a derivative nowhere

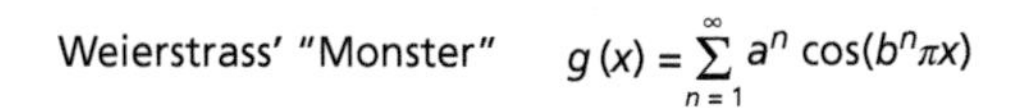

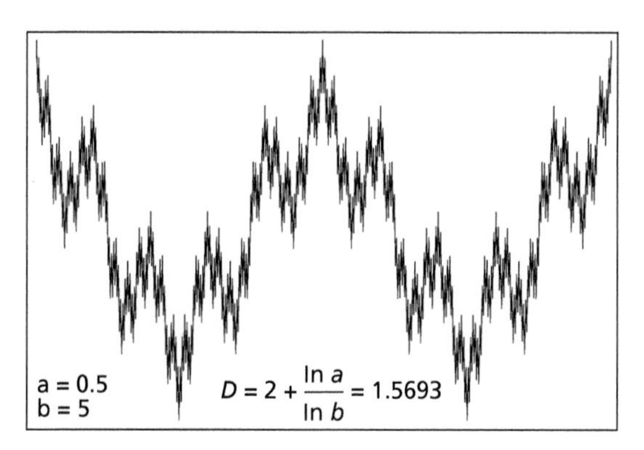

ig. 5.2 Weierstrass' "Monster" (1872) with a = 0.5, b = 5. This continu-
us function is nowhere differentiable. It is a fractal with fractal dimension
$= 2 + \ln(0.5)/\ln(5) = 1.5693$

e Fig. 5.2). This pathological function, called a "Monster" by Charles rmite, is now called the Weierstrass function.[16] Partially inspired Weierstrass' discovery, a young German mathematician began tear- g down the concepts of dimension while building up the concepts infinity.

Georg Cantor (1845–1918) was born in Russia, but the family had ved to Germany while Cantor was still young. In 1863, he enrolled he University of Berlin where he sat on lectures by Weierstrass and onecker. He received his doctorate in 1867 and his Habilitation in 1869, oving into a faculty position at the University of Halle and remaining here for the rest of his career. Cantor published a paper earlier in 1872 n the question of whether the representation of an arbitrary function y a Fourier series is unique. He had found that even though the series night converge to a function almost everywhere, there surprisingly ould still be an infinite number of points where the convergence failed. Originally, Cantor was interested in the behavior of functions at these points, but his interest soon shifted to the properties of the points themselves, which became his life's work.

In a radical and groundbreaking publication in 1874, Cantor began to lay down the foundations for a new field of mathematics known today as set theory. The paper was titled *Über eine Eigenschaft des Inbegriffes aller reallen algebraischen Zahlen* (On a Property of the Collection of All Real Algebraic Numbers)[17] where he proved that there is more than one "size" of infinity. For instance, the algebraic numbers (irrational numbers that are the roots of polynomials with integer coefficients) constituted a countable infinity, as do the rational numbers. A countable set is one that stands in a one-to-one correspondence to the natural numbers. An infinitely large set that is countable is one type of infinity, and Cantor assigned it a cardinal number designated as $\aleph_0$ to denote this type of infinity as the first transfinite cardinal number. Cantor then showed that there were other transfinite numbers, such as the number of reals, which is an uncountable infinity of type $\aleph_1$. Using the new concepts of set theory, Cantor devised a method for generating transcendental numbers (the irrational numbers that are not algebraic) that was different than a technique introduced by Liouville in 1844, and he devised a new proof of one of Liouville's theorems stating that there are infinitely many transcendental numbers in any interval.

Transfinite numbers introduced a radical new level of mathematical abstraction that could be used as a new tool in the mathematical toolbox to address questions of continuity and infinitesimals that are closely tied to the foundations of differential and integral calculus. However, many in the mathematics world were mortified by transfinite mathematics, and there was strong opposition from eminent mathematicians and philosophers. Despite his hopes that his discovery would be the springboard to launch him to heights of success and recognition, Cantor's work became his albatross, and he spent much of his time explaining and defending his work. In the process, Cantor refined his ideas, devising clear and intuitive examples of infinite sets. One famous example is Cantor's ternary set that he published in 1883 in *Grundlagen einer allgemeinen Mannigfaltigkeitslehre* (Foundations of a General Theory of Aggregates).[18] The set generates a function that has zero derivative almost everywhere, yet whose integral is equal to unity,[19] It is a striking example of a function that is not equal to the integral of its derivative! Cantor's ternary set is illustrated in Fig. 5.3, showing its iterative construction as well as the integral over the set that generates the Cantor staircase. Cantor demonstrated that the size of his set is $\aleph_1$, which is the cardinality of the real numbers, but whereas the real numbers are uniformly distributed, Cantor's set is "clumped." This clumpiness is an essential feature that distinguishes it from the one-dimensional number line, and it raises important questions about dimensionality.

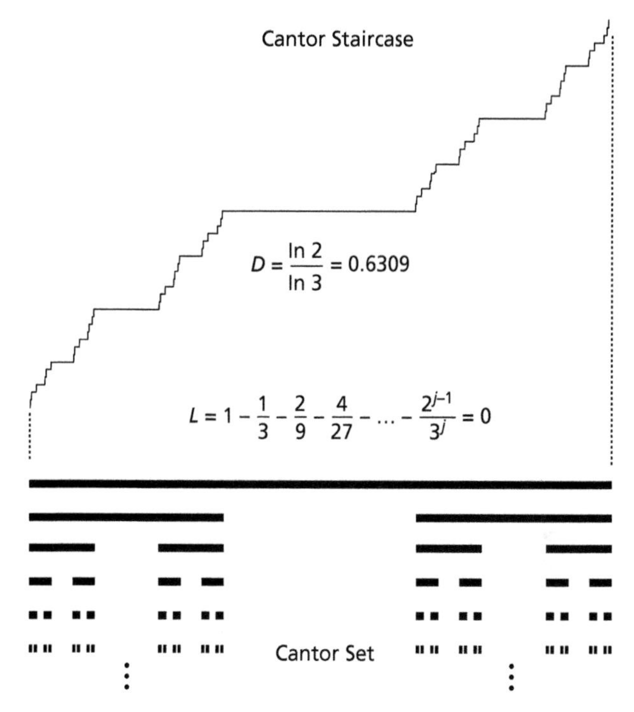

Fig. 5.3 The 1883 Cantor set (below) and the Cantor staircase (above, as the indefinite integral over the set)

But this was just the first of many assaults on the age-old concept of dimensionality, and it was beginning to crumble.

In 1878, in a letter to his friend Richard Dedekind, Cantor went further and showed that there was a one-to-one correspondence between the real numbers and the points in any n-dimensional space. He was so surprised by his own result that he wrote to Dedekind "I see it, but I don't believe it." The solid concepts of dimension and dimensionality were dissolving before his eyes. What does it mean to trace the path of a trajectory in an n-dimensional space, if all the points in n dimensions were just numbers on a line? What could such a trajectory look like? A graphic example of a plane-filling path was constructed in 1890 by Peano, who was a peripatetic mathematician with interests that wandered broadly across the landscape of the mathematical problems of his day—usually ahead of his time. Only two years after he had axiomatized linear vector spaces, Peano constructed a continuous curve

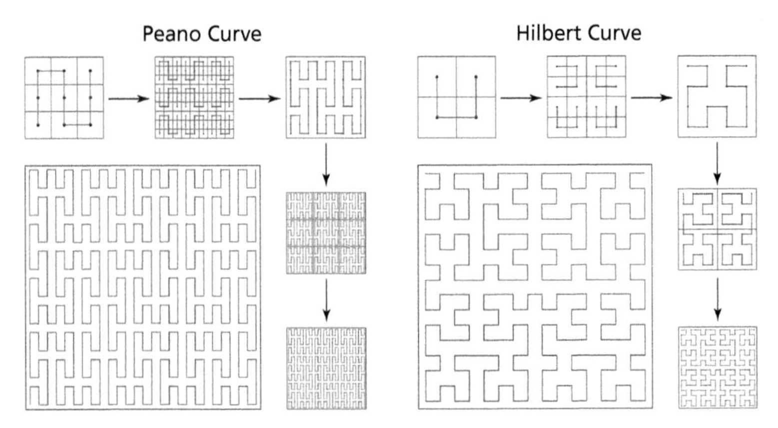

Fig. 5.4 Peano's (1890) and Hilbert's (1891) plane-filling curves. When the iterations are taken to infinity, the curves approach every point of two-dimensional space arbitrarily closely, giving them a dimension D = 2

that filled space, published in *Sur une courbe, qui remplit toute une aire plane* (On a curve, which fills an entire flat area).[20]

The construction of Peano's curve, shown in Fig. 5.4, proceeds by taking a square and dividing it into nine equal sub squares. Lines connect the centers of each of the sub squares. Then each sub square is divided again into nine sub squares whose centers are all connected by lines. At this stage, the original pattern, repeated nine times, is connected together by eight links, forming a single curve. This process is repeated infinitely many times, resulting in a curve that passes through every point of the original plane square. In this way, a line is made to fill a plane. Where Cantor had proven abstractly that the cardinality of the real numbers was the same as the points in n-dimensional space, Peano created a specific example. This was followed quickly by another construction, invented by David Hilbert in 1891, that divided the square into four instead of nine, simplifying the construction, but also showing that such constructions were easily generated.

The space-filling curves of Peano and Hilbert have the extreme properties that a one-dimensional curve approaches every point in a two-dimensional space. This ability of a one-dimensional trajectory to fill space mirrored the ergodic hypothesis that Boltzmann relied upon as he developed statistical mechanics (see Chapter 6). These examples by Peano, Hilbert and Boltzmann inspired searches for continuous

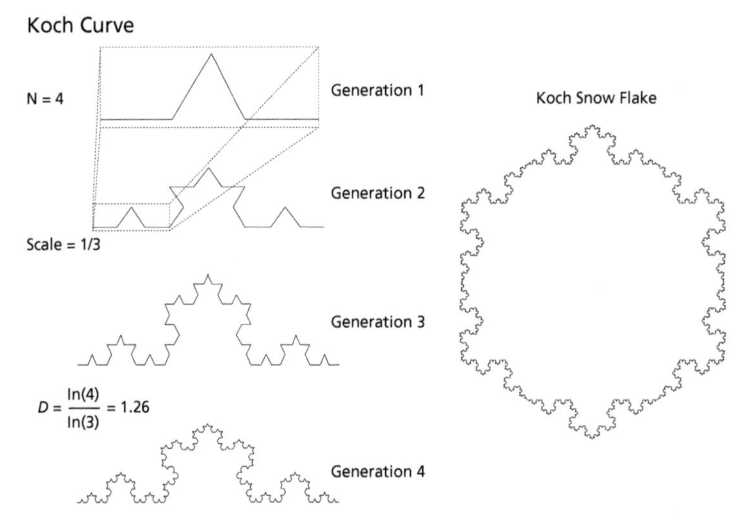

Fig. 5.5 Construction of the Koch curve and Koch Snowflake from line segments (1904). The portions of the curve are self-similar

curves whose dimensionality similarly exceeded one dimension, yet *without* filling space. Weierstrass' Monster was already one such curve, existing in some dimension greater than one but not filling the plane. The construction of the Monster required infinite series of harmonic functions, and the resulting curve was single valued on its domain of real numbers. An alternative approach was proposed by Helge von Koch (1870–1924), a Swedish mathematician with an interest in number theory. He suggested in 1904 that a set of straight line segments could be joined together, and then shrunk by a scale factor to act as new segments of the original pattern. The construction of the Koch curve is shown in Fig. 5.5. When the process is taken to its limit, it produces a curve, differentiable nowhere, which snakes through two dimensions. When connected with other identical curves into a hexagon, the curve resembles a snowflake, and the construction is known as "Koch's Snowflake" (see Fig. 5.5.).

The work by Cantor, Peano and von Koch created a crisis in geometry. Previously, geometric objects were separated into inviolable categories based on their dimensionality—points moving through space create lines, lines moving through space create planes, planes moving through space create volumes—and after Riemann, volumes moving through n-dimensional space create hyper volumes, etc. Now, Cantor and Peano had erased the solid walls between these categories. Over the following quarter of a century, mathematicians struggled to rescue concepts of dimensionality. An important byproduct of that struggle was a much deeper understanding of concepts of space, especially in the hands of Felix Hausdorff.

Felix Hausdorff (1868–1942) had a split personality, of sorts, with the two sides of his personality leading separate lives—one as Felix Hausdorff the brilliant mathematician, and the other as Paul Mongré the successful author and playwright—and many who knew him in one role were unaware of his other. He was born in Breslau, at that time a part of Prussia, and the family moved to Leipzig where he was educated. Hausdorff was a talented musician, and he originally wished to pursue a career in music, but his father dissuaded him, suggesting a more practical path in the sciences. At the university in Leipzig, Hausdorff studied physics and astronomy and wrote his doctoral dissertation in mathematical physics on atmospheric refraction of light, later writing his habilitation on atmospheric absorption. He did not have much impact on the field, but his academic skills secured him a teaching position at the university. In his early years at Leipzig, when most young professors are supposed to immerse themselves in their academic careers, he instead was active as an author, and he adopted the pseudonym Paul Mongré under which he published a growing repertoire of poetry and plays. He spent most of his free time circulating in the creative and exciting world of artists and poets in Leipzig, but wisely never gave up his day job.

Hausdorff obviously had broad interests, and philosophical curiosity attracted him to set theory. In 1901, he gave one of the first courses in set theory and began making original contributions to the field. This work became so successful that he finally devoted his full creative energies to it. In 1914, on the eve of World War I, he published a definitive textbook on the topic, *Grundzüge der Mengenlehre* (Foundations of Set Theory),[21] which also contained new results, setting the stage for much of the future work by others. Hausdorff was spared the personal carnage that many other mathematicians experienced during the war, and he continued his work in set and measure theory. In 1918, as the war was ending, he published a little paper *Dimension und Äuseres Mass* (Dimension and Outer Measure).[22] This paper was written in response to a paper by Konstantin Cathodory who had shown how to construct a p-dimensional measure in a q-dimensional space. Hausdorff realized that it was possible to extend Cathodory's measure to cases when p was not an integer and could take on fractional values.[23] Thus, he introduced what today is called the Hausdorff measure, and from it generalized the notion of fractional dimension.

Hausdorff's first concrete example of a set with a fractional dimension was the Cantor ternary set. He showed that the outer measure of the Cantor set would go discontinuously from zero to infinity as the fractional dimension increased smoothly. The critical value where the measure changed its character became known as the Hausdorff dimension. For the Cantor ternary set, the Hausdorff dimension is exactly $D = \ln(2)/\ln(3) = 0.6309$. This value for the dimension is less than the Euclidean dimension $E = 1$ of the support (the real numbers on the interval [0, 1]), but it is also greater than $E = 0$ which would hold for a countable number of points on the interval.[24] Koch's Snowflake has a fractal dimension of $D = \ln(4)/\ln(3) = 1.26$, and Weierstrass' monster has a fractal dimension of $D = 2 + \ln(a)/\ln(b)$. The work by Hausdorff became well known in the mathematics community, but remained largely unknown to wider audiences until nearly sixty years later when Benoit Mandelbrot coined the term *fractal* to describe geometric objects like the Cantor set.

Benoit Mandelbrot (1924–2010) was born in Warsaw Poland, but emigrated with his family to Paris in 1936 to join his uncle who was an academic. The family was Jewish, and was barely able to avoid being sent to a concentration camp when France was occupied by the Nazis. Hausdorff and his family had not been so lucky, being ordered to a concentration camp in 1942. Hausdorff, his wife and his wife's sister escaped this fate only by committing collective suicide. After the war, Mandelbrot studied under Gaston Julia and Paul Lévy at the École Polytechnique and received his Master's degree from Cal Tech in 1949 and his PhD from the University of Paris in 1952 in mathematical science. He joined the IBM Thomas J. Watson Research Center in Yorktown Heights, New York in 1958.

At the IBM research center Mandelbrot gained access to powerful computers that enabled him to study Julia sets, named after his teacher in Paris, and to visualize them in great detail using the latest advances in computer visualization. He was fascinated by the self-similar characters of the Cantor and the Julia sets. In 1967, he published a paper titled *How long is the Coastline of Britain?* in Science magazine,[25] where he showed that the length of the coastline diverged with a Hausdorff dimension equal to D = 1.25. Later, in 1975, Mandelbrot coined the term *fractal* to describe these self-similar point sets, and he began to realize that these types of sets were ubiquitous in nature, ranging from the structure of trees and drainage basins, to the patterns of clouds and mountain landscapes. He published his highly successful and influential book *The Fractal Geometry of Nature* in 1982, introducing fractals to the wider public and launching a generation of hobbyists interested in computer-generated fractals.[26] The rise of fractal geometry coincided with the rise of chaos theory that was aided by the same computing power. For instance, important geometric structures of chaos theory, known as strange attractors, have fractal geometry.

The concepts of trajectories and of functions share much in common. Clearly, the physical trajectory of a mass must be continuous, like Descartes's first construction of a curve, or else unphysical behavior ensues. Even as dynamical spaces expand in dimension and abstraction, continuity of trajectories must remain inviolate if the future is to be deterministic. After Weierstrass created his Monster, continuity momentarily lost its footing, with serious questions raised about the nature and existence of continuity. Mathematicians like Dedekind, Cantor and Hausdorff rescued it, giving it a firm foundation, preparing the ground for the advent of chaos theory a hundred years in the future, which we will explore in Chapter 9. But a final step remains in this geometric evolution from simple trajectories to abstract, waiting for a final nudge to bring it into existence, requiring the concept of a unique state trajectory in the geometric space of dynamics known as phase space.

6

The Tangled Tale of Phase Space

Hamiltonian Mechanics is geometry in phase space.
V. I Arnold (1978)[1]

In our journey to understand the sources of modern dynamics, phase space is a central abstraction, like an attractor that we cannot resist, towards which we irresistibly have been drawn. Modern dynamics lives in phase space. The diverse dynamics of economics and evolution, of gravity bending light and of storms sweeping across the oceans, of chaos and networks—these all are expressed as dynamical flows through phase space. Here, finally, is the logical conclusion of Galileo's trajectory, begun as a simple arc, one among many, but now seen as a low-dimensional projection, the tip of the iceberg, of a complete dynamical scaffold, invisible behind Galileo's simple parabola, capable of supporting dynamical behavior of great diversity, imagined or unimagined.

Yet, the origins of the concept of phase space, as well as its name, are historically obscure—which is surprising in view of the central role it plays in so many aspects of modern physics.[2] The historical origins have been obscured further by overly generous attribution. In virtually every textbook on dynamics, classical or statistical, the first reference to phase space is placed firmly in the hands of the French mathematician Joseph Liouville, usually with a citation dated to his 1838 paper[3] in which he supposedly derived the theorem on the conservation of volume in phase space. However, Liouville makes no mention of *phase space* in his paper, let alone dynamical systems. Liouville's paper is purely mathematical, on the behavior of a class of solutions to a specific kind of differential equation. Though he lived to a ripe old age (he died in 1883), he was apparently unaware of its application to statistical mechanics by Boltzmann[4] even within his lifetime. Therefore, Liouville's famous paper, cited routinely as the origin of phase space by even the most rigorous textbooks, and by the most noted chroniclers of the history of mathematics,[5] surprisingly is not it!

Galileo Unbound. David D. Nolte, Oxford University Press (2018).
 DOI: 10.1093/oso/9780198805847.001.0001

As we explore the origins of phase space, who imagined it, who struggled with it, and ultimately who named it, we will be led to a now obscure encyclopedia article published in 1911 that had etymological side effects not fully intended by its author. Along the way to answering these questions, our search for the origins of phase space will bring us along a path that passes by key historical events in the development of Galileo's trajectory, such as the discovery of entropy and the development of statistical mechanics. These topics give a first glimpse of the deeper role that phase space plays in modern dynamics, a role that will be explored in later chapters. The tangled tale of the present chapter begins with Joseph Liouville, the most renowned French mathematician of the early part of the nineteenth century during the golden age of differential calculus and the beginning of differential geometry.[6]

Liouville's Theorem

Joseph Liouville (1809–1882) was born in the small city of St. Omer, France, not far from Calais, with roots back to the Middle Ages. He was the son of an army captain, from whom he inherited a dashing appearance and a head of flowing hair. The family had the means to provide him a modest education, with the goal of training him for the practical career of an engineer. However, Liouville fell in love with mathematical analysis, eventually displaying a virtuoso mastery of topics ranging from number theory and complex analysis to differential geometry and topology. His energy and industry knew almost no bounds as he launched, and managed as its editor, a key French mathematics journal that still exists today. He also found time for politics after the revolutions of 1848 swept across Europe, getting himself elected to the Constituent Assembly, driven by his democratic zeal, yet getting himself voted out of office the very next year.

Liouville is known among mathematicians and physicists for several mathematical theorems, most notably the Sturm-Liouville theory of integral equations. The motivation for much of his work came from physical problems such as celestial mechanics, but also electrodynamics and the theory of heat. The properties and solutions of differential equations of multiple variables (many dimensions) were among his main areas of interest, and it was in this context that he worked on the solution of differential equations with constant integrals. Liouville's et se réduit simplement à

$$u \frac{dP_1}{dx_1},$$

parce que les sommes

$$\Sigma\left(\pm \frac{dx_2}{da} \cdot \frac{dx_2}{db} \cdots \frac{dx_n}{dc}\right), \ldots \Sigma\left(\pm \frac{dx_n}{da} \cdot \frac{dx_2}{db} \cdots \frac{dx_n}{dc}\right)$$

sont nulles en vertu d'une propriété bien connue. La valeur complète de $\frac{du}{dt}$ est donc

$$\frac{du}{dt} = u\left(\frac{dP_1}{dx_1} + \frac{dP_2}{dx_2} + \ldots + \frac{dP_n}{dx_n}\right).$$

Fig. 6.1 Liouville's theorem of 1838[7]

now famous paper of 1838 (the paper attributed as the proof of the conservation of volume in phase space (see Fig. 6.1) for a facsimile of the key equations in the paper) appeared only a few years after William Rowan Hamilton's publication of his dynamics of 1834 and 1835,[8] and yet Liouville made no mention of the application of his results to dynamics, possibly because he was unaware of Hamilton's work.[9] The connection with mechanics was made a few years later by the Prussian mathematician Carl Jacobi.

Carl Gustav Jacob Jacobi (1804–1851) was born in Potsdam, Germany, to wealthy parents who ensured a formal and rigorous education for their bright boy. Enrolling at the University of Berlin at the age of sixteen, he decided to devote himself to mathematics despite his easy mastery of classic languages and philosophy. After completing his undergraduate degree in three years, he submitted his doctoral thesis in his fourth and began lecturing at the University (after converting to Christianity from Judaism as a step necessary to take the post). Despite his rapidly rising star, there was no permanent position for him in Berlin, so he took an assistant professorship at the University of Königsberg. He became a guiding force in the mathematics department, his energy and enthusiasm infectious to the students and faculty who attended his lectures in thrall, lectures that extended over a dozen hours across several days on the topics of his latest mathematical researches. Jacobi soon ranked among the first-rate mathematicians in Europe—until he was suddenly stricken in his thirty-nineth year with debilitating diabetes.

Because the harsh winters of Königsberg were harmful to his health, he moved to Berlin where he, like his friend Liouville, was caught up by the democratic fervor of 1848. As young Bernhard Riemann stood a nervous night guarding the royal palace, Jacobi was speaking too freely at the Constitutional Club, coming under suspicion by the royalists. His teaching stipend from the University was withdrawn. When he threatened to take a position at the University of Vienna, the royalists relented, but within a few more years, Jacobi succumbed to mounting assaults on his health.

In the winter semester of 1847–1848 at Berlin, in the midst of the political turmoil, Jacobi gave a series of lectures on applications of differential equations to dynamics, demonstrating that the conditions imposed upon the system of differential equations that Liouville had studied in 1838 were satisfied by Hamilton's equations.[10] Jacobi's work represents the first mathematical application of phase space, and he is clearly the one who discovered the conservation of volume in phase space (see Fig. 6.2 for a facsimile of Jacobi's published version of his lectures). Building upon the mathematical work of Liouville (explicitly referencing Liouville's 1838 paper), Jacobi was the first put Liouville's theorem into a mechanical context. What Jacobi did not do, and could not do in his time, was to recognize phase space as a space. In the early 1840s, there

Daher lässt sich der vollständige Differentialquotient von $\lg R$ nach x unter der merkwürdigen Form

$$(2.)\quad \frac{d\lg R}{dx} = \frac{\partial X_1}{\partial x_1} + \frac{\partial X_2}{\partial x_2} + \cdots + \frac{\partial X_n}{\partial x_n}$$

darstellen, wo

$$R = \Sigma \pm \frac{\partial x_1}{\partial \alpha_1} \frac{\partial x_2}{\partial \alpha_2} \cdots \frac{\partial x_n}{\partial \alpha_n}.$$

Nach vollendeter Integration des Systems (1.) findet man also R aus der Gleichung (2.) durch eine Quadratur nach x. Aber es giebt Fälle, in welchen die Determinante R vor allen Integrationen angegeben werden kann, nämlich wenn sich die Summe $\frac{\partial X_1}{\partial x_1} + \frac{\partial X_2}{\partial x_2} + \cdots + \frac{\partial X_n}{\partial x_n}$ mit Hülfe des Systems (1.) in einen vollständigen Differentialquotienten nach x transformiren lässt, oder, was ein noch einfacherer Fall ist, wenn X_1 kein x_1, X_2 kein x_2 u. s. w. X_n kein x_n enthält. Alsdann ist $\frac{\partial X_1}{\partial x_1} + \frac{\partial X_2}{\partial x_2} + \cdots + \frac{\partial X_n}{\partial x_n} = 0$; daher

$$\frac{d\lg R}{dx} = 0,$$

$$R = \text{Const.}$$

Fig. 6.2 Jacobi's derivation of Liouville's theorem[11]

was no concept of *space* beyond the three dimensions of our physical space. This was the time before Arthur Cayley,[12] Hermann Grassmann[13] and Bernhard Riemann[14] and their new notions of multidimensional manifolds. For Jacobi, there was no *space*, only products of differentials of many variables. And there were no *trajectories* through the *phase space*, only the physical trajectories of individual particles. Therefore, Jacobi may be the originator of the analytical treatment of dynamical systems of many variables, but he cannot be designated as the originator of *phase space* or its name. The time was not right. First, the concept of multidimensional spaces had to enter the psyche of nineteenth-century scientists. Furthermore, science of the mid-1800s needed to come to grips with a strange new physical principal known as entropy. It was in the search for the meaning of entropy that the first multidimensional dynamical spaces would emerge.

Clausius' Transformation

The industrial revolution, with its insatiable need for power, was built upon the steam engine. The first practical steam engine was constructed by the English inventor Thomas Newcomen in 1712, and was later improved upon significantly by the Scottish engineer James Watt. These engines provided power that far surpassed the labor of man or even of horses, but they remained badly inefficient. In 1820, the French engineer and physicist Sadi Carnot (1796–1832) became intrigued by these engines, and sought to find out how efficient they could be in ideal cases. His research lead to his short book *Réflexions sur la puissance motrice du feu et sur les machines propres à développer cette puissance* (Reflections on the Motive Power of Fire) published in 1824 that contained the seeds of the science of thermodynamics. Unfortunately, Carnot was still working within the outdated framework of heat as a substance, known as *caloric*, and no one noticed Carnot's publication until Émile Clapeyron (1799–1864) reworked Carnot's views in 1834. Even then, it took a little over decade before it was taken up by Rudolf Clausius, who was able to isolate Carnot's flashes of genius from his moments of confusion.

Rudolph Clausius (1822–1888) was Prussian physicist, receiving his primary education at the Gymnasium at Stettin (where Hermann Grassmann would replace his own father and finish his career a few years after Clausius had left for a degree in Berlin) and a doctorate from the University of Halle in 1847. Clausius was a passionate imposing man, with high forehead and deep piercing eyes, harboring strong feelings for the emerging German empire forming under the expert manipulations of Chancellor Bismark. When Bismark helped fabricate the Franco-Prussian war of 1870 as a pretext to unify the North German Federation with the southern German states, Clausius had recently arrived at the University of Bonn after spending most of his career away from his beloved Germany at the University in Zürich. Clausius organized a student ambulance corps, serving behind the advancing German troops as they swept across France, participating in the vicious battle of Gravelotte where the Germans suffered 20,000 casualties and the French 13,000. Clausius was awarded the Iron Cross for his service, but was wounded during the conflict, an injury that he carried with difficulty but as a badge of honor for the remainder of his time at Bonn, where he rose to the position of Rector of the University in his last years.

Clausius is principally responsible for the First and Second Laws of Thermodynamics. He was unknown to the scientific world when he published his momentous paper in 1850 *Über die Bewegende Kraft der Wärme und die Gesetze, welche sich daraus für die Wärmelehre selbst arbleiten lassen* (On the moving force of heat and the laws regarding the nature of heat itself which are deducible therefrom). This paper contains the essential ideas behind the First and Second Laws of Thermodynamics that he constructed to reconcile Carnot's theory with Hermann von Helmholtz' recent discovery of the conservation of energy that was published in 1847. Clausius' First Law captures the conservation of energy in the forms of heat and work, while his Second Law concerns the physics of irreversibility and the thermodynamic arrow of time, or that heat cannot flow unaided from cold to hot bodies. He followed this work in 1854 with more exact definitions of the First and Second Laws and more rigorous mathematical derivations of their consequences. In this second paper, a quantity appears that had the unexpected property that it always increased. The quantity remained unnamed for over a decade until Clausius finally called it *Entropy* in 1865. Clausius' choice of word was made as a parallel to the word *Energy*. Entropy comes from the Greek word τροπη that means "transformation," and it was in the transformations between heat and work that entropy appears in Clausius' theory.

From the beginning of Clausius' studies in the science of heat, he had realized that work and heat, which seemed like such different entities at the large scale, must have a common origin at a microscopic scale. This realization is part of what helped him see through Carnot's problems, aided also by the equivalence of work and heat that had been defined so accurately by James Prescott Joule[15] in 1845. Clausius remained quiet on the microscopic origins of heat through his 1850 and 1854 papers, but this changed abruptly in 1856, when the German physicist August Kronig proposed a simple kinetic theory of gases. In Kronig's theory, gases are composed of material points that obey the laws of physics governing kinetic energy and work. Clausius immediately saw that the concepts of Kronig's paper paralleled his own and he published in 1857 his paper *The Nature of Motion that We Call Heat* in which he laid out several macroscopic consequences of the motions and collisions of microscopic molecules.[16] He initially assumed that molecules were point particles, but by considering the diffusion of fragrance through air, he concluded in a paper in 1858 that molecules must have a finite size and hence travel a finite distance, called the mean free path, between collisions with one another. These papers, with precedents from earlier authors, inaugurated the new field of *kinetic theory*.

One of the avid readers of Clausius' papers was the Scottish physicist James Clerk Maxwell (1831–1879), later to become famous for his development of statistical mechanics and the theory of electrodynamics. He was an unknown young faculty member of Aberdeen University in Scotland when a mathematics prize was announced in 1857 by St. John's College, Cambridge, to be awarded to the best mathematical treatment of the physics of the rings of Saturn. The problem of the stability of Saturn's rings had eluded physicists ever since Christiaan Huygens first discovered that they were rings of material orbiting the planet, as opposed to Galileo's "handles." Maxwell realized that to apply Newton's deterministic laws to each particle in the ring would be impossible because the number of equations would be uncountably large, so he adopted a probabilistic approach in which he derived how groups of particles would behave on average under the influence of Saturn's gravity and under their mutual collisions. This approach was radical at the time. No one had attempted to use the somewhat seedy mathematics of probability, which had previously been restricted to problems of dice and gambling, to a fundamental physical problem. His approach was fruitful, and in 1859, he was awarded $130 and the Adams Prize for his solution—a solution that was ultimately validated in the 1980s by the Voyager flybys of Saturn.

Because of his bold use of probability theory in the solution to the stability of Saturn's rings, Maxwell was in a perfect position to appreciate Clausius' papers on kinetic theory. Clausius had assumed that all molecules moved at the same average molecular speed. Maxwell realized that the molecules must have a distribution of speeds defined by a probability. He published his first paper on the topic in 1860, unveiling what is today known as the Maxwell velocity distribution.[17] The topic continued to keep his attention up through 1866 (in addition to his foundational work during this same time on electrodynamics) as he expanded on his initial theory and succeeded in deriving thermodynamic properties of gases based on the velocity distribution. However, one problem continued to give him trouble, which was the well-known decrease of air temperature with increasing height.

This same year 1866 saw the arrival of a young and gifted theorist on the scientific stage—Ludwig Boltzmann (1844–1906), an Austrian physicist who studied at the University of Vienna under Joseph Stefan. His PhD dissertation was on the kinetic theory of gases, which he published in 1866. Boltzmann adopted Maxwell's probabilistic approach to kinetic theory, but based his theory on molecular energy rather than on molecular speed. Despite this seemingly nuanced difference, it provided him with a new set of tools and results, and in 1868, he showed how the gas distribution laws could include potential energy as easily as kinetic energy. This difference allowed him to prove that it was not only the temperature that decreased with height for a gas in thermal equilibrium, but also the molecular density that decreased. This work introduced what is called *the Boltzmann factor* to the gas probability equations. The Boltzmann factor has much broader applicability than simply describing the properties of gases varying with altitude, appearing today in virtually every textbook on chemistry, solid-state or condensed matter physics, bioenergetics and more.

With his 1866 thesis on kinetic theory, Boltzmann had embarked on his life's work that would consume him until the end. The program that he set himself was no less than to derive all the consequences of thermodynamics from kinetic theory—including Clausius' troublesome Second Law of Thermodynamics. On the journey to his goal, Boltzmann uncovered many fundamental aspects of physics that echo today in the fields of statistical mechanics and chaos theory, including the concept of phase space and the system trajectory.

Boltzmann's Phase

As Boltzmann built upon Maxwell's work on kinetic theory, he discovered the conservation of volume in phase space. He published two papers that included the theorem in early 1871.[18] In the first paper he appears to have been unaware of Jacobi's original work and the connection with Liouville, but in the second, Boltzmann makes explicit reference to Jacobi's derivation that was related to the conservation of phase space volume, although Jacobi had called it the theorem of the last multiplier for technical reasons, not recognizing it as a conservation of volume. Despite Jacobi's reference of Liouville's theorem in his *Vorlesungen über Dynamic* (1842), Boltzmann makes no mention of Liouville at this time. In these seminal 1871 papers, there is no language of *phase* or *space*, although in Boltzmann's hands the conservation of phase space for a conservative dynamical system appears for the first time in its mathematically modern form.

The second of these two 1871 papers is where Boltzmann first made the analogy[19] between physical trajectories of particles in two-dimensional space and what are called Lissajous figures (although he did not refer to Lissajous). Lissajous figures (also sometimes known as Bowditch figures after the scientist, navigator and America's first insurance actuary Nathanial Bowditch), are two-dimensional patterns that arise when two harmonic time series are plotted against each other—as best experienced in physics labs using an oscilloscope and two function generators. When the two harmonic frequencies are rational fractions, periodic patterns occur. But when the frequency ratio is irrational, then the system trajectory visits all points on the plane bounded by the signal amplitude. This is Boltzmann's first description of what later became his ergodic hypothesis,[20] which states that a dynamical system samples all parts of its dynamical space. In Lissajous figures, the relative phase between the harmonic signals determines the instantaneous configuration, and the point on the figure is referred to as the *phase point*. Only a year later in 1872 Boltzmann uses the term *phase* for the first time in his paper "On further studies of the equipartition theory of gas molecules."[21]

Boltzmann's paper of 1872 is one of his most famous. This was where he applied probability theory to the microscopic distributions of states in the gas and derived his famous *H-theorem*. Boltzmann defined a quantity he called H (Greek *eta* or E) that was based on mechanical principles of colliding gas molecules. He discovered that the quantity H always decreased during any irreversible process, and he boldly made the connection that H was the negative of entropy (beginning with German E). What Boltzmann had achieved with his H-theorem was to take a phenomenological law, the Second Law of Thermodynamics, and place it in on a mechanistic footing using only Newton's laws, thereby cementing the importance of statistical mechanics in the history of physics. Despite this brilliant success, a nuance was buried deep in Boltzmann's derivation of the H-theorem that threatened to overthrow the fragile new science he had invented. This mathematical nuance dealt with the thermodynamic arrow of time and the eventual fate of the Universe.

Throughout his career, Boltzmann had a knack for getting into arguments with his contemporaries. This was driven in part by the blunt and self-important manner in which he communicated with others. But it also stemmed from the difficulty of what he was attempting to accomplish, and the difficulty others had in trying to comprehend his theories. One of the most fruitful arguments was with Joseph Loschmidt, who was a colleague and friend at the University of Vienna. Loschmidt was well versed in kinetic theory by the time he and Boltzmann began their famous dialogue. Shortly after Maxwell published his statistical extension of Clausius' kinetic theory, Loschmidt recognized that Maxwell's theory made it possible to estimate the size of gas molecules if one knew the number density of molecules in a volume of gas. He found a good estimate based on the volume of an equal mass of molecules in a liquid state and published in 1865 his estimate of the molecular size of air as one nanometer, the year before Boltzmann's thesis.

Much later, in 1876, Loschmidt saw a paradox in Boltzmann's derivation of his H-theorem that connected it to a controversy that had been simmering for years within the inner circles of physics. This controversy related to the paradox between the microscopic reversibility of Newton's laws and the macroscopic irreversibility of thermodynamics. Reversibility means that physics looks the same if run forwards or backwards through time. This is clearly true for ideal collisions between perfectly elastic balls. If you make a movie of colliding billiard balls and run it forwards or backwards, you cannot tell which is which (if you don't run the movie long enough to see friction slow things down). However, the Second Law of Thermodynamics states that the entropy of a system must either increase with time or stay the same, but never decrease. Gases obviously exhibit irreversible behavior, because if a gas were made initially to occupy only half of a container, when released it would quickly expand to fill the entire container. The reverse process is never observed: if one starts with gas filling an entire container, it is never observed to spontaneously evacuate half of the container at some future time. This dichotomy between the reversible physics of mechanistic systems and the irreversible physics of statistical systems is called *the reversibility paradox*. It was raised as early as 1867 in communications between Maxwell and his friends Thomson and Tait, which had led Thomson to adopt a probabilistic view of irreversibility as being statistically the most probable, but not a law.

Loschmidt used the reversibility paradox in 1876 to question Boltzmann's H-theorem of 1872. He was motivated by a desire to prove that the future *heat death* of the Universe, predicted by Clausius based on increasing entropy, was not a certain end to all things. Loschmidt said that he wished ". . . to destroy the terroristic nimbus of the second law, which has made it appear to be an annihilating principle for all living beings of the universe."[22] Loschmidt argued that if the atoms of a gas began with an initial configuration that was low in entropy, and was allowed to evolve in time to a higher entropy configuration, and then if all velocities were exactly reversed, the system would evolve from the higher entropy configuration to a low one, apparently violating (and hence invalidating) the conclusions of Boltzmann's H-theorem. Loschmidt remarked that this would "open up the comforting prospect that mankind is not dependent on mineral coal or the sun for transforming heat into work, but may have available forever an inexhaustible supply of transformable heat."

Of course, this is not true, and Loschmidt had missed the mark, but his criticisms spurred Boltzmann on to perhaps his greatest accomplishment. In two papers published in 1877, he made the important distinction between microstates of a system (positions, velocities and numbers of molecules) and macro states (temperature, pressure and entropy).[23] The crucial point was that one could define the relative probabilities of all the different microstates which could contribute to a given macro state. With this approach, Boltzmann was able to derive a simple equation that relates the entropy of a macro state to the probability of all the different microstates of which it was composed. The German word for *probability* is *Wahrscheinlichkeit*, and Boltzmann's famous equation relating entropy S to probability W is $S = k \log W$, which is carved on the headstone of his final resting place. In this way, Boltzmann's argument with Loschmidt turned out to be highly productive, and frankly gave Boltzmann one of his greatest professional successes, while Loschmidt has largely been forgotten.

With his work on the dynamical state of a kinetic system, Boltzmann came very close to making the first significant expansion of concepts of the trajectory since Galileo, by representing the *multiple* trajectories of *many points* in a *single* three-dimensional space as the *single* trajectory of a *single* point moving in a *multi*dimensional space. If he had used the language of the system trajectory in his original papers from 1871, it would have been astounding, and he would have been far ahead of his time. However, this revolutionary step in abstraction was to be taken by a rising mathematician, contemporary to Boltzmann in the 1890s, who had a keen mind for visualizing complicated motions of dynamical systems—Henri Poincaré (1854–1912).

Poincaré was not a revolutionary at heart, but he became a revolutionary in spite of himself. He was born in the city of Nancy in eastern France, trained at the École Polytechnique, the top technical school in France, studied under Charles Hermite, and received post-graduate training at the École des Mines, becoming a mining engineer with the Corps des Mines. This was a practical occupation, with an emphasis on mechanics, and it suited Poincaré's aptitudes, which were strongly mechanical-visual, instilling in him a reliance on intuitive mechanistic thinking. Poincaré believed that physical intuition was more important than mathematical rigor—that conceptual models were more important than complex equations seeking to capture every detail of a problem. Part of the outlook he developed over his years of technical training was his conviction that mechanical systems tended towards stability, towards equilibrium, and this conviction was the origin of one of the most fortuitous mistakes in the history of complex dynamics.

The Three-Body Problem

In 1885, the Swedish mathematician Gösta Mittag-Leffler was looking forward to the sixtieth birthday of King Oscar of Sweden and sought his patronage to support a mathematics prize that would increase European attention on the emerging scientific and mathematical activities of Sweden. Mittag-Leffler sought the advice of Karl Weierstrass in Berlin to help frame the problem of the competition. At that time, Weierstrass had been working without success on the long-term stability of planetary orbits. He had taken up the problem because it was widely believed that Dirichlet had already proven the fact thirty years earlier, but had died before he could publish his results. The competition was announced in July 1885, and the prize was to be awarded in January of 1889 on the King's birthday. The problem was:

> Given a system of arbitrarily many mass points that attract each other according to Newton's law, under the assumption that no two points ever collide, try to find a representation of the coordinates of each point as a series in a variable that is some known function of time and for all of whose values the series converges uniformly. (King Oscar II Award announcement: *Acta Mathematica*, vol. 7 1885–1886.)

In other words, it tasked the entrants to find stable solutions to an n-body gravitational problem. The physics of two mutually gravitating bodies is completely solvable in closed form, but the physics of three bodies is not. Calculations can trace the time-evolution of three interacting bodies, but no general function of the orbits can be obtained. The three-body problem had defied the efforts of the world's most renowned mathematicians, including Newton and Euler.

Euler had investigated simplified versions of the three-body problem around 1760 and wrote several memoires treating the problem of a body attracted to two fixed centers of gravity and moving in the plane.[24] He was able to solve it using elliptic integrals. Although the Earth and Sun are not fixed, the solution to this problem provides a simple model that approximates the physical case when the two fixed centers are viewed in a coordinate frame that is rotating with the Sun-Earth system. In 1762 he tried another approach where he considered a massless Moon attracted to a massive Earth orbiting a massive Sun, again all in the plane. This model of the three-body problem is called the restricted three-body problem. Euler could not find general solutions to this problem, but he did stumble on an interesting special case when the three bodies remain collinear throughout their motions in a rotating reference frame.

When Poincaré read the competition announcement he was attracted by the similarity between the stated goal of the prize and a topic upon which he was already working. In his position at the Corps des Mines (during which time he witnessed and was assigned to investigate one of the worst mine explosions in France), Poincaré pursued his PhD in mathematics under Hermite at the

Polytechnique in Paris. His thesis studied the properties of functions defined by difference equations. A difference equation is a discrete iterated mapping that defines a point in terms of a previous point, taking the form $x_{n+1}=F(x_n)$ in a simple one-dimensional case, or $\vec{x}_{n+1} = \vec{F}(\vec{x}_n)$ in the multidimensional case. Poincaré became skilled at using difference equations, developing ever-deeper intuition into their behavior, and how they might be applied to solve practical problems. After receiving his PhD in 1879, Poincaré published several papers that explored increasingly sophisticated properties of groups of solutions of differential equations. He was not so interested in obtaining specific solutions, or single trajectories, but sought instead to understand and classify the broader behavior of entire groups of trajectories.

These papers were the beginnings of the major shift towards the general definition and use of phase space, ushering in a far-reaching new perspective on Galileo's trajectory. The new approach was a global view, asking questions that were qualitative, such as whether trajectories remained bounded within finite regions, or whether they were drawn towards or diverged away from particular positions. This method departed from the normal practices of the day that emphasized numerical accuracy in the positions of planets and asteroids. Poincaré's work was accurate in the sense of mathematical proofs and theorems, but not numerically. He was asking the big questions rather than focusing on the details, and the stability of the solar system was perhaps the biggest question of all.

Poincaré formulated King Otto's prize problem in terms of Euler's restricted three-body problem, and he seemed to prove the absolute stability of the three-body problem. Poincaré was awarded the prize on 21 January 1889, and as part of the prize process he began to write up his essay for publication in *Acta Mathematica*. The paper was already through proofs and initial printing late in 1889 when he checked one of his most important conclusions and discovered an error. He had originally shown that, if the motion of the small body were perturbed slightly, it would remain arbitrarily close to the original motion. But a Swedish mathematician who was sent the manuscript for review could not reconcile this conclusion of the recurrence theorem with a more detailed analysis of Poincaré's difference equations. As Poincaré attempted to respond to this reviewer's concerns, he became concerned himself, then alarmed, and then mortified as he realized that his grand conclusion of the stability of the solar system was false. He wrote a hasty letter to Mittag-Leffler imploring him not to print the essay, but it was too late. Mittag-Leffler subsequently wrote hasty letters to his colleagues around Europe who had received advance copies, asking them to destroy or return them. Weierstrass was furious. What, he demanded to know, remained of the prize-winning essay if stability was no longer assured? Poincaré would take nearly two months to answer this question, and in the process, would discover *chaos*.

Poincaré worked feverishly from December 1889 to January 1890 to correct the manuscript, and then paid for the reprinting of the journal volumes out of his own pocket.[25] His mistake arose at special points in phase space known as homoclinic points. A homoclinic point is like the saddle of a mountain pass. Along the direction of a trail through the pass, the elevation of the trail rises from one valley to a maximum and then falls again into the adjacent valley. In the orthogonal direction, transverse to the trail through the mountain pass, the elevation rises on both sides to the adjacent mountain peaks. If a ball is placed at the very center of the mountain pass, it will perch there precariously forever. This is known as a fixed point—in this case a homoclinic fixed point. If the ball is slightly displaced along the trail, no matter how little, it will fall away to one valley or the other. The path of the ball in this case is known as the unstable manifold. Conversely, if the ball is displaced perpendicular to the trail, rising slightly towards one mountain peak or the other, the ball will return to the center of the pass. The path of the ball in this case is known as the stable manifold. In terms of dynamical systems, points on the stable manifold move towards the fixed point, while points on the unstable manifold move away. However, the homoclinic fixed point is a position of unstable equilibrium, and the slightest deviation of the system state from this point eventually must fall away. The instability of the homoclinic points in the three-body problem was the source of difficulty in Poincaré's approach to the stability of the system. The question was how to analytically describe the trajectories passing near these points (see Fig. 6.3).

Poincaré used his expertise in difference equations to convert the complicated differential equations of the three-body problem into easier-to-analyze difference equations. He achieved this by envisioning an infinite plane stretched across phase space. Every time the system trajectory passes through this plane, the point of intersection is recorded, and successive iterations of the trajectory form a sequence of points. This procedure creates a discrete map, also known as a *Poincaré*

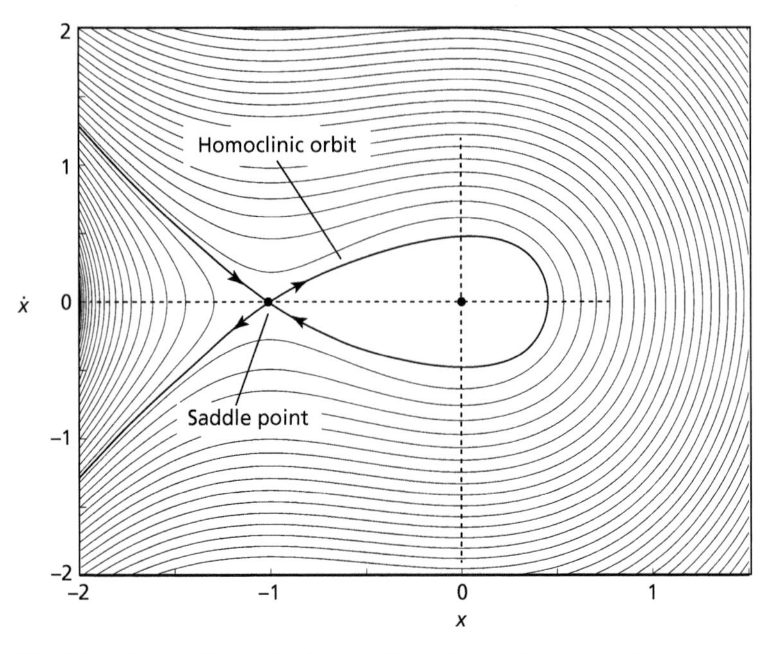

Fig. 6.3 A homoclinic orbit in two-dimensional phase space for a system with one degree of freedom. The black trajectory is the homoclinic orbit. The neighboring trajectories are either unbounded (outside the homoclinic orbit) or periodic (inside the homoclinic orbit). The homoclinic orbit crosses itself at the homoclinic point that has stable and unstable manifolds. One unstable manifold turns back on itself and becomes the stable manifold. When one more degree of freedom is added to this problem, the homoclinic point becomes the source of chaotic behavior[26]

first-return map, and the plane of intersections is called a *Poincaré section* (see Fig. 6.4). Poincaré introduced the concept of a limit cycle when a trajectory converges onto a steady-state orbit of the full phase space. In this case, the successive points of the first-return map approach a limiting point that is the fixed point of the discrete mapping. If the limit-cycle orbit is stable in the full phase space, then small deviations will relax back to the stable orbit. But if the limit-cycle orbit is unstable, then small deviations will move the successive points away to successively larger distances. Some orbits have mixed behavior, with some points approaching and some points diverging from the fixed point—this describes a homoclinic point in the Poincaré section.

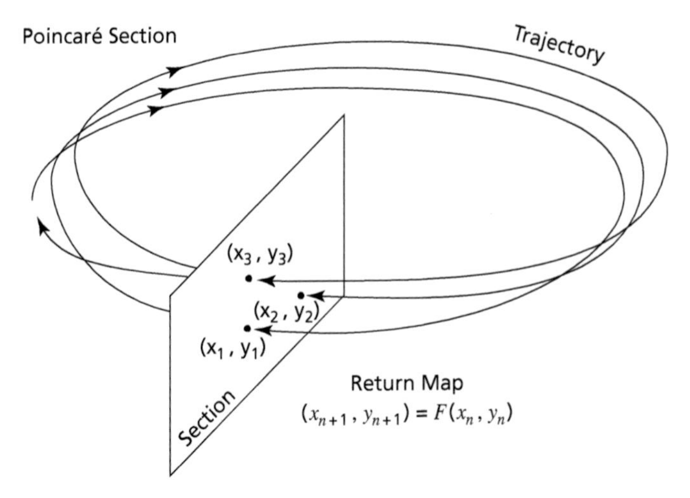

Fig. 6.4 View of a Poincaré section in 3D phase space. A system trajectory intersects the plane at discrete points x_1, x_2, x_3, ... that reduce the dimensionality of the dynamics and can be analyzed using difference equations

Poincaré traced the sequence of points falling away from the homoclinic point along trajectories on unstable manifolds, as well as converging towards the homoclinic point along stable manifolds. What he discovered in the three-body problem was that unstable manifolds and stable manifolds never ended, so that stable manifolds and unstable manifolds had to coexist within the same regions of the Poincaré section. Furthermore, the unstable manifold from one homoclinic point, as it approached the homoclinic point along the direction of the stable manifold, crossed the stable manifold an infinite number of times, executing ever wilder swings to and fro across the section. Some of these swings would cross the unstable manifold near the original homoclinic point, creating new homoclinic points on the section that would generate new unstable and stable manifolds with their own wild swings, and so on.

This "tangle" was generating the arbitrarily large response to small changes in initial conditions that he had discovered. He was amazed by his own findings and stated that:

> [I]f one seeks to visualize the pattern formed by these two curves and their infinite number of intersections... these intersections form a kind of

> lattice-work, a weave, a chain-link network of infinitely fine mesh; each of the two curves can never cross itself, but it must fold back on itself in a very complicated way so as to recross all the chain-links an infinite number of times. . . . One will be struck by the complexity of this figure, which I am not even attempting to draw. Nothing can give us a better idea of the intricacy of the three-body problem, and of all the problems of dynamics in general.[27]

It was clear to him that he had discovered a fundamentally new aspect of dynamical motion. This was the discovery of sensitivity to initial conditions, which is at the heart of chaos theory.

Part of this same publication was the first clear exposition of the idea of *the* system trajectory: a single trajectory (of a many body system) through a multidimensional space with its many axes defined by all the dynamical variables of the many bodies. He stated, "if there are p degrees of freedom, the situation of the system can be represented by the position of a point in the space of 2p-1 dimensions."[28] Poincaré's use of the word "situation" is equivalent to Boltzmann's "phase." Poincaré also generalizes the word "space" to describe the multidimensional dynamical space of the dynamical system, no longer reserving that word only for our physical three-dimensional space[29]. This expression of *the* system trajectory came in the section titled *Tentative Generalizations* of his Acta publication and is dated 1888, which is when he finished his submission to the prize committee rather than when the paper appeared in print.

Poincaré expanded his studies of dynamic motion and published his results in three volumes *New Methods of Celestial Mechanics.* Much of the work was carried out in phase space, introducing new tools and geometric approaches that have become workhorses of modern dynamics, such as Poincaré maps and fixed-point classifications. The third volume contained the material introducing chaotic motion arising from the homoclinic tangle. Along the way, Poincaré derived the theorem on the conservation of phase space volume, that he called an integral invariant, unaware of Boltzmann's derivation of it, who in his turn had at first been unaware of Jacobi's original derivation.

Boltzmann's Atoms

Several years after Poincaré's triumph, Boltzmann collected his thoughts and contributions to kinetic theory into his book *Lectures on Gas Theory* (1896). In it, he refers to n-fold integrals over n-fold regions. In the *Lectures*, Boltzmann defines *phase* for the first time in its modern context as the collective state of the gas defined by the positions and momenta of all the gas molecules:

> Instead of saying that the values of the variables lie within the region g_1 for a certain system, we shall often use the expression: this system has the phase *pq*. We can therefore say also: the expression gives the number of systems that have phase *pq* at time t.[30]

Boltzmann takes a step closer to describing a single trajectory of the system through its dynamical space:

> When one wishes to discuss any curve whose equation contains an arbitrary parameter, it is customary to consider simultaneously all the curves obtained by giving this parameter all its possible values. We are now dealing with a mechanical system (characterized by given equations of motion) whose motion depends on the values of the 2μ parameters P, Q.[30]

This view of a dynamical system as a single trajectory is made by analogy to a curve rather than made explicitly. Boltzmann still does not seem to be able to take that final step of speaking of a single trajectory through a multidimensional space. All the mathematics are in place, and he makes the analogies, and uses the geometric language of n-fold regions, but he implies the trajectory rather than stating it explicitly. This inability is clearly related to the fact that he never uses the word *space*, nor did he take the last step to call it *phase space*.

Given Boltzmann's invention of phase space and his discovery of conserved volumes within it, one wonders why Liouville gets all the credit today. The answer is that Boltzmann himself is to blame through his generosity. Although by 1871 Boltzmann already knew of Jacobi's application of the last multiplier, in which Jacobi referenced Liouville's theorem, it was only later in his *Lectures* of 1896 that Boltzmann generously and definitively placed Liouville's name on the conservation theorem in a way that stuck.[31] Had it not been for Jacobi's reference to Liouville in his Vorlesungen, Boltzmann would likely never have known of Liouville's paper. In turn, had Boltzmann not given the credit to Liouville, then what could reasonably have been called Boltzmann's theorem on the conservation of phase-space volume, is called Liouville's theorem because of Boltzmann's own generosity. Ironically, his naming it "Liouville's Theorem" is one of the chief reasons for the obscurity of Boltzmann's own role in the discovery and use of phase space.

Boltzmann's stature in the history of physics is assured by the importance of so many of his original contributions. Some stand alone, such as his derivation of the kinetic theory of entropy. Others attracted new supporters like Einstein with his explanation of Brownian motion. Echoes of Boltzmann's ideas were to be heard during the development of modern physics, as in Einstein's proof of the existence of the quantum of electromagnetic energy, later named the photon. The reason why Einstein drew so heavily from Boltzmann was because of the power of Boltzmann's methods. But all of this came too late for Boltzmann, who took his life in 1906 by the shores of the Adriatic Sea. Boltzmann had succumbed to his lifetime battle against depression—a depression brought about partly by his anguish at seeing his life's work nullified by his colleagues and his inability to master the necessary philosophical rigor to counteract them. He thought all that he had fought for had been lost, but he was wrong, because it was all about to begin.

Eherenfest's Legacy

The American theoretical physicist J. Willard Gibbs (1839–1903) was contemporary with Boltzmann. Gibbs began his professional career at Yale University in New England in 1871 a year before Boltzmann published his paper on his H-theorem. Despite being in a scientific backwater that, at that time, had no serious scientific culture (the United States), Gibbs' work had great impact internationally. He was a clear and careful thinker, and had a knack for seeing new things for what they were, and naming them, as others groped through the fog.

Gibbs' first scientific paper in 1873 was on the graphical representation of thermodynamic properties and the phase diagrams of phase transitions (the use of *phase* here is in the sense of a phase of matter and not the phase of a point in phase space). In England, Maxwell was so taken with the idea that he personally constructed a three-dimensional clay model of Gibbs' thermodynamic surface and sent Gibbs a plaster cast of it. Gibbs followed this initial paper with a three-hundred-page treatise *On the Equilibrium of Heterogeneous Substances*, published in two parts in 1875 and 1878 that helped establish the foundations of thermodynamics and extended thermodynamics to systems of many components. Gibbs' tome is considered the "*Principia*" of chemical thermodynamics and established many of the equations and notations in thermodynamics, not the least of which is the equation of state[32] used by every student of thermodynamics today that combines the first law of thermodynamics (energy conservation) with the second law (entropy) while introducing the new concept of chemical potential.

Towards the end of the 1890s, Gibbs turned his keen eye to the works of Boltzmann and the topic of statistical mechanics. Boltzmann had discovered and established most of the important features of statistical mechanics, but he did not have Gibbs' talent for clarity. Much of Boltzmann's work has an anachronistic look and feel to it. Boltzmann's *Lectures on Gas Theory* of 1895 was his crowning achievement, but it was a work firmly rooted in the sensibilities of the nineteenth century. Gibbs took a fresh look at the subject, giving it its modern name "statistical mechanics" and simplifying many of the complicated theoretical devices and physical arguments of Boltzmann.

Gibbs' last published work was his *Elementary Principles in Statistical Mechanics*, published in 1901. Here we see an explicit reference to the trajectory of the phase point in a high-dimensional space, expressed as a footnote in his textbook, stating "if we regard a phase as represented by a point in space of 2n dimensions, the changes which take place in the course of time in our ensemble of systems will be represented by a current in such space."[33] However, even in Gibbs' *Elementary Principles* there remain anachronisms and hesitations. Although he rederives the conservation of phase space volume, he does not use the word "volume," but instead uses the term "extension" borrowed from Grassmann. Even more telling is that the only place where Gibbs uses the word "space" is in a footnote, as if the notion of a trajectory in a high-dimensional space was an aside or an analogy, literally a footnote, rather than a fundamental principle. Even Gibbs was not immune to the prejudices about space at the turn of the century. Although Gibbs was a great inventor of terminology, giving us "statistical mechanics" and "ensemble" and establishing the modern nomenclature of vector analysis, he did not invent the phrase "phase space." That was to come a decade later.

The "space" aversion at the end of the 19th century quickly evaporated by the first decade of the twentieth century, especially with the advent of relativity and the growing conception of Minkowski's four-dimensional space time.[34] Boltzmann by this time was dead (by his own hand), but one of his students, Paul Ehrenfest (1880–1933), was asked by Felix Klein to write a review of Boltzmann's work for the Encyclopedia of Mathematical Sciences. Paul Ehrenfest, with his physicist wife Tatyana, published the encyclopedia article in 1911.[35] They approached the subject systematically, seeking to make precise definitions. This was partly in response to the controversies that had raged during the latter part of Boltzmann's life on proofs or dis-proofs of the ergodic nature of gas systems. Therefore, the Ehrenfests took great pains to define a rigorous name for the multidimensional dynamical space... and used the term *Γ-space* where the instantaneous state of the system was the *Γ-point*.

There is an irony here. By this time, the stigma of using the expression *space* for *n*-dimensions had disappeared, and so the Ehrenfests were comfortable using *space* to define Boltzmann's *n*-dimensions. But they dispensed with the term *phase*, possibly because of its obscurity. Yet, at the very beginning of the encyclopedia article, in order to set the context for the definition of *Γ-space*, they referred back to Boltzmann's usage of *phase*, briefly mentioning *Phasenraum* (phase space) in the article to set the stage, and then dispensed with it. This was not the first use of *Phasenraum* in print, which appears for the first time in a paper by Paul Hertz in 1910 on the foundations of statistical mechanics.[36] Hertz and the Ehrenfests carried on an extensive correspondence during the five years that the Ehrenfests were developing their encyclopedia article, so it is possible that the term *Phasenraum* came originally from the Ehrenfests. However, it is also possible that the term had been used informally in the hallways, so to speak, among Boltzmann's followers.

Encyclopedia articles in the Ehrenfests' day were widely read, like Reviews of Modern Physics today, and the Ehrenfests' article was no exception. And here is the irony: what stuck in readers' minds was the toss-away phrase *phase space*, while virtually everyone ignored his *Γ-space*. Within two years of the Ehrenfests' article of 1911, two papers appeared in the same issue of *Annalen der Physik* using the expression *phase space*. These were papers in 1913 on ergodic theory by Artur Rosenthal[37] and Michel Plancheral,[38] applying the new measure theory to the question whether ergodic systems, the systems that Boltzmann relied upon so heavily for his derivations of entropy and the H-theorem, could exist. The unequivocal conclusion of both authors was that ergodic systems were impossible! Had Boltzmann still been alive, he would surely have felt struck to the core, as he always did when his work was questioned. However, the mathematicians were concerned with systems that visited *exactly* every point in phase space, while Boltzmann's arguments were based on systems that approached as closely as needed to every point in phase space. This subtle difference allowed Boltzmann's results to survive Plancheral and Rosenthal. The result of these papers was that the usage of the term *phase space* stuck, first appearing in a journal paper title in 1918,[39] and becoming increasingly common after that.

New Abstractions

The story of phase space fits neatly within the bookends of the nineteenth century, beginning with Lagrange and his generalized coordinates at the turn into the century and ending with Boltzmann and Gibbs at the turn out of the century. During those hundred years, a major shift occurred in the concept of the trajectory. Lagrange had freed particles and bodies from the rigid spatial dimensions of x and y and z, allowing their dynamics to be described by any parameters whose variations captured the degrees-of-freedom of the dynamical system, but this new freedom came with a price. Geometry disappeared from dynamics, replaced by the cold analytic equations of rational mechanics. Lagrange was genuinely proud that he had banished geometry from dynamics, because in his view, the spatial trajectory no longer mattered.

When Jacobi combined Liouville's theorem with Hamilton's dynamics, he took the first step towards rehabilitating geometry as a worthy context of dynamics. Jacobi was one of the first mathematicians of the nineteenth century to begin to generalize the idea of dimensions to higher levels, including expressions for higher-dimensional surfaces and volumes. He did not do this with explicit language other than the language of mathematics and functional analysis, but the mathematical forms were high-dimensional integrals, yielding expressions that are easily interpreted today as four-dimensional volumes and beyond. But Jacobi was still of the old school of rational mechanics and did not change his language or his outlook. He had created phase space without realizing it, and so he could not put a name to it because he did not recognize it.

The urgency to enlist phase space as an important new tool in dynamics was driven by the extreme demands placed on statistical mechanics to make sense of the physical interactions of an Avogadro's number of particles. Boltzmann extended Jacobi's work into these extreme conditions, and multidimensional volumes within phase space became his weapons to combat the elusive vagaries of entropy, as well as to deflect the annoying arrows of his foes. He made Liouville's name famous, recognized by any student of physics today, by showing that it was only after integrating over both positions and momenta that invariant properties of the dynamics emerged—the conservation of phase-space volume. His personal generosity towards Liouville doomed the memory of his own central role in the tangled tale of phase space. But Boltzmann could not put the true name to his creation. It took Poincaré, fully versed in the new methods of multidimensional spaces of the later nineteenth century, to conceive of the system trajectory, finally expressed in explicit language by Gibbs and Ehrenfest. In this language, a single point moving through a high-dimensional phase space represents the trajectory of a system of many particles—even a huge number of particles equal to Avogadro's number. Here is the central abstraction, the major expansion beyond Galileo's trajectory!

Along the way to this grand new view of physics, the seeds of later revolutions were sown. The strange uncertainty of Clausius' entropy prophesied the future uncertainties of Heisenberg in quantum theory, and the tangled homoclinic orbits of Poincaré presaged Lorenz' butterfly in chaos theory. By the beginning of the twentieth century, classical physics had expanded to fill vast new abstract spaces, but even greater abstractions were about to challenge the logical structure of reality. Quantum physics and chaos theory would condemn us to fundamental ignorance while freeing us of our fate. Meanwhile, lurking in the shadows was gravity, the oldest and weakest of the physical forces, where trajectories are not only described by geometry but are caused by it.

7

The Lens of Gravity

> [The Star,] like the Cheshire cat, fades from view. One leaves behind only its grin, the other, only its gravitational attraction.
>
> JOHN ARCHIBALD WHEELER, December 1967.[1]

To fall freely is an odd and rare experience. The first few moments after jumping off a low wall give only a fleeting feeling of free fall. Even if you jump out of an airplane (with a parachute), there are only a few seconds before air friction begins to dominate and you approach terminal velocity. To truly experience free fall, you need to take a ride on NASA's "vomit comet," the research jet that repeatedly makes parabolic dives to match Galileo's Law of Fall. Even better, you can become an astronaut, or a high-paying future customer of Virgin Space.

Free fall is force-free motion—at least it is free of forces in your freely falling reference frame. In spite of gravity's incessant tug on you during free fall, you personally experience no forces. The strange absence of all those action-reaction pairs, on all the components of your body pushing against each other, induces free-fall sickness. Astronauts in the International Space Station (ISS), at an altitude of only 400 km above the surface of the Earth, are subject to a force of gravity that is only a little weaker than on the surface of the Earth (by about 11 percent), yet they experience no forces at all as they fall freely in their nearly circular orbit. This is not a matter of perception or psychology, because an accelerometer on the ISS measures zero g's. The circular motion of the orbit is not necessary to draw this conclusion. Even an accelerometer falling directly to Earth from space, accelerating at the acceleration of gravity, measures zero g's. Free-fall acceleration under gravity is acceleration free! The other word for this is *inertial*—the body in free fall is in an inertial frame—but what has happened to gravity?

Galileo Unbound. David D. Nolte, Oxford University Press (2018).
 DOI: 10.1093/oso/9780198805847.001.0001

What Stays the Same if Everything is Relative?

At the turn of the twentieth century, the place to be in mathematics was at the University of Göttingen (see Chapter 5). It had a long tradition of mathematical giants that included Carl Friedrich Gauss, Bernhard Riemann, Peter Dirichlet, Felix Klein and David Hilbert. Under the guidance of Felix Klein, Göttingen mathematics was undergoing a renaissance. Klein had attracted Hilbert from the University of Königsberg in 1895, and in 1902 succeeded in bringing Minkowski from Zürich (where he had been one of Einstein's teachers). Minkowski and Hilbert had been raised as children in or around Königsberg and had become friends when they met at the University.[2] While Minkowski was still a student at Königsberg, he published a manuscript on the theory of quadratic forms, for which he received the Mathematics Prize of the French Academy of Sciences in 1883. A quadratic form is a function of squares, for instance the Pythagorean theorem $a^2 + b^2 = c^2$ for a right triangle where the sum of the squares of the lengths of the sides equals the square of the length of the hypotenuse. Quadratic forms are easily extended into multiple dimensions, and can take on unexpected properties when applied to distances in Riemannian geometry. Of particular importance in the theory of quadratic forms are invariants whose values remain unchanged when the coordinates are altered by a coordinate transformation.

Shortly after Minkowski arrived at Göttingen, the relativity revolution broke, and he was perfectly situated to apply his theory of quadratic forms and invariants to the Lorentz transformations derived by Poincaré and Einstein. Although Poincaré had published a paper in 1906 that showed that the Lorentz transformation was a generalized rotation in four-dimensional space,[3] Poincaré continued to discuss space and time as separate phenomena, as did Einstein. For them, simultaneity was no longer an invariant, but events in time were still events in time and not somehow mixed with space-like properties. Minkowski recognized that Poincaré had missed an opportunity to define a four-dimensional vector space filled by four-vectors that captured all possible events in a single coordinate description without the need to separate out time and space.

In 1908, at the age of 44, Minkowski presented a paper at the eightieth *Assembly of German Natural Scientists and Physicians* (21 September 1908). In his opening address, he stated that[4]:

> The views of space and time which I wish to lay before you have sprung from the soil of experimental physics, and therein lies their strength. They are radical. Henceforth space by itself, and time by itself, are doomed to fade away into mere shadows, and only a kind of union of the two will preserve an independent reality.

This audacious claim was not mere grandstanding. He was well versed in Riemann's metric description of manifolds, and his work on quadratic forms included non-Euclidean spaces that had hyperbolic coordinates. With this background, Minkowski was able to construct a valid four-vector in what is now called *Minkowski space*.

To illustrate his arguments Minkowski constructed the most recognizable visual icon of relativity theory—the space-time diagram in which the trajectories of particles appear as "world lines," as in Fig. 7.1. On this diagram, one spatial dimension is plotted along the horizontal axis, and the value *ct* (speed of light multiplied by time) is plotted along the vertical axis. In these units, a photon travels along a line oriented at 45 degrees, and the world line (the name Minkowski gave to trajectories) of all massive particles must have slopes steeper than this. For instance, a stationary particle, that appears to have no trajectory at all, executes a vertical trajectory on the space-time diagram as it travels forward through time. With this new formulation by Minkowski, space

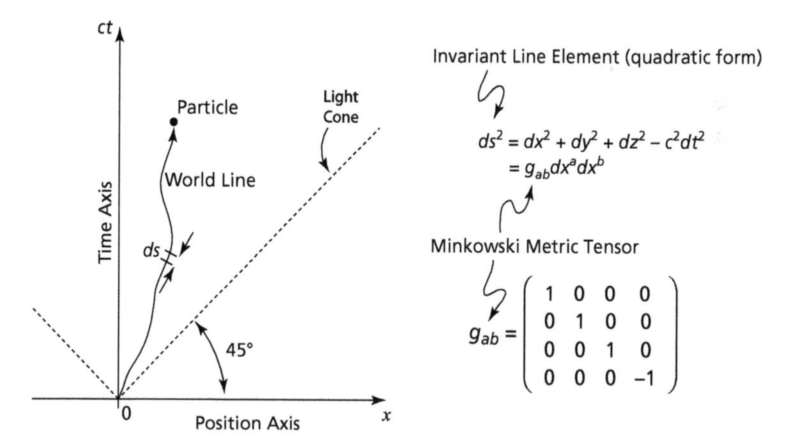

Fig. 7.1 The world line trajectory in Minkowski space. Light travels at an angle of 45 degrees. The world line of a massive particle must always travel inside a sequence of light cones attached to each point on the trajectory

and time were mixed together in a single manifold—space-time—and were no longer separate entities. This audacious step makes it an important landmark on the historical trajectory from Galileo's simple Law of Fall to today's complex dynamics in high-dimensional spaces.

The forward progress of physics has often been driven by the search for invariants, from Aristotle and his teleology, through Galileo and his Law of Fall, Newton and his Universal Gravitation, and Maupertuis and his least action. With space-time, Minkowski unveiled a powerful new form of invariant with which to tame the confusion of shifting frames—quadratic forms. Quadratic forms on a manifold are invariant to transformation in coordinates. This is just a simple statement that a vector is an entity of reality that is independent of how it is described. The length of a vector in our normal three-space does not change if we flip the coordinates around or rotate them, and the same is true for four-vectors in Minkowski space subject to Lorentz transformations. In relativity theory, this property of invariance becomes especially useful, because part of the mental challenge of relativity is that everything looks different when viewed from different frames. How do you get a good grip on a phenomenon if it is always changing, always relative to one frame or another? Invariants are anchors that we can hold on to as reference frames shift and morph about us.

For instance, the mass of a particle in its rest frame becomes an invariant mass, always with the same value. In earlier relativity theory, even in Einstein's papers, the mass of an object was a function of its speed. How is the mass of an electron a fundamental property of physics if it is a function of how fast it is traveling? The construction of invariant mass removes this problem, and the mass of the electron becomes an immutable property of physics, independent of the frame. Invariant mass is just one of many invariants that emerge from Minkowski's space-time description. The study of relativity, where all things seem relative, became a study of invariants, where some things never change. In this sense, the theory of relativity is a misnomer. Ironically, relativity theory became the motivation of post-modern relativism that denies the existence of absolutes, even though relativity theory as practiced by physicists is all about absolutes.

When Minkowski first proposed his theory in 1908, mathematicians received it favorably, but physicists were aghast. He was tinkering with the "real" space in which all physicists worked. Even Poincaré, who was a leading philosopher of physics as well as a mathematician, and who was one of the first to put forward a principle of relativity, believed that three-dimensional space was a part of objective reality—a view called his doctrine of physical space.[5] Furthermore, physics in that day was grounded on the principle that space was Euclidean. Minkowski's space-time, on the other hand, was not Euclidean, nor did it subscribe exactly to Riemannian geometry, because space had no separate reality from time, and distances in space-time could be negative. To physicists, even mathematical physicists, this looked like mathematical wizardry. Just because Lorentz' equations could be mapped onto a four-dimensional non-Euclidean pseudo-Riemannian geometry did not make it real—it was just a mathematical analogy. Several leading physicists criticized Minkowski's grand proposal, and Einstein himself was one of his harshest critics.

Minkowski was probably taken aback by the hostile reception among physicists, even his former students, of what he considered to be the apex of his life's work, and what he hoped was a momentous contribution that would provide a new world view. He immediately began to tone down the more unappetizing aspects of his theory, emphasizing "world" events and "the world" in place of space-time events and space-time, and he played down aspects of non-Euclidean geometry. He probably would have continued to woo physicists to his theory, but he died suddenly of a burst appendix in 1909. This sad event did not end the prospects for space-time, because Arnold Sommerfeld (who went on to play a central role in the development of quantum theory, as presented in Chapter 8) took it up, and he systematized it in a way that was palatable to physicists. Then Max von Laue extended it while he was working with Sommerfeld in Munich, publishing the first physics textbook on relativity theory in 1911, establishing the space-time formalism for future generations of German physicists. Further support for Minkowski's work came from his distinguished colleagues at Göttingen (Hilbert, Klein, Wiechert, Schwarzschild) as well as his former students (Born, Laue, Kaluza, Frank, Noether). With such champions, his work was immortalized in the methodology of physics, representing one of the crowning achievements of the Göttingen mathematical community.

Einstein, at first hostile to space-time theory because he felt it was overly complicated, came around to embrace it as he took a deeper look at the acceleration of masses in free fall. Galileo (and Stevin) observed that all bodies fall at the same rate regardless of their weight, and

Newton derived this fact simply by equating inertial mass with gravitational mass. But there was no explanation of why these two masses—the mass m that appears in the formula $F = ma$, and the other mass m that appears in the formula $F = GmM/R^2$—should be the same. As Einstein was about to tackle this long-standing problem of physics—the equivalence of inertial and gravitational mass—he needed Minkowski's space-time to succeed.

Warping Space-Time

Despite the power and success of relativity theory, it still had not answered some of the Big Questions. It had not gone beyond Newton to answer why gravity acts at a distance. Furthermore, the theory by early 1907 only applied to reference frames in uniform motion—to inertial frames. How would the theory of relativity treat frames that were not in uniform motion, for instance, frames that were uniformly accelerated? This was the situation when Einstein had an epiphany in November of 1907 as he was working on a review article on relativity theory for a German journal of radioactivity and electronics.[6] He called it the "glücklichste Gedanken meines Lebens" (the happiest thought of my life) "*Because for an observer falling freely from the roof of a house there exists*—at least in his immediate surroundings—*no gravitational field.*"[7]

This poor observer, if he lets go of a hammer, will fall at the same rate as the hammer, as if he and the hammer, or anything else he let go of, were in an inertial and force-free frame. Here is D'Alembert's principle again, the one that Lagrange used with such effectiveness, stating the apparent tautology that $F - ma = 0$. With this epiphany, Einstein immediately knew that "the relativity postulate has to be extended to coordinate systems which, relative to each other, are in non-uniform motion."[8] Relativity theory was not just a special case (hence the *Special* Theory of Relativity) but was more general (hence the *General* Theory of Relativity). Einstein had always suspected this, but until 1907 had not seen the path forward. Now he did.

Einstein turned the equivalence principle—that equates gravitational with inertial mass—into an extension of the relativity principle. He made this argument not by appealing to free fall, but by the opposite. He envisioned two frames. One far from any gravitating body that is accelerating uniformly with the acceleration g, and another frame that is stationary but in a gravitational field with acceleration $-g$. Einstein's new equivalence principle stated that no local experiments, performed by the observers in these two very different situations, could tell them apart. The uniformly accelerated frame was equivalent to the stationary frame in the gravitational field. Therefore, to discover the effects of gravity upon observations, all one needed to do was calculate those effects for the uniformly accelerating frame.

In rapid succession, Einstein applied his new equivalence principle to derive two important consequences of accelerating frames. The first was the gravitational redshift of photons escaping from a star—they must be slightly redder when they are collected at a large distance than when they were first emitted. The second and most dramatic effect was the bending of a light path by gravity. Einstein grasped these effects within a single month of his "happiest thought," and derived approximate theoretical expressions for both of them.

Einstein's derivation of the gravitational redshift considered that as light climbed out of the gravitational potential of the star, the wavelength would slightly increase, pushing it slightly towards the red end of the spectrum. Einstein derived this fact not by dealing with the gravitational potential, which was impossible because there was no theory of gravity and light yet, but by considering three frames of reference: one an inertial frame at rest, another non-inertial frame that is uniformly accelerating, and a third frame that is in uniform motion that, for a moment, is co-moving at the same speed as the accelerating frame. By comparing the first to the third frame (both inertial), he derived the redshift of light. There was no gravity in the derivation. Einstein merely applied the principles of special relativity to the two inertial frames, and then used the equivalence principle to translate the problem back to the gravitational situation.

Einstein's second derivation using the equivalence principle was the bending of light by gravity. A popular description today explains this simply by considering a light ray that enters a uniformly accelerating elevator at right angles to the elevator motion, as shown in Fig. 7.2. The light ray executes a perfectly straight path, but the elevator is moving upward with increasing speed. Therefore, in the time it takes the light to traverse the width of the elevator, the elevator has moved upwards, and the light exits the elevator at a spot on the wall below the entry level. The path of the light ray, as observed by someone in the elevator, is a parabola—Galileo's parabola. Yet the light ray in free space was a perfect

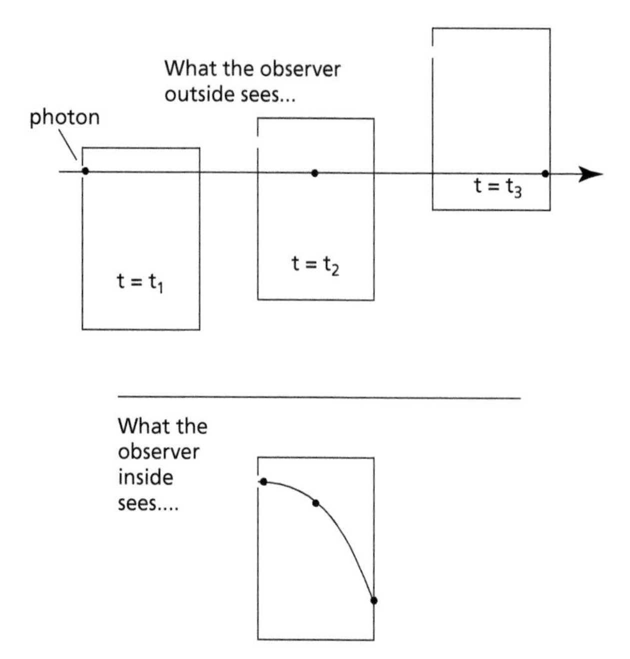

Fig. 7.2 The apparent deflection of a photon traveling through an elevator accelerating upwards at acceleration equal to g. The inside observer feels as if a uniform gravitational field is present, and observes the photon executing Galileo's parabolic trajectory[9]

straight line. Therefore, the observer in the elevator, who cannot tell the difference between uniform acceleration and a gravitational field, assumes that the light is bent by a gravitational field.

This simple elevator argument, attributed to Einstein, came years later. In 1907 Einstein took a different approach that yielded a result that caused him pause. Einstein naturally began with Maxwell's electromagnetic equations for the propagation of light. By applying the usual Lorentz transformation to the two frames using the equivalence principle, he found that Maxwell's equations had the same form as always, but with an unexpected side effect. The speed of light appearing in them was no longer a constant! This was a shock. The very cornerstone of special relativity is the constancy of the speed of light in all frames for all observers. But in the accelerating frame, the speed of light depends on the gravitational potential (after applying the new equivalence principle).

Without hesitation, Einstein boldly accepted this unexpected result and immediately understood that it would cause light to be deflected by a gravitational potential. His courage and confidence in the consequences of his own calculations are awe inspiring, but even *he* understood that he had taken a dangerous step beyond the ordinary, and beyond what many of his colleagues were prepared to accept. Though confident in his own conclusions, he was nervous about the reaction of the physics world. He need not have worried, because few of his colleagues read the review article, and fewer understood the arguments. Furthermore, Einstein himself was drawn into more challenging terrain by the quantum theory, which kept him away from his investigation into general relativity for nearly four years until 1911.

From 1905 through 1911 Einstein was engaged in a lively defense of the quantum theory of light that he had developed in 1905, and it took most of his time and attention. The photon was at the heart of wave-particle duality, which remains to this day a difficult problem of epistemology. Also during these years, Einstein's career was going through many changes. In the spring of 1908, he was given the position of *vienia docendi*, giving him the dubious right to teach without salary at the University of Bern, and a year later in 1909 he finally became an associate professor of theoretical physics at the University of Zürich, receiving a salary that allowed him finally to resign from the patent office. In early 1911, he was offered a full professorship at the German University in Prague, where he arrived in March to take up the post. Finally attaining some security, in terms of his position as well as international reputation, Einstein was ready to take up again the consequences of the equivalence principle.

The non-constancy of the speed of light in the presence of a gravitational potential must have been in the back of his mind ever since he first conceived of it in 1907. It was the most surprising and most troubling result of his initial foray into the consequences of the equivalence principle. Therefore, this was the first topic that he tackled in his continuing development of a general theory of relativity. As he grappled with the problem, he recognized that numbers were invariant properties amidst relative frames. If there are N quantities in one frame, then there must be N quantities in any other, regardless of how they moved or what potentials they might be in. In the emission of light from a source, the number of wave crests or troughs were just numbers,

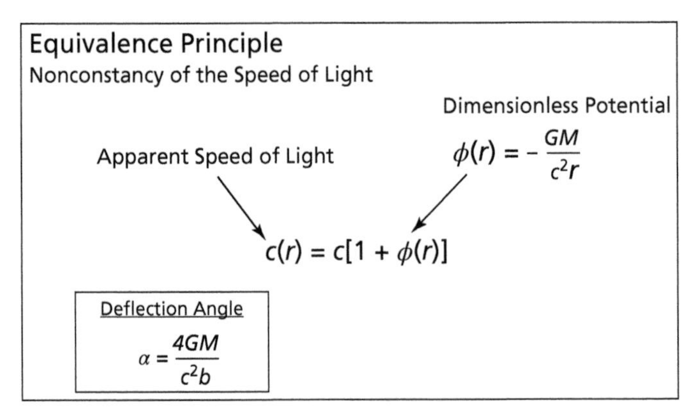

Fig. 7.3 The position-dependent apparent speed of light derived from the Equivalence Principle. The angular deflection of light by the Sun for a light ray with impact parameter *b*, derived from the Equivalence Principle of 1911 was half the correct value of 1915 that included the warping of space

and hence they must be the same for all observers. Using this simple approach, he was able to show that the speed of light in a gravitational potential depended on the gravitational potential difference between two locations. Because the speed of light changes depending on its position (see Fig. 7.3), Einstein further recognized that this is nothing other than a problem of the refraction of light. In a single night, using Huygens's principle,[10] Einstein calculated the deflection of light caused by refraction when a light ray passes near a 1/r potential. He published this result in 1911 and challenged astronomers to observe this deflection in the positions of stars. For light passing by the limb of the Sun, the angle he calculated would have been just barely measurable given the equipment at the time. Of course, it would need to take place during a total eclipse in order to see the stars' positions, but it was within the realm of possibility.

In his paper of 1911, Einstein had pushed the equivalence principle as far as it could go using only the Lorentz transformations. Yet, Einstein was fully invested in seeking a general theory of gravitation, so he pushed beyond the equivalence principle and beyond the Lorentz transformations. In early 1912, he published two papers that extended his results of the non-constancy of the speed of light in the presence of a gravitational potential. He expanded this idea into a *field theory*, in which the speed of light becomes a field that fills all space, and in the 1912 papers, the erstwhile constant *c* became a static field.[11] Einstein derived a scalar field equation in which the spatial variation of the field depended on the sum of mechanical and electromagnetic energy as well as on the density of the field, making it a nonlinear equation in which the calculation of a property depends on the property itself.

This nonlinear equation had the disturbing feature that it removed the general validity of the equivalence principle. Just as he once had abandoned the constancy of the speed of light, the rock upon which the special theory was built, now he was moving beyond the very equivalence principle that had opened his eyes to all the possibilities of gravity. Apparently, Einstein would build an entire theory, and then tear it all down again in his inexorable pursuit of the answers he sought. But this had the side effect that it would cast him adrift, where once there had been anchors. He had succeeded in constructing a scalar field theory of gravity, which yielded some fruit, but he knew it was not correct. No sooner had he published his 1912 papers, than he already was looking for ways to supersede them, and he suspected he knew the direction to look. He noted "the laws of Euclidean geometry most probably do not hold in a uniformly rotating system in which, because of the Lorentz contraction, the ratio of the circumference to the diameter should be different from π if we apply definitions of lengths."[12] Here is the first statement that Euclidean geometry was not sufficient to capture the complete laws of gravity. Non-Euclidean geometry would be needed. The problem was that Einstein had no background in non-Euclidean geometry.

It was then that Einstein received a fateful letter from Marcel Grossmann (1878–1936), an old friend at the ETH in Zürich. Einstein's reputation had continued to climb, and he was now a highly sought-after candidate for faculty positions across Europe. Because he was Swiss, and had attended the ETH for his graduate studies, ETH held the best cards, and Grossmann was enquiring whether Einstein would be interested in a professorship at ETH. His reply was immediate and affirmative, and he and his family arrived back in Zürich in August of 1912. As soon as he arrived, Einstein already knew what he wanted to ask Grossmann. Einstein had decided that the key to finding a new way forward to a general theory of relativity lay in the variational principle applied to the line element, the same line element that Jacobi had geometrized with his principle of least action. Although Einstein did not have the math background to make progress, he knew who would—Grossmann.

Grossmann was a mathematician, and although he did not do research in non-Euclidean geometry, he was acquainted with the topic. At one of their first meetings, Einstein discussed with Grossmann whether non-Euclidean geometry might provide a fruitful basis for a theory of gravity. Between 10 August and 16 August 1912, in only a single week,[13] it is clear from Einstein's letters that his entire thinking shifted to the tensor theory of Gregorio Ricci-Curbastro (1853–1925). In short succession Einstein also became aware of Riemann's work and especially the central role played by the metric tensor and the line element $ds^2 = g_{ab}dx^a dx^b$. One can surmise that it must have been with at least a little regret when Einstein realized that Minkowski, whom he had severely criticized for his four-dimensional space-time, had hit on the correct basis from which to begin an attack on gravitation. Minkowski's space-time was the "flat" manifold that the equations of gravity must relax to in the limit of large distances. However, closer to a gravitating body, Minkowski's flat space-time would need to warp. The next step was to find out precisely how this should happen.

An intense period of collaboration began between Einstein and Grossmann that spanned about six months of work. This type of detailed mathematical work was not natural for Einstein, who was famous for stating simple postulates that led to relatively simple results that usually needed only straightforward math. With Riemannian geometry and Ricci's tensor calculus, he struggled, even with Grossmann's help. A famous quote from Einstein expresses this, written in a letter to Sommerfeld in October of 1912[14]:

> ... in all my life I have labored not nearly as hard, and I have become imbued with great respect for mathematics, the subtler part of which I had in my simple-mindedness regarded as pure luxury until now. Compared with this problem, the original relativity is child's play.

They forged ahead, applying tensor calculus to the invariant line element $ds^2 = g_{ab}dx^a dx^b$ subject to the condition of stationarity—the same form of stationarity developed by Lagrange in his variational calculus. In the hands of Einstein and Grossmann, the variational principle produced the simple equation $\delta \int ds = 0$. whose solution is a geodesic. In other words, the path of fall in a gravitational potential is a geodesic—the shortest and straightest path through the warped space-time. In addition, Einstein imposed the condition of a new and more powerful form of the equivalence principle. Einstein demanded that, for a valid theory of gravitation, a special coordinate transformation must exist that converts, at least locally, the metric tensor into the Minkowski metric tensor of flat space-time. This coordinate transformation represents the frame of free fall—the inertial force-free frame of his "happiest thought." By early 1913, after grueling sessions working together and long hours working independently, they arrived at a new formulation of gravity. Einstein's earlier scalar field equations were replaced by tensor field equations that defined how energy density warped space-time and how warped space-time determined the trajectories of particles. This to-and-fro between matter and space is captured in the teleological statement often made about General Relativity that "Matter tells space how to curve, and space tells matter how to move."[15]

The paper they wrote together and published in 1913 had two parts, one by Grossmann and one by Einstein. Grossmann restricted himself to the mathematics and Einstein provided the physics. They each separately read oral papers at the annual meeting of the Swiss Physical Society to great acclaim, but also to great incomprehension. They had moved beyond what physicists could handle. Later that same year they published their second and last paper together that added some observations about the properties of their field equation. By that time, Planck had wooed Einstein away from Zürich to Berlin. Although he was ambivalent towards Germany—the country where he was born, but also the country he had resolutely rejected when he became a Swiss citizen—he could not resist moving to the center of the physics universe at that time. Once in Berlin, Einstein continued working to solve several important shortcomings that still existed in the theory. The most important shortcoming was the lack, in the 1913 form of the theory, of what is known as general covariance.

General covariance for a law of physics requires that it must be independent of choice of coordinate frame, even frames that are in non-uniform motion. For example, Maxwell's equations, when written in their space-time differential forms with gradients and curls, has general covariance because no explicit coordinates are used to write them, and the equations look the same in all reference frames. When Einstein was still in Prague, before he had begun working with Grossmann, he already was seeking a theory of gravity that had general covariance. But after all the hard work with Grossmann, the theory had not achieved this goal. Driven by his conviction that general covariance *must* be obeyed by a correct theory of gravity, Einstein started over, going back to Riemannian geometry, and corresponding with Levi-Civita, a former student of Ricci's, who pointed out some errors in the 1913 paper.

Finally, after an additional year of hard effort, Einstein constructed a theory of gravity that had general covariance. The paper was published at the end of 1915 in the form that is now recognized as Einstein's general theory of relativity. Armed with the final theory, Einstein made three calculations, some of which he had presented in preliminary form before, but now presented in their complete form: 1) the precession of the perihelion of Mercury; 2) the gravitational redshift; and 3) the bending of light. The first calculation, on the precession of the perihelion of Mercury, was in close agreement with known observations. Here was the first dramatic success of the general theory of relativity. This precession had been known for many years to be beyond Newton's theory of gravity. It was also known *not* to be a consequence of special relativity. In Einstein's general theory, a slight deviation from Newton's $1/r$ potential is anticipated, which allows an elliptical orbit to precess, and the amount of precession agreed beautifully with the theory. Einstein's second calculation, on the gravitational redshift, agreed with his scalar theory and was not modified by the new theory. The gravitational redshift was beyond experimental spectroscopy capabilities at that time. In the third calculation, the bending of light from the scalar theory was now seen to have been only half correct. Light certainly is bent by gravity, but the complete theory modifies both space and time, and the angular deflection is twice as big as earlier prediction. This angle was within experimental capabilities, and Einstein had already challenged astronomers to observe the deflection of starlight by the Sun. None had yet done so, but in the midst of World War I, when most British scientists were cut off from German science, Arthur Eddington (a British astronomer and pacifist) was keenly following Einstein's work.

Arthur Eddington (1882–1944) was born in the countryside of northern England, the son of Quaker parents. He excelled at school, attended Trinity College at Cambridge and received his first position at the Royal Observatory in Greenwich in 1906. He returned as a Fellow to Cambridge and became the director of the Cambridge Observatory by 1914. As secretary of the Royal Astronomical Society during the war, he was one of the first to receive papers submitted for publication on topics related to Einstein's theory of General Relativity. In March of 1917, he and the Astronomer Royal Frank Dyson began planning a scientific

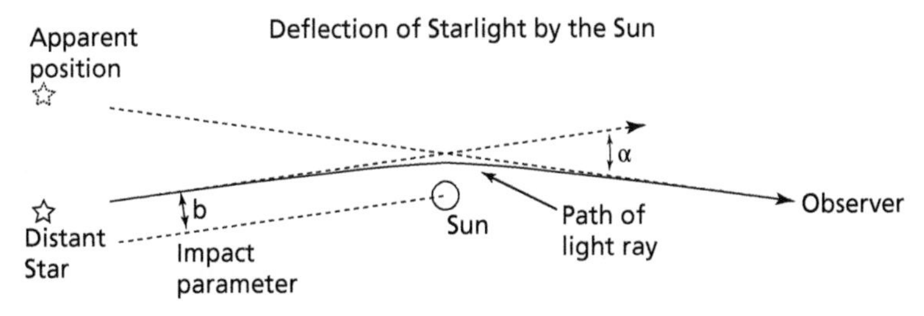

Fig. 7.4 Deflection of starlight by the Sun

expedition to the island of Principe off Africa, and to the town of Sobral in Brazil, which would be excellent viewing sites during a total solar eclipse on 29 May 1919. Stars near the limb of the Sun would only be observable when the Sun's disc was fully occluded by the Moon. At that moment, by comparing the positions of stars just beyond the radius of the Sun against their known positions in the sky, the displacement of the apparent positions could be measured (see Fig. 7.4). Eddington was well suited for this job, because his first position at the Royal Observatory was the study of stellar parallax caused by the motion of the Earth around the Sun. Parallax involves minute shifts in the apparent position of stars through the year, and Eddington had devised precise techniques to measure these deflections. He was now poised to use these same techniques during the eclipse of 1919.

Preparations for the expedition were well under way when Eddington was called up by the draft board in early 1918. The war had stalemated in the ugly trench warfare of northern France. Hundreds of thousands of British soldiers were being maimed or killed and reinforcements were necessary. Because of his Quaker upbringing, Eddington was a conscientious objector at a time when this was not recognized by the government and was punishable by imprisonment. Dyson attempted to intercede for Eddington on the grounds that the preparations for the scientific expedition were a national priority and that Eddington was indispensable to the project. The draft board probably understood neither the need nor the argument, but they did agree to postpone Eddington's enlistment for a short time. Fortunately for Eddington, the war ended in August of 1918, and by spring of 1919 they had devised portable observatories that could be installed in a matter of days—they were ready to travel. Both the Principe and Sobral expeditions left Liverpool on 8 March, going their separate ways at

Madeira. Eddington landed with his expedition on the Isle of Principe on 23 April, a little more than a month before the date of the eclipse.

Principe is a small island in the Gulf of Guinea about 120 miles off the west coast of Africa. It was a possession of Portugal and covered with cocoa plantations. Despite the years of planning and calibrations of their instruments, the success of the expedition ultimately hinged on the fickle weather. A tone of concern can be noted in the official publication of the expedition:

> The climate is very moist, but not unhealthy. The vegetation is luxuriant, and the scenery is extremely beautiful. We arrived near the end of the rainy season, but the bravado, a dry wind, set in about May 10, and from then onwards no rain fell except on the morning of the eclipse. We were advised that the prospects of clear sky at the end of May were not very good, but that the best chance was on the north and west of the island. The days preceding the eclipse were very cloudy. On the morning of May 29 there was a very heavy thunderstorm from about 10 a.m. to 11.30 a.m.—a remarkable occurrence at that time of year. The Sun then appeared for a few minutes, but the cloud gathered again.[16]

Minutes before the total eclipse, the Sun was still mostly behind clouds, but as totality approached, the Sun could be seen continuously, even if at times through a thin veil of clouds. The duration of totality was only five minutes, but many photographic plates were successfully acquired. The expedition to Sobral had a similar experience with clouds, but they too were able to make clear photographs during the five minutes of totality. With the data in hand, both parts of the expedition returned to England to begin the analysis.

The eclipse team had two reasons to be extraordinarily careful in the analysis of the data to make sure the results were correct beyond any doubt. The first, and most obvious, was the historic character of what the results would mean scientifically. The precession of the orbit of Mercury, explained by general relativity, had been the first definitive step beyond Newton in over 200 years, but it was merely a correction. The bending of light by gravity, on the other hand, was caused by no force acting at a distance across space, but by space itself. The Universe we live in would be seen in a new light, so there was no room for mistakes. The second reason the team needed to be confident about their results was their symbolic character. The horrific carnage of the First World War was not yet two years in the past, and the damage to open scientific exchange persisted between Germany and much of the rest of the world. The symbolic act of a British team verifying the theory of a German scientist could begin a rapprochement.

On 6 November 1919, the long-anticipated paper was read before the Royal Society and published in the Philosophical Transactions on 27 April 1920. The measurement of the deflection of the starlight was found to be consistent with the predictions of Einstein's theory. In a rare instance of worldwide scientific awareness, major newspapers latched onto the announcement and spread the news of the success. Overnight, Einstein became a celebrity, the first and most lasting scientific superstar of modern times.

This was the birth of a new law of fall, a law that transformed Galileo's trajectory from the short arc of a circular path into a multidimensional geodesic curve threading its way through warped space-time. The new law removed dynamics from motion. There was no action at a distance, no force of gravity accelerating masses according to Newton's Second Law. Mass and energy warped space-time, and mass and energy moved freely through the warped topography of their own making. By the time the dust settled after the 1920 announcement of the bending of light by gravity, everyone was suddenly living in an altered world. The Sun still rose each day in the East, as it always had done through the history of the solid Earth, but the causes had changed.

Einstein may owe his greatest debt to Galileo. The "happiest thought of his life," of the workman falling from the roof at the same rate as his hammer, was the same thought Galileo had when he removed enough friction to see that all objects fall at the same rate. Certainly, the perspective had changed. Einstein was grappling with what it meant to be in a force-free reference frame, while Galileo was struggling even to formulate the notion of a force. Nevertheless, Einstein's (and our) debt to Galileo is eternal. The rock thrown from the cliff moves by the same cause as the Moon around the Earth and the Earth around the Sun. But hidden inside Einstein's equations are causes that go beyond rocks and Earth and Moon to places where gravity becomes so strong that our familiar world vanishes, replaced by bizarre behavior that challenged even Einstein's keen intuition.

Schwarzschild's Radius

In 1915, when Einstein published the full theory of general relativity, Germany was embroiled in World War I, and even physicists and mathematicians were at the front, including a promising astrophysicist—Karl Schwarzschild (1873–1916). Schwarzschild had received his PhD in 1896 in theoretical physics and subsequently taught at the famous mathematics department at Göttingen from 1901 to 1909 as a colleague of David Hilbert and Hermann Minkowski. Schwarzschild's work concentrated on astrophysical problems, and in 1909 he moved to the German Astrophysical Observatory outside of Berlin. When war broke out he enlisted in the army and rose to the rank of lieutenant of artillery.

In the moments of calm between artillery bombardments on the Eastern Front, Schwarzschild retired to his bivouac where he read the scientific news from home. In December of 1915, he received a copy of Einstein's latest paper, containing the full theory of general relativity, including the nonlinear field equation. The strongly nonlinear character of the field equations suggested that exact solutions might be difficult or impossible to find. Yet, within only a month of publication, Einstein received a letter from the Eastern Front. Schwarzschild had found a simple metric solution when space had perfect spherical symmetry. Einstein was thrilled, and he communicated the letter to the Berlin Academy where it was published in January 1916. Schwarzschild continued writing papers despite his chaotic environment, and his final paper was published on the day he died on the Russian Front from a rare autoimmune disease. One of these papers solved Einstein's equations around a spherically symmetric mass (see Fig. 7.5), but there was a troubling feature in Schwarzschild's solution that contained an infinity at a finite distance away from the center of the massive object. This critical value came to be called the Schwarzschild radius, and the divergence to infinity is called the Schwarzschild singularity.

Physics at the Schwarzschild singularity is strange but eerily recognizable. Lengths of objects at the Schwarzschild radius would be infinitely contracted along the radial direction, and clocks would stop ticking. The critical radius has many of the attributes of a reference frame traveling at the speed of light. For instance, photons emitted from the critical radius would be infinitely red shifted and hence would have no energy to be detected. The star would be dark.

Einstein was a highly instinctual physicist, which was his greatest strength, but in this case, his intuition told him that the Universe would not allow the Schwarzschild singularity to come into existence. The density of matter that would be required to create the singularity was beyond anything ever experienced, and beyond anything that could be

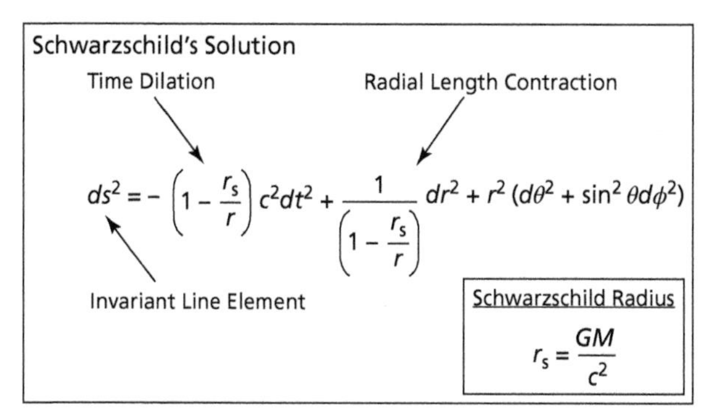

Fig. 7.5 Schwarzschild's solution to Einstein's field equations. The invariant line element ds^2 is related to metric tensor components that include a time-like term and space-like terms. Only radial lengths are contracted, which is why an in falling spaceship would appear flattened into a pancake

conceived. Even if a star became exceedingly dense at its core (which stars do), he believed that the critical radius could be approached but not passed. Some physical process, possibly unknown, would arise that would limit the density to physically reasonable values.

Through a quarter of the twentieth century, Schwarzschild's singularity lurked in the background, a nagging doubt on everyone's minds. Although the Schwarzschild solution was unassailable and was used routinely by physicists, the obvious divergence in the metric formula was nervously avoided. Although some accepted the consequences of the equations and the possible existence of an ultra-compact object from which even light could not escape, the consensus from 1915 through the 1950s was that some physical process would intervene and prevent such massive objects from forming. The simplest mechanism was that dying stars blew off considerable mass during their collapse, forming planetary nebula. Perhaps enough mass was always ejected to prevent the stars from collapsing to the singularity.

J. Robert Oppenheimer (1904–1967), an American physicist with appointments at Berkeley and The California Institute of Technology (Cal Tech), had other ideas. By 1938, he was convinced that a neutron star (a star with the mass density of a nucleus whose theoretical feasibility had been proven by Subrahmanyan Chandrasehkar in 1930), as it collapsed, would compress to the Schwarzschild radius and possibly beyond. But what would this look like to an outside observer far away? Even more fundamentally, what would this look like to an observer on the surface of the compressing star, riding it all the way down through the Schwarzschild radius? Would something catastrophic happen at the critical radius, perhaps space-time stretching to infinite curvature, tidal forces ripping the observer apart? To answer these questions, he enlisted one of his graduate students at Berkeley, Hartland Snyder, to undertake the formidable calculations.[17] Hartland Snyder was a tenacious student, not easily distracted, with the no-nonsense attitude of a kid from Utah, who reveled in hard work and hard mathematics. Even so, the calculations were tricky, and it was necessary to make gross simplifications to arrive at answers.

Fortunately, Oppenheimer had an impeccable feel for what was essential and what could be left out of a calculation. For instance, he immediately saw that they could use a theorem proven a decade earlier by George Birkhoff (1884–1944) at Harvard that showed that the Schwarzschild solution to Einstein's field equations was valid outside any spherical non-rotating mass, even if the mass were changing size and density. Therefore, an imploding star would just be a sequence of solutions of successively smaller and more compact objects. This was easy! And it yielded one of the first fundamental answers to a problem that had plagued Einstein and others ever since Schwarzschild had communicated his calculations from the trenches on the Eastern Front: does the curvature of space-time become infinite at the critical radius? No, answered Synder and Oppenheimer, the curvature remains finite and hence the laws of physics, even under ultra-strong gravity, remain finite at the Schwarzschild radius. There was nothing catastrophic, nothing to fear, and nothing abhorrent about the critical radius that had repelled so many physicists.

Armed with the confidence afforded by this new insight, Snyder embarked on the calculation of the collapse of a star. To make the problem tractable, Snyder and Oppenheimer needed to simplify the star properties to the extreme, making it perfectly spherical, without angular momentum, and without internal pressure. Only gravity would remain untouched by these simplifications. With the help of Oppenheimer as well as Richard Tolman, a professor at Cal Tech, Snyder finally had his answer. He constructed a complete solution, on paper, of the collapsing star that could answer any question: what it looked like from far away, from the in-falling surface, or even from inside the star. As usual with problems of relativity, the answers to these different questions, all asked about the same physical process, but asked from different relative perspectives, could not have looked more different.

For instance, to an outside observer far from the imploding star, the speed of collapse initially accelerates, just like Galileo's free fall accelerating under gravity. But as the surface of the star nears the critical radius it slows down, decelerating because of gravitational time dilation, shrinking ever more slowly as it approaches the limit. Photons from the star at the same time become progressively red-shifted because of the strong gravity, and just as the star's surface approaches the critical radius, the rate of collapse stalls, and the photons become so red-shifted that they are undetectable. In short, the in-falling star "freezes" just as it reaches the Schwarzschild radius, and at the same time the star goes dark.

From this distant viewpoint, it might look as if the collapse, which so many physicists believed to be impossible, had been avoided. However, Snyder's same solutions could be interpreted from the point of view of an observer on the surface of the shrinking star, riding the collapse downward like a child on an amusement park ride. For this observer, nothing special happens at the critical radius. The curvature of space-time remains finite, just as Snyder and Oppenheimer had concluded using Birkhoff's theorem, without infinite tidal forces ripping the observer apart, and time does *not* stop, as local clocks keep ticking away. The surface of the collapsing star passes the critical radius without event, the laws of physics remaining intact. Snyder and Oppenheimer published their results in 1939, convinced that they had solved the problem of collapsing stars, removing objections to the Schwarzschild limit. The broader physics community, on the other hand, was not so easily convinced, and controversy still raged, partly because of the extreme simplifications that Oppenheimer and Snyder made to arrive at their answers.

The world had to wait for more detailed analysis. On the day that their paper was published in the *Physical Review* on 1 September 1939, Germany invaded Poland. With the advent of World War II, intellectual life in Europe was thrown into chaos, and the United States initiated a government-backed scientific effort, focused on the war, at a scale never before seen in the history of science, consuming the time and energy of virtually every key physicist in the country, especially Oppenheimer. Because in that same 1 September issue of the *Physical Review*,

Niels Bohr (1885–1962) in Copenhagen published a paper on a model for the nuclear fission of Uranium. Fission had been glimpsed a year earlier by Otto Hahn shortly after his close collaborator Lise Meitner had fled Germany to Sweden. Bohr already had suspected that the nucleus could be deformed, and Meitner working with her nephew Otto Frisch correctly argued that if the deformation were large enough, that the nucleus could split. Bohr quickly set to work refining Meitner's calculations and published his work in that fateful 1 September issue of the *Physical Review*. The fission of Uranium could provide an energy source of unparalleled power, including cataclysmic destructive power that could be harnessed for weapons. Not long afterwards, a general from the US Army, General Groves, was knocking on Oppenheimer's door, changing his life forever. Neutron stars and black holes would have to wait, not just for the end of World War II, but also for the Cold War with Russia to stabilize. Two decades were lost before these problems again caught the attention of the world's scientists.

Black Hole Trajectories

The co-author on Bohr's fission paper was a young scientist visiting on a fellowship from America by the name of John Archibald Wheeler (1911–2008). Wheeler was born in Florida into a family of librarians. His parents were librarians and his siblings were librarians—*he* became a physicist. After receiving his PhD in 1933 from Johns Hopkins, filing a dissertation on the application of quantum mechanics to the optical properties of helium, he received a fellowship that allowed him to go to New York to work with the nuclear theorist Gregory Breit at New York University from 1933–1934, and then to Copenhagen to work with Niels Bohr from 1934–1935. He became an assistant professor at the University of North Carolina at Chapel Hill, but then moved to Princeton University in 1938, where he remained until his retirement in 1976. Wheeler was short in stature, but that just seemed to compress his energy. He was open and enthusiastic, an optimistic advisor who helped launch the careers of important physicists including Hugh Everett, Richard Feynman, Charles Misner, Kip Thorne, and William Unruh, among others.

Oppenheimer and Wheeler were matter and antimatter. Where Oppenheimer was a rigorous thinker, following the path down which mathematics and established physical principles logically led, Wheeler was open-minded and speculative, jumping ahead and looking for ways the laws broke down. Where Oppenheimer was politically liberal and a borderline communist, recognizing nuances and degrees, Wheeler was politically conservative and tended to see things in black and white. Where Oppenheimer was a dove (despite having developed the first nuclear weapon) and opposed the development of the hydrogen bomb, Wheeler was a hawk and was convinced of the evil of communist Russia and the need for the H-bomb project directed by the Hungarian-American physicist Edward Teller (1908–2003). And initially, where Oppenheimer believed that the ultimate fate of massive stars was gravitational cut-off (i.e., black holes), Wheeler was opposed to such an extreme fate, believing that the gravitational energy of collapse must somehow be dissipated to prevent gravitational cut-off. In 1958, Oppenheimer and Wheeler were nearly neighbors at Princeton. Oppenheimer was the director of the Institute for Advanced Study, and Wheeler was in the Physics Department. They traveled separately to a scientific meeting in Brussels where Wheeler presented a wide-ranging attack on the Oppenheimer-Snyder calculations. His central argument was based on the extreme simplifications used in the calculations, especially the lack of spin, since collapsing stars would spin faster as they shrank, like an ice skater pulling in arms and legs. Oppenheimer was the first to take the floor after Wheeler had finished, and quietly and methodically explained to the attentive audience why Wheeler was wrong. Wheeler was not fazed, and the issue was not resolved that day.

The collapse of realistic stars with spin and equatorial bulges and turbulent flow was far too complex to calculate on paper, but spin-offs of Edward Teller's H-bomb project included new computers and computer codes that could deal with these kinds of problems. In the early 1950s, as the super bomb development approached its first operational success with the Ivy Mike explosion on 1 November 1952, a young nuclear physicist with a degree from Cornell, Stirling Colgate (1925–2013), was hired to develop computational diagnostics of thermonuclear explosions. Colgate was an heir to the Colgate toothpaste fortune, and by odd coincidence had been sent to an elite, but little-known boarding school in Los Alamos, New Mexico, called the Los Alamos Ranch School. In 1942, in his junior year, a secretive military delegation, that included Robert Oppenheimer and E. O. Lawrence, arrived for a tour of the school, and shortly afterwards the school was suddenly closed, and the students sent packing with rushed diplomas but no explanation. The US Army had just appropriated the Ranch

School site for the upcoming Manhattan Project. Ten years later, Colgate was back, working on computer simulations of implosions. The codes and computers steadily increased in power, until by the late 1950s Colgate realized that they could be used to simulate the implosion of stars. With encouragement from Teller and working with Richard White and Michael May, he developed increasingly sophisticated simulations that showed by the early 1960s that the qualitative aspects of the Oppenheimer-Snyder calculations were correct. Stars could indeed collapse to gravitational cutoff. Wheeler was wrong.

Actually, Wheeler was thrilled. Kip Thorne remembers, as a student at Princeton, when Wheeler rushed into the graduate course on gravitation he was teaching with the news about Colgate's results, filling blackboard after blackboard with drawings and equations.[18] By this time, Wheeler and the broader theory community had already been primed to accept the Oppenheimer-Snyder results and to finally shed the aversion to the Schwarzschild singularity. A paper in 1958 by David Finkelstein was published in the *Physical Review* that reported a new set of coordinates that could stand in for Schwarzschild's coordinates—and the singularity simply disappears. This was the mathematical proof that the Schwarzschild singularity is merely a coordinate singularity, an artifact of the choice of coordinates that carries no physical significance, or at least no physical divergence. The Schwarzschild radius in Finkelstein's new coordinate frame was still the location inside which nothing, not even light, can escape. But nothing diverged at that radius. Oppenheimer and Snyder had been right, and now Wheeler was a convert too. Not only was Wheeler a convert, but like many a convert, he became an evangelist, enlisting his considerable energy and his cohort of students to pursue the deeper physics of "gravitational cut-off." He even got rid of that unwieldy expression, replacing it with the more memorable and descriptive "black hole" in a 1967 talk at a conference on pulsars.

Wheeler, with his students and visiting scientists at Princeton, as well as a growing international community, ushered in a so-called golden age of general relativity (GR) as they explored a widening array of physics under conditions of strong gravity. Most of the early work in GR had concentrated on weak-field limits, like those encountered in the solar system. With new perspectives on physics in ultra-strong fields, as well as computers to help when analytical work failed, nothing was off limits. In addition to tackling problems of increasing complexity, such as spinning and charged black holes, they also worked through many types of trajectories in the vicinity of black holes. And what they studied, and what they found, was essentially a new law of fall, extending Galileo's simple rule of distance versus squared time in the weak-field limit, to a bizarre new menagerie of trajectories in strong-field limits.

Finding trajectories in the new law of fall was simple, even if the behavior of the trajectories was strange. Trajectories and orbits around black holes are geodesic paths, shortest and straightest distances through space-time warped to extremes in the vicinity of black holes. The weak bending of light that Eddington observed, and that launched Einstein to stardom, becomes more extreme around black holes. Light passing by a black hole can be captured if it gets too close, as in Fig. 7.6. Even more surprising, light can whip around a black hole and still escape to infinity, or light can orbit a black hole in a perfect circular orbit at a radius 1.5 times the Schwarzschild radius on a surface called the photon sphere. Each of these paths is a null geodesic in the vicinity of

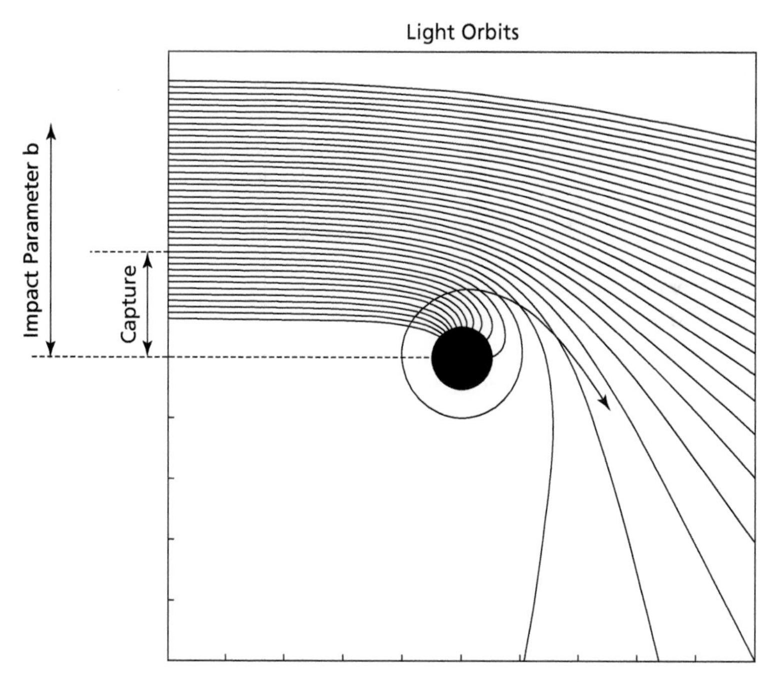

Fig. 7.6 Light trajectories in the vicinity of a black hole.[19] Some photons are deflected, others are captured, but some can orbit the black hole

a black hole. Each is simply in free fall, but in a much more complicated geometry than Galileo had ever imagined.

Once black holes became mainstream objects of interest, the search was on for observational evidence of their existence. One of the first candidates was the x-ray source Cygnus X-1 that was proposed in 1972 to be a star caught in a binary orbit with a black hole. As the black hole sucked the gases out of the star, the increasing acceleration of the inward spiraling gases emitted x-rays. By working out the orbital period, it became convincing that this system contained a black hole. It is now known that supermassive black holes occur at the centers of most galaxies, and smaller black holes abound throughout our own galaxy and are common enough that two black holes can be in orbit around each other. This was demonstrated dramatically on 14 September 2015 when the dual LIGO (Laser Interferometer Gravitational-Wave Observatory) detectors in Louisiana and Washington detected the gravitational waves generated by the merger of two black holes of around 30 stellar masses. This was the first observation of gravitational waves predicted originally by Einstein in 1916, but he considered them too weak to ever be detectable. The Nobel Prize in Physics for 2017 was awarded to Kip Thorne, Barry Barish and Rainer Weiss for their visionary work establishing the LIGO technology and proving that Einstein was right in theory, but uncharacteristically short on optimism.

Galileo Bound

Near the dark horizon of a dead star, Galileo's trajectory has wandered far from his simple parabola. The warping of space-time twists the parabola into knots of precessing orbits, wild loops of photons and the pile-up of gas into compact accretion rings. Yet all this pales in comparison to the fate of the soul who dips beneath the Schwarzschild radius that lurks as the eternal prison from which Galileo's trajectory can never escape. Although the coordinate singularity causes no problem for a local observer falling through that radius, it still holds a special place as the event horizon. Up until the last second, while the observer stays outside that radius, there is hope of escape to the outer Universe. But once that radius is passed, the only future is the true singularity at the core of the black hole. No amount of energy or rocket thrusts can prevent that eventuality. All world lines, even the world lines of photons, terminate on the central singularity. There has been no end to speculation about what happens at the origin. The warp of space becomes so extreme that the laws of physics must break down. Some believe that the center of a black hole can connect, through an Einstein-Rosen wormhole, to distant points in the Universe.

From the extremes of black holes emerges a connection of the astronomically large to the microscopically small. The event horizon and the central singularity challenge ideas about energy and matter and make contact with the role of quantum mechanics in the physics of gravity. For instance, Hawking showed, in his famous paper of 1974, that pairs of particles can pop into existence straddling the event horizon.[20] One particle falls to its fate at the central singularity, while the other particle escapes from the prison of the black hole, returning to life in the Universe. In this way, fluctuations slowly erode a black hole through Hawking radiation. Furthermore, the entropy of a black hole is quantized on its surface into discrete units that meld gravity with another fundamental constant of nature: Planck's constant. Therefore, just as Galileo's trajectory becomes irretrievably bound as it crosses the event horizon, it springs forth with new life into the quantum realm.

8

On the Quantum Footpath

To those of us who participated in the development of atomic theory, the five years following the Solvay Conference in Brussels in 1927 looked so wonderful that we often spoke of them as the golden age of atomic physics.

(Heisenberg, 1969)[1]

Reality is a hard thing to pin down. What we think we know about the world often gets tangled up with our own perceptions and our consciousness. But pragmatic physicists tend to reject this line of thought, replacing *senses* with *sensors* whose physical performance is governed by physical law and not by subjective interpretation. The Universe, and everything that takes place in it, exists whether we are there to sense it or not. This pragmatic approach has been unfailing in pinning down the object behavior of the Universe, independent of the need for overt observation—except when it comes to quantum phenomena.

Einstein was a pragmatist. He loved to create thought experiments to shed light on concepts that, at first, seem counterintuitive or subjective, and then he would use simple arguments to help crystalize them into clear understanding. Yet even Einstein was stumped by quantum mechanics. Everyone is stumped by quantum mechanics. As scientists, we instinctively try to make sense of things as cause and effect acting through mechanisms, such as local mechanisms that affect trajectories. Along the trajectory of a classical gas atom, like a billiard ball, the atom collides with other atoms, making local contacts and exchanging momentum with the other atoms, each atom affecting the trajectories of the others. In contrast, identifying cause in quantum interactions can lead to paradoxes, paradoxes that Einstein tried to use, in the famous Solvay Conference debates with Bohr in 1927 and 1930, to show that quantum mechanics was missing something. The debate hinged on concepts of reality and on what the act of observation causes to happen in quantum systems. Ultimately Einstein failed, but not before

Galileo Unbound. David D. Nolte, Oxford University Press (2018).
 DOI: 10.1093/oso/9780198805847.001.0001

he framed the questions and established the language that we use today when we talk about quantum phenomena.

As observers were elevated in stature to become causal agents and integral participants in quantum evolution, trajectories were demoted to a standing only slightly above nonexistence. In the time-evolution of quantum states, Lagrange's formerly continuous trajectories evaporate into dust. Even the two rival founders—Heisenberg and Schrödinger—of the two rival quantum theories—quantum mechanics and wave mechanics—agreed on this. Bohr's electron orbits of 1913, those enduring icons of the atomic age, dissolved into fuzzy balls in the competing theories of Heisenberg and Schrödinger between 1925 and 1927. Trajectories of quantum particles, whether through configuration space or through phase space, no longer existed as sacred constructs of reality—at least not in the old sense.[2]

Matrix Mechanics

When Werner Heisenberg (1901–1976) joined the theory group of Arnold Sommerfeld in Münich in the fall of 1920, he immediately was taken under the wing of Wolfgang Pauli (1900–1958), another of Sommerfeld's students who was two years Heisenberg's senior. Heisenberg had originally intended to work in relativity theory, but Pauli steered him clear,[3] telling the impressionable young student that all the really important problems were in atomic theory. Heisenberg took Pauli's advice to heart and consumed as much quantum theory as he could, learning most of it from Sommerfeld's seminars.

Werner Heisenberg was an intensely ambitious young scientist with barely concealed delusions of grandeur, but he was clever and worked like the devil. He was also likable and had many friends and champions throughout his life. He was born in Wurzburg in 1901, the son of a gymnasium teacher who would rise through the German university system to become the only extraordinary professor of Medieval and modern Greek in Germany.[4] Heisenberg's relationship with his father was not easy, his father's frequent disapproval serving as a source of tension as well as a possible source of Heisenberg's ambition through his early life. His actions may have been as much to impress as to rebel.

Sommerfeld suggested that Heisenberg should try something with the Zeeman effect, a splitting of atomic spectral lines under a magnetic field. Presented with a chance to shine, he snatched up the challenge and set to work. Having had no indoctrination into the quantum theory of the day, he felt no qualms in devising a solution to the Zeeman effect by assuming it was governed by half-quantum numbers. To the old quantum school this was heresy! Everything that had happened in quantum theory, from Planck's first hypothesis in 1900 through Bohr's correspondence principle of 1920 (that recovered classical behavior for large quantum numbers) obeyed the sacred mandate that quantum numbers were integers. Even Einstein's questionable (and not yet proven by that time) quanta of light came in integer numbers. Integer quantum numbers were inviolable—but Heisenberg was immune to all this. His half-integer quantum numbers simply worked!

At first taken aback and highly skeptical, Sommerfeld was intrigued by Heisenberg's solution. It was not a specious whim, but a carefully considered assessment of the best experimental data. Together, Heisenberg and Sommerfeld worked to polish the arguments, and the paper was submitted for publication at the end of 1921. Its reception by the broader community was not warm. Bohr was appalled at Heisenberg's half quantum integers, and if Bohr was not on your side, no one was. Pauli too, was highly critical, as he always was to everyone and everything. He had a habit of telling Heisenberg, "You are a complete fool."[5] But Pauli and Heisenberg, stark opposites in personal character—Heisenberg up early every day hard-working and diligent, Pauli carousing in cabarets, working all night and sleeping until noon—had mutual respect and close feelings for each other. Pauli's criticisms were not mean, but forced Heisenberg to refine his ideas and his arguments.

A major event in the German quantum physics community was planned for the summer of 1922 in Göttingen, when Bohr, that year's new Nobel Laureate, was to make a special visit and deliver several lectures over two weeks—an event that has since come to be called Göttingen's *Bohr Festspiele*. Sommerfeld sent his young student to attend with the hopes that Heisenberg would learn more physics and perhaps get deeper insights into the Zeeman effect. At the end of one of Bohr's early lectures, during the question period, Heisenberg, a mere student, and to the mortification of the collected audience, stood to question and criticize a technical point in the lecture.[6] Bohr handled the question with tact, but later vectored in on the brash 20-year-old, suggesting they take a walk together. They climbed the Hainberg outside of town and talked long and deeply about more than atomic physics. This meeting had a fundamental and lasting effect on Heisenberg and launched a close scientific and personal relationship that would endure until their fateful meeting in Copenhagen in 1941. But at that time, they respectfully disagreed, and Bohr sharply criticized Heisenberg's half-integer quantum numbers in one of his later lectures at the *Bohrfest*. Heisenberg returned to Münich unabashed. Furthermore, Max Born at Göttingen had been favorably impressed by Heisenberg and arranged with Sommerfeld to have Heisenberg visit Göttingen while Sommerfeld was away on sabbatical in the United States for a year working with Norbert Wiener at the Massachusetts Institute of Technology (MIT).

Max Born (1882–1970) had taken up the position of Director of the Physical Institute in Göttingen, replacing Peter Debye (1884–1966) in 1921. He first arrived in Göttingen in 1904 to work for his doctorate, and quickly became acquainted with David Hilbert, Felix Klein and Hermann Minkowski, who were there at that time. By 1922, Born was established in Göttingen as a leading researcher in quantum theory, attracting some of the best young minds in Germany to be his assistants. Pauli, fresh from Sommerfeld's supervision, was Born's assistant from 1921 to 1922, although they did not get along. Pauli's work habits were antithetical to Born's conservative character, and it galled Born to need to send a maid to roust Pauli out of bed at 10:30 every day.[7] Heisenberg arrived just as Pauli left to spend a year in Copenhagen working with Bohr. Although originally skeptical about Heisenberg's half-integer quantum number approach to the atom, Born and the Göttingen School were won over by Heisenberg shortly after he arrived. Born and Heisenberg used the half-integers, much to the chagrin of Bohr and Pauli in Copenhagen, to model helium and also to partially explain Bohr's "building up principle" that described the periodic table as a filling of electron shells.

However, by early 1923 progress in atomic model building had come nearly to a standstill. Atoms with more than two particles remained stubbornly outside the Bohr-Sommerfeld theory. In a 1923 review article on the state of quantum theory, Born was forced to admit "not only new assumptions in the usual sense of physical hypotheses will be necessary, but the entire system of concepts of physics must be rebuilt from the ground up."[8] He called for a new theory of "quantum mechanics" that would replace the existing models. It was a broad, though respectful, repudiation of the Bohr-Sommerfeld model. It was also prescient, and Heisenberg was in the middle of it.

Born was anxious to retain the talents of the wunderkind that Heisenberg had become and offered Heisenberg a position in Göttingen upon his graduation from Münich with his doctorate. Heisenberg returned to Münich at the end of the spring semester in 1923 upon Sommerfeld's return to complete his doctoral degree on a topic of hydrodynamics that Sommerfeld had inexplicably assigned to him, despite Heisenberg's deep immersion and successes in quantum theory. When it was time to defend his dissertation, one of his examiners was the eminent experimental physicist Wilhelm Wien, the winner of the 1911 Nobel Prize in Physics for his famous displacement law of black body radiation that had helped inspire Planck's quantum hypothesis and Ehrenfest's adiabatic principle.[9] Heisenberg had taken an experimental lab course with Wien, but had not put in much effort because his interests lay elsewhere. His barely veiled boredom with experimental arts had irked Wien. Now that it was time to stand and defend, Wien had an opportunity to put this prima donna in his place. The questions Wien posed were surprisingly outside Heisenberg's experience. They began at a difficult level, and when Heisenberg faltered, he asked progressively simpler questions. Finally, in exasperation, Wien asked Heisenberg to derive the resolving power of a simple optical microscope—an undergraduate problem that every physics student should know. Heisenberg, by this time completely flustered, was unable to do it.

Wien turned to a shocked Sommerfeld and pronounced a failing grade—an F—on the examination. Only after the kind of negotiation that can take place in academia between faculty colleagues was it agreed that Heisenberg would receive a C on his doctoral exam.[10] It was a passing grade, but Heisenberg was embarrassed and dismayed. He packed his bags and left town in disgrace that night, arriving on Born's doorstep the next morning in Göttingen fully two months in advance of the agreed-upon start date. He told Born of his debacle and asked if Born would still offer him a position. Perplexed, Born asked for the complete story, which Heisenberg miserably supplied. To Born's credit, he realized that the fate of a brilliant theoretical physicist hung upon the whims of an aging experimentalist. He confidently reiterated his offer to Heisenberg. Ironically, Heisenberg's struggle with the resolving power of a microscope would raise its ugly head again in the near future.

Just as Sommerfeld's sabbatical had enabled Heisenberg to be in Göttingen for two semesters in 1922–1923 to work with Born, Born's sabbatical in 1924–1925 enabled Heisenberg to visit Bohr's Institute for two semesters, again with the promise to return after Born was back. Heisenberg's visit to Copenhagen proved inspirational. In March of 1925, Pauli visited from Hamburg, and the four great physicists (Bohr, Pauli, Heisenberg and Bohr's assistant Hendrik Kramers) worked intensely together for several weeks from 9 a.m. to midnight, often in loud bouts of (friendly) arguments. Then it was time for Pauli to leave and for Heisenberg to return to Göttingen, as promised, to prepare for the summer semester.

Heisenberg returned to Göttingen on 27 April 1925 with all the wild discussions of Copenhagen fresh in his mind. His thinking shifted progressively further away from orbits and visual models of electrons to rest upon only what can be observed in an experiment. The optical transition energies measured in the laboratory became the rock of his reality—not hypothetical paths that could not be observed. He began to construct a new type of algebra of transition amplitudes that required the amplitude of one step in a process to be multiplied, not just by the amplitude of the succeeding step, but multiplied by a sum over all possible alternative steps between the initial and the final state. A disturbing side effect of his new algebra for multiplying the amplitudes was the strange property that the order of multiplication mattered. He got different answers depending on the sequence of multiplications. In modern terminology, the quantities did not *commute*, such that $A \cdot B \neq B \cdot A$. Nonetheless, Heisenberg knew he was close to a new description of quantum phenomena—he was just missing a final key element.

In a widely recounted story, Heisenberg came down with a bad case of hay fever and retreated for relief to the barren island of Helgoland off the northern coast of Germany. One evening, standing alone on the shore, as the Sun set over the Atlantic, he had a flash of inspiration that allowed him to use energy conservation to restrict the too-numerous sums of transition amplitudes to only those that conserved energy between initial and final state. He worked feverishly through the night, calculating the energy levels of quantum oscillators and quantum rotators, finishing in time to watch the sun rise. Looking back, it is hard to understand why energy conservation would have been the crucial link, but mid 1925 was only a few years after Compton had first shown that photons conserved energy and momentum in electron scattering. It had only been a few months earlier in 1925 that laboratory experiments had confirmed that energy and momentum conservation also held in light scattering from bound electrons in atoms, putting the nails in the coffin of a Bohr-Kramers-Slater model that had rejected energy conservation in a last attempt to salvage the ephemeral electron orbits. Heisenberg quickly wrote a draft paper of his new quantum theory, and when he returned to Göttingen, he asked Born to look it over to see if he thought it had merit. In a letter to Pauli around that time, he said: "My entire meager efforts go toward killing off and suitably replacing the concept of the orbital paths that one cannot observe."[11]

Around the same time, Born suddenly realized that Heisenberg's strange multiplication properties were the multiplication properties of matrices. While Heisenberg was temporarily away in Copenhagen, Born and his new assistant Pascual Jordan wrote a paper on the matrix approach to quantum mechanics. They submitted their paper in the September 1925 issue of *Zeitschrift für Physik* with the unremarkable title "On a quantum-theoretical reinterpretation of kinematic and mechanical relations."[12] The three of them (Born, Heisenberg and Jordan) then wrote a definitive paper on the new matrix approach to quantum mechanics that they submitted in November 1925 to *Zeitschrift*,[13] a paper that is universally referred to as the *Dreimännerarbiet* (Three-man Work). With the theory out in the open, Pauli in Hamburg and Dirac at Cambridge used the new quantum mechanics to derive the transition energies of hydrogen,[14] while Lucy Mensing and J. Robert Oppenheimer in Göttingen extended it to the spectra of more complicated molecules.

Heisenberg's achievement was a radical breakthrough. Quantum theory had been stuck for nearly a decade, and now Heisenberg had thrown wide the doors with an approach that could treat broad new classes of problems that the old quantum theory could not touch. Ironically, the matrix approach to quantum mechanics relies exclusively on the particle properties of quanta, yet it completely abandons the idea of particle trajectories. Particles were the foundation of the new physics, but they only existed as mathematical entries—observables—in matrices. The age-old orbits and trajectories—Galileo's orbits and trajectories—had been abolished. But this new view of reality was not welcomed by all. A relatively unknown Austrian physicist was at work on a radically different approach to quantum systems that would restore, at least partially, the picture of electrons evolving in space and time, and with properties that challenged the particle view.

Wave Mechanics

Erwin Schrödinger (1887–1961) had been teaching as a mid-career professor at the University of Zurich since 1921. He had yet to distinguish himself in physics, though he was known for his breadth of knowledge, as if waiting for the right idea to come along. Schrödinger was not enthusiastic about Heisenberg's abstract new quantum mechanics that rejected visualization for the electrons. A cornerstone of Heisenberg's approach was the reliance on experimental observables, rejecting questions about the configurations of the electrons not only as irrelevant, but as ill posed. Heisenberg believed that electronic configurations were unobservable and hence were not real. Schrödinger believed the opposite. For him, the electronic configurations were what gave the electronic states their properties.

In late 1925, Schrödinger read a paper by Einstein that mentioned in a footnote a result of the thesis of Louis de Broglie (1892–1987) that connected mechanical momentum to wavelike properties as a general new principle in quantum theory. De Broglie's doctoral dissertation had been published in 1924 under the direction of Paul Langevin, and de Broglie had communicated with Einstein because of the close connection of de Broglie's hypothesis with Einstein's theory of the photon. De Broglie's theory postulated that all particles behaved as waves, not only the massless photon, but also massive particles. This new theory of matter waves explained the quantized Bohr orbits as standing waves: the principle quantum number in Bohr's theory was simply the number of standing-wave nodes. Schrödinger obtained a copy of de Broglie's thesis and was asked by Peter Debye to present a seminar on 23 November describing the new idea to the physics theory group in Zürich. At the end of the seminar, Peter Debye was unimpressed by the qualitative nature of the wave description, and remarked that a proper wave theory must have a wave equation.

A month later, over Christmas, Schrödinger took a trip to the Italian Alps with Debye's remark on his mind and a mistress on his arm (the identity of whom has remained a mystery to this day).[15] Whether inspired by his illicit companion, or in spite of her, Schrödinger had an epiphany in a manner similar to Heisenberg's on Helgoland, enabling him to connect de Broglie's momentum operator to a mathematical eigenvalue problem. An eigenvalue problem is a differential equation, applied to a function, which yields a stationary state (an eigenstate) with a defined scalar property (an eigenvalue). In the case of de Broglie's matter waves, the function was the matter wavefunction and the eigenvalue was the quantum energy of the quantum state. Once Schrödinger had his wave equation for the electron in a hydrogen atom, he needed to solve it to find the eigenstates and their associated eigenvalues. By the time he returned to Zürich, Schrödinger had already worked out the solution to the angular variables. With the help of his colleague Hermann Weyl, he soon had the radial solution as well. The startling and satisfying result of the wave equation approach was that quantum numbers, that had to be postulated in the old quantum theory and which were buried in Heisenberg's quantum mechanics, emerged directly from the discrete solutions to the wave equations. In January of 1926, the same month that Pauli and Dirac submitted their solutions of hydrogen using Heisenberg's quantum mechanics, Schrödinger submitted his paper on the eigenvalue solution of hydrogen using his new wave equation (see Fig. 8.1).[16]

The timing could not have been more inconvenient for Heisenberg. He had just reached a pinnacle of his career, inventing a powerful new quantum mechanics that physicists across the globe had begun to use to solve long-standing problems, when a challenger appeared to steal the spotlight. Heisenberg's quantum mechanics was radically abstract and used difficult mathematics. In contrast, Schrödinger's wave equation was mathematically familiar—everyone was comfortable with differential equations. Furthermore, the wavefunction was something that could be visualized with a spatial pattern that could be plotted and that could change in time, reminiscent of a trajectory. Visualizability questions about quantum processes had been one of the key reasons Heisen-

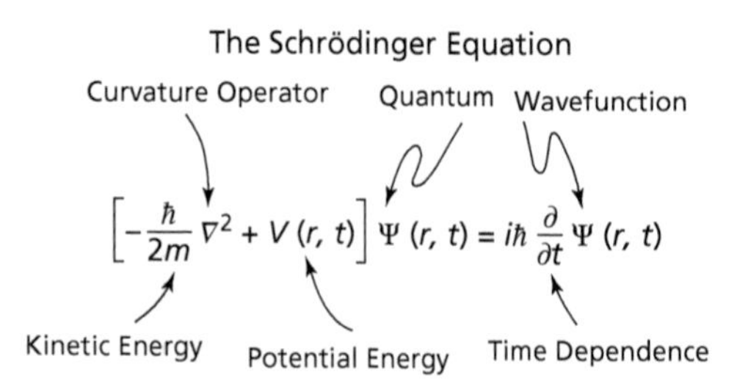

Fig. 8.1 The time-dependent Schrödinger Equation is a differential equation in space and time whose solution is the quantum wavefunction that depends on the potential

berg had rejected orbits and trajectories, going so far as to deny them any real existence. Just when Heisenberg thought he had succeeded in exorcising them from quantum theory, Schrödinger slipped them back in. What made it worse was that physicists immediately embraced Schrödinger's wave mechanics for precisely the reason that the wave functions were easy to visualize. Within months of the publication of Schrödinger's paper, Heisenberg's newborn quantum mechanics was in danger of being supplanted.

Schrödinger himself worked to solidify his wave mechanics as the central theory of quantum processes over Heisenberg's abstract quantum mechanics. Part of his agenda was to reestablish visualizability and reality into quantum theory. To this end, he (incorrectly) interpreted the square of the wavefunction to be the physically real electron density around the hydrogen nucleus. He furthermore rejected the discontinuous jumps of quantum mechanics in favor of smooth time evolution dictated by the time-dependent version of his differential equations. These interpretations by Schrödinger were in direct conflict with Heisenberg's positions and became the first battle in a never-ending war for the heart and soul of quantum theory—the battle over wave-particle duality.

Even Born, Heisenberg's ally, turned to the new wave mechanics to solve the problem of a free electron scattering from an atom. Schrödinger's wavefunction approach promised to be ideal for this problem, because the free electron had a continuum of energies that did not easily fit into Heisenberg's discrete quantum matrices. Yet Born immediately encountered a problem with his calculations. When he used Schrödinger's squared wavefunction as a real electron density, the theoretical results made no sense. In a flash of insight (for which he later was awarded the Nobel Prize in physics), Born realized that if the squared wavefunction were interpreted as a probability density for finding the location of a discrete electron, then the scattering calculations agreed perfectly with experiment.[17] The consequence of this interpretation was that experiments should be performed many times, using an ensemble approach with many identical systems, each one possessing discrete particles and discrete events, but with each of these elements occurring randomly according to the probability density. The experimental results would be the average over many repetitions. Born had swept away Schrödinger's real charge density and replaced it with quantum probability, bringing with it all the randomness and uncertainty that had dogged quantum theory from the beginning. Schrödinger was neither pleased nor convinced. Nor was Einstein, who was growing increasingly dissatisfied with the direction that quantum theory was going. Nonetheless, Bohr and his school in Copenhagen were immediately receptive of Born's probability interpretation, and they incorporated it as a central pillar of what became the Copenhagen interpretation after one final important piece was put into place, but not before Heisenberg and Bohr were to argue bitterly over it.

Uncertain Physics

In April 1926, Bohr's longtime assistant Kramers accepted a faculty position at the University in Utrecht, and Bohr immediately offered the assistant position to Heisenberg who was happy to accept. He arrived in Copenhagen in May just as Schrödinger published an astounding paper showing the mathematical equivalence of wave mechanics and quantum mechanics.[18] Rather than smoothing over the differences between the two camps, this paper had the opposite effect. If the two approaches were mathematically equivalent, then whatever approach someone chose came down to "taste," that is to say, "philosophy," and philosophical disagreements are always the ugliest. Schrödinger on one side and Heisenberg on the other began to attack each other in lectures and in print. The diatribes unsettled Bohr, and he sought to mediate between them, although he was clearly on the side of Heisenberg. Therefore, he invited Schrödinger to visit Copenhagen in late September 1926 to give a lecture on his wave mechanics and to engage him in friendly discussion.

The friendly discussion between Schrödinger and Bohr became a heated debate that raged from early morning until late at night, Bohr insisting on the fundamental discontinuities of quantum jumps, Schrödinger accepting only continuous change. Bohr's intensity was so great that he dogged Schrödinger until he became ill, and even Bohr's wife had to intercede on his behalf as Schrödinger took to his bed.[19] Heisenberg, by accounts of this meeting, did not participate directly, but stayed on the sidelines, watching and listening. When Schrödinger decamped from Copenhagen after several fraught days, he had conceded nothing, but Bohr's defense of quantum mechanics was not a defense of Heisenberg's extreme point of view. Bohr was already seeking a middle ground that admitted both Heisenberg's quantum mechanics and Schrödinger's wave mechanics. Over the months following Schrödinger's departure from Copenhagen, Bohr and Heisenberg continued the debate, each searching for a resolution of the conflict. Bohr was convinced that both waves and particles must be relevant elements of reality, while Heisenberg admitted physical reality only to measurements of discrete quanta. The arguments grew tiresome, each side entrenched, and finally the two men, so close as friends and collaborators, drew apart. Remarkably, this time apart was a formative time for them as they each developed a deeper understanding, in their separate ways, of quantum theory, creating principles that lie at the heart of modern quantum interpretation.

Heisenberg was strongly affected by Schrödinger's visit, becoming aware that quantum theory was missing a guiding principle, much like the situation of relativity before Einstein, when the Lorentz transformations were being discussed in terms of classical space and time. The equations worked, but there was no consistent perspective that led to deeper understanding, until Einstein flipped the problem by assuming what Lorentz was trying to prove—the constancy of the speed of light. Once Einstein established his postulates, the full structure of relativity and space-time crystallized. Likewise, both quantum mechanics and wave mechanics worked, and were even mathematically equivalent, as Schrödinger had shown, but a key perspective was missing—something that went beyond the squabbles over wave or particle viewpoints. At this time Heisenberg began to put mathematical expressions to an idea that he had been playing with since 1924 when the classically sacred electron orbit was rejected in quantum systems because it made no sense to talk about the simultaneous position and momentum of the electron in such an orbit.

Heisenberg revisited this problem of simultaneous determination of position and momentum and realized that he had prematurely discarded the concept of a quantum trajectory. Within the laws of quantum mechanics, it was possible to make successive measurements of position and momentum, and if the time between successive measurements was reduced to zero, then these measurements could be viewed as simultaneous. The question then was: what would be measured? To answer this question, he turned to the probabilistic theory of Born. Ironically, the Born interpretation was linked directly to Schrödinger's unpalatable (to Heisenberg) wave mechanics, but it did provide mathematical transformations between spatial representations and momentum representations of quantum systems. By assuming a Gaussian probability distribution for position measurements with an uncertainty in position given by δx, Heisenberg applied a new probabilistic transformation theory developed by Dirac and Jordan to derive the probability distribution for the momentum measurements on the same system. He was fascinated and thrilled to find that the momentum distribution also was a Gaussian distribution, and the width, or uncertainty, in the momentum distribution δp was inversely proportional to the uncertainty in position. Moreover, the product of the uncertainties in position and momentum was proportional to Planck's fundamental constant. What Heisenberg had derived was the famous relation known today by his name—the Heisenberg Uncertainty Relation (see Fig. 8.2).

This key result was obtained while Bohr was away in Norway on a ski trip. Heisenberg had already shared it with Pauli in a long letter, and had been encouraged by Pauli's rare enthusiastic approval to turn the letter into a paper, which he had already submitted to the *Zeitschrift* on 22 March 1927 before Bohr's return and without the traditional approval

also je genauer der Ort bestimmt ist, desto ungenauer ist der Impuls bekannt und umgekehrt; hierin erblicken wir eine direkte anschauliche Erläuterung der Relation $pq - qp = \frac{h}{2\pi i}$. Sei q_1 die Genauigkeit, mit der der Wert q bekannt ist (q_1 ist etwa der mittlere Fehler von q), also hier die Wellenlänge des Lichtes, p_1 die Genauigkeit, mit der der Wert p bestimmbar ist, also hier die unstetige Änderung von p beim Comptoneffekt, so stehen nach elementaren Formeln des Comptoneffekts p_1 und q_1 in der Beziehung

$$p_1 q_1 \sim h. \tag{1}$$

Daß diese Beziehung (1) in direkter mathematischer Verbindung mit der Vertauschungsrelation $pq - qp = \frac{h}{2\pi i}$ steht, wird später gezeigt werden. Hier sei darauf hingewiesen, daß Gleichung (1) der präzise Ausdruck für die Tatsachen ist, die man früher durch Einteilung des Phasenraumes in Zellen der Größe h zu beschreiben suchte.

Fig. 8.2 First presentation of the uncertainty principle by Heisenberg (1927).[20]

of the institute director. When Bohr did return, with a fresh new idea of his own, he was not pleased with Heisenberg's rush to publish. Bohr had become convinced that a fundamental wave-particle duality was at the core of quantum phenomena, in which neither could stand apart from the other. Which character—wave or particle—emerged from an experiment depended on what experiment was performed. In this sense, the experimentalist was a participant in the quantum phenomenon. Bohr called this his Complementarity Principle. Heisenberg agreed with Bohr on this point—as they agreed on most of the features of quantum theory—but there remained a fundamental disagreement on whether the wave nature of matter was an integral part of the uncertainty relation. Bohr was certain that it was, and pressed Heisenberg to withdraw his paper. Heisenberg was certain that it was not, and refused. In fact, Bohr had found a flaw in one of Heisenberg's physical arguments in the submitted manuscript, and this flaw held the key to the importance of wave phenomena in the interpretation of the uncertainty relation.

To provide physical insight and intuition into the uncertainty relation that he derived mathematically using the Dirac-Jordan transformation theory, Heisenberg used a simple thought experiment to illustrate the role of measurement in the uncertainty principle. He envisioned a gamma-ray microscope that measured the position of an isolated electron by scattering a gamma photon from the electron and detecting the position of the photon in a detector. To measure the position of the electron to ever-greater accuracy, the wavelength of the photon would be made ever smaller. However, through the Compton effect, the scattered photon imparts momentum to the electron in the measurement process. Because the photon momentum increases as the wavelength decreases, greater position accuracy requires greater momentum transfer and hence greater uncertainty in the electron momentum. Voila! There is a tradeoff between knowledge of position and knowledge of momentum.

Not so fast, warned Bohr, because Heisenberg had forgotten to include the collecting lens of the microscope in his analysis. A lens must collect the scattered photon to project it onto the detector. In the theory of Compton scattering, there is no uncertainty in the position or the momentum of the electron or photon during the process. However, there is an uncertainty in the experimental measurement, because the photon can enter anywhere within the lens aperture, which is directly related to the resolving power of the microscope. Therefore, Bohr insisted, the origin of the uncertainty in the thought experiment is the resolving power of the microscope, and *that* is a wave phenomenon. The resolving power of a microscope! This was exactly Wien's question three years before when Heisenberg had failed to explain it during his exams, nearly costing him his doctoral degree. Heisenberg was abashed, and though he stubbornly refused to withdraw the paper, he did finally, after a month of arguments that at times lead to tears,[21] agree to add Bohr's observation about the resolving power of the microscope in a postscript of the paper proofs.

The publication of Heisenberg's paper in the *Zeitschrift* in April 1927 was the capstone to the edifice that Bohr and his band had been building for the past five years. What eventually became known as the Copenhagen Interpretation of quantum theory combined Born's probabilistic interpretation of Schrödinger's wavefunction with Heisenberg's uncertainty principle and Bohr's complementarity. The Copenhagen interpretation provided a conceptual framework and a language to help think about and talk about enigmatic quantum phenomena. Later that year, at the fifth Solvay Congress, Bohr and Heisenberg presented a united front, despite their earlier disagreement, after which the Copenhagen interpretation of quantum physics became the approach adopted by most physicists.

Despite the crowning success of Heisenberg's uncertainty principle paper, and the subsequent adoption of the Copenhagen interpretation, lurking beneath the surface were serious conceptual difficulties. For instance, Heisenberg's paper contained a radical new viewpoint of quantum phenomena, as radical and as transformative as Einstein's relativity theory was twenty years earlier. The radical new viewpoint was that the trajectory of a quantum particle does not exist in reality until measurement brings it into existence. He wrote to Pauli in a letter that "the path only comes into existence through this: that we observe it."[22] This radical new view concerning the quantum trajectory has come to be called wavefunction collapse. John von Neumann provided the axiomatic underpinnings of wavefunction collapse in 1932, and it became an integral part of the Copenhagen interpretation. However, wavefunction collapse causes severe problems when the wavefunction describes two or more particles that are separated by large distances. This was a point that Einstein made with high drama in a last-ditch effort to bring quantum theory back to reason.

Entanglement

The Solvay congresses were unparalleled scientific meetings of their day. They were held about every three years, always in Belgium, supported by the Belgian chemical industrialist Ernest Solvay. Attendance was by invitation only, and invitations were offered only to the top scientists concerned with the selected topic of each meeting. The first meeting, held in 1911, was on radiation and quanta. The fifth meeting, held in 1927, was on electrons and photons, focusing on the recent advances in quantum theory. The old quantum guard was invited—Planck, Bohr and Einstein. The new quantum guard was invited as well—Heisenberg, de Broglie, Schrödinger, Born, Pauli and Dirac.

Heisenberg and Bohr joined forces to solidify the Copenhagen interpretation of quantum physics. The chief conclusion that Heisenberg and Bohr impressed on the assembled attendees was that the theory of quantum processes was complete, concluding that unknown or uncertain characteristics of measurements could not be attributed to lack of knowledge or understanding, but were fundamental and permanently inaccessible. Einstein was not convinced by the argument, and he rose to his feet to say so after Bohr's informal presentation of his complementarity principle. Einstein insisted that uncertainties in measurement were not fundamental, but were caused by incomplete information, that, if known, would accurately account for the measurement results. Bohr was not prepared for Einstein's critique and brushed it off, but what ensued in the dining hall and the hallways of the Hotel Metropole in Brussels over the next several days became enshrined as one of the most famous scientific debates of the modern era, known the Bohr–Einstein debate on the meaning of quantum theory.

The spirit of this back and forth encounter between Bohr and Einstein is caught dramatically in the words of Paul Ehrenfest who witnessed the debate, partially mediating between Bohr and Einstein, both of whom he respected deeply.

> Brussels-Solvay was fine! ... BOHR towering over everybody. At first not understood at all ..., then step by step defeating everybody. Naturally, once again the awful Bohr incantation terminology. Impossible for anyone else to summarize. ... (Every night at 1 a.m., Bohr came into my room just to say ONE SINGLE WORD to me, until 3 a.m.) It was delightful for me to be present during the conversation between Bohr and Einstein. Like a game of chess, Einstein all the time with new examples.

> In a certain sense a sort of Perpetuum Mobile of the second kind to break the UNCERTAINTY RELATION. Bohr from out of philosophical smoke clouds constantly searching for the tools to crush one example after the other. Einstein like a jack-in-the-box; jumping out fresh every morning. Oh, that was priceless. But I am almost without reservation pro Bohr and contra Einstein. His attitude to Bohr is now exactly like the attitude of the defenders of absolute simultaneity towards him. . . .
>
> Ehrenfest[23]

The debate gently raged night and day through the fifth congress, and was renewed three years later.

By the time the sixth Solvay Congress was convened in 1930, quantum physics had made several significant steps that were best understood in the context of the Copenhagen interpretation. Therefore, Einstein was at a greater disadvantage than the previous conference because the orthodox interpretation was becoming accepted by fiat. Even if aspects of the theory were objectionable on philosophical grounds, the theory described experimental reality with great accuracy. Nonetheless, this time he had a thought experiment whose interpretation hinged on his particular strength: relativity. It involved the measurement of the time of emission of a photon from an enclosed box. Einstein sought to show that the product of the energy uncertainty of the photon and the time uncertainty for emission was much smaller than Planck's constant, apparently violating Heisenberg's precious uncertainty principle.

Bohr was stopped in his tracks with this challenge. Although he sensed immediately that Einstein had missed something (because Bohr had complete confidence in the uncertainty principle), he could not put his finger immediately on what it was. That evening he wandered from one attendee to another, very unhappy, trying to persuade them and saying that Einstein could not be right because it would be the end of physics. At the end of the evening, Bohr was no closer to a solution, and Einstein was looking smug. However, by the next morning Bohr reappeared tired, but in high spirits, and he delivered a masterstroke. Where Einstein had used *special* relativity against Bohr, Bohr used Einstein's own *general* relativity against him to show that the products of uncertainties were precisely as required by Heisenberg's principle.

Einstein graciously admitted defeat, and afterwards in his writings he no longer tried to refute the uncertainty principle. He accepted that it was an integral part of measured quantum behavior. However, he did not accept that quantum mechanics was a complete theory of reality. When the Solvay conference reconvened in 1933 to discuss the structure of the nucleus, it seemed that the Bohr-Einstein debate was over; and that Bohr had won. But the most important salvo in the battle was to come only two years later.

In May 1935, Einstein published a paper in the Physical Review with two coauthors, Boris Podolski and Nathan Rosen. The title of the paper was "Can quantum-mechanical description of physical reality be considered complete?" and the topic of the paper has come to be called the EPR paradox—an acronym formed by the author's names.[24] The EPR paradox remains today one of the deepest and most subtle paradoxes of physics. The paper was a reprise of the 1927 and 1930 Solvay debates on whether quantum mechanics provides all possible information about quantum systems, but with a more sophisticated approach that used the joint wavefunction of two particles.

Consider a quantum particle, or an excited state, that decays into two photons that travel to two far away detectors. The physical reality of the detection process consists of pairs of measurements (x_1, x_2) or (p_1, p_2) for positions or momenta. Detection of x_1 uniquely defines x_2 and vice versa, or detection of p_1 uniquely defines p_2 and vice versa. Even if the detectors are separated by a large distance, the detection of one part of the pair instantaneously determines the measurement of the other part of the pair, implying transmission of information at a speed faster than the speed of light—something not allowed by relativity and causality. Therefore, argued Einstein, there must be hidden information in the two-particle system, not accessible to conventional quantum theory, which removes the need for transmission faster than the speed of light. In other words, there is a deeper store of information beyond that which can be measured quantum mechanically, and hence quantum mechanics is not complete.

When Bohr at his institute in Copenhagen was alerted to the EPR paper, he stopped what everyone was doing until they could respond to the paper. The New York Times had already run an article claiming that quantum mechanics was in peril, so Bohr meant put a stop to such nonsense as soon as possible. By then he was an expert at defusing Einstein's attacks, but this one was different, far subtler, so he had to work much harder to get to the key weakness. It took Bohr months, and in the process, he had to abandon one of his key beliefs, but finally he published his reply in October 1935, five months after the original EPR paper, using the exact same title.[25]

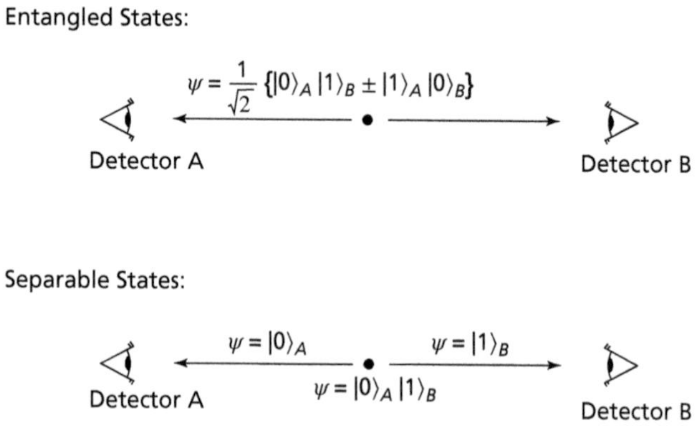

Entangled States:

$$\psi = \frac{1}{\sqrt{2}}\{|0\rangle_A |1\rangle_B \pm |1\rangle_A |0\rangle_B\}$$

Detector A

Detector B

Separable States:

$\psi = |0\rangle_A$

$\psi = |1\rangle_B$

Detector A

$\psi = |0\rangle_A |1\rangle_B$

Detector B

Fig. 8.3 Entangled states compared to factorizable states. The measurement of one of an entangled pair defines the measured properties of the other, while for separable states, measurement of one has no influence on the measured properties of the other. The symbols for the wavefunctions are Dirac's "bra-ket" notation

Bohr identified the central premise in the EPR argument that set up the paradox (see Fig. 8.3). A decision must be made at detector A whether to measure position or momentum. If position is measured to arbitrary accuracy, then all momentum information is lost, and vice versa. Because B is so far away, the decision whether to measure position or momentum at A cannot be known, so the decision at B is selected independently of the decision at A. This is where Bohr reintroduced his concept of complementarity, one of the pillars of the Copenhagen interpretation, upon which he centered his counterattack. When A and B happen to measure the same variable, position or momentum, then their results agree perfectly. However, when they choose to measure opposite variables, then no correspondence between their measurements will be found, in complete agreement with the uncertainty principle. Therefore, the measurement apparatus plays a central role in the description of reality. One could say that reality is defined by what one measures. This was the point that Einstein and his colleagues had overlooked.

On the way to his conclusion, Bohr had to give up his belief that a measurement apparatus interacts only with its local particle. Central to his response is the fact that the measurement at A causes a collapse during the measurement process, as established by John von

Neumann in his axoimatization of quantum mechanics in 1932,[26] of the joint wavefunction across space of arbitrary distance, as illustrated in Fig. 8.3. No information is transmitted, and hence there is no violation of relativity, and no need to look for a more complete theory, but there is also a nonlocal influence. For Bohr, this was a willing sacrifice, while for Einstein, this was "spooky action at a distance" that he abhorred. The physics community at large thought that Bohr had put the matter to rest. Einstein put the matter behind him, at least publicly. But Schrödinger was not satisfied. The Copenhagen interpretation was becoming enshrined as truth, but it still contained absurdities that he was determined to expose.

In late 1935, Schrödinger published a series of papers discussing the current situation in quantum mechanics.[27] He had been corresponding with Einstein through the summer, and as each letter arrived, back and forth across the Atlantic, it raised the stakes in their discussions, refining their arguments. They agreed that the Copenhagen interpretation required too much suspension of disbelief, but they differed in their own interpretations. Schrödinger remained a realist. Although Born had established almost ten years before that the quantum wavefunction was not related to the physically real electron density, Schrödinger continued to ascribe to it a level of physical reality. To reduce the Copenhagen interpretation to absurdity, he took Bohr's measurement argument to extremes.

Schrödinger proposed placing a cat in a sealed box with a radioactive substance known to have a half-life of an hour. If the particle decays, a Geiger counter in the box releases a hammer that smashes a vial of poison, killing the cat. At the end of an hour, the chance the cat is alive or dead is 50/50. In the Copenhagen interpretation, because all elements in the box consist of quantum mechanical wavefunctions, and no overt measurement is made, then the cat must be in a quantum superposition of both alive and dead states. Now if an observer opens the box to look inside, the wavefunction of the cat collapses into one state or the other. To Schrödinger, such a state of affairs was absurd, clearly illustrating the limitations of the Copenhagen interpretation that required wavefunction superpositions to remain unmolested if no overt measurement takes place.

The Schrödinger cat paradox became sensationally famous in ways that Schrödinger himself had not intended. He had fashioned the example to combat what he saw was an unrealistic acceptance of the

Copenhagen interpretation of quantum mechanics. While succeeding in showing that quantum mechanics was bizarre, he had hoped to show that it was wrong, or at least missing something. To his chagrin, there was general acceptance of the dead-or-alive superposition, at least in principle. Schrödinger had called the mutual influence of quantum states upon each other as *entanglement*, and the word stuck. The discussions elicited by the cat paradox side-stepped the philosophical problems of quantum mechanics and focused instead on the physical differences between microscopic states, where quantum superposition held sway, versus macroscopic states, where classical physics emerges.

Schrödinger's cat caused other problems as well, as the dynamics of quantum systems no longer executed clear trajectories in phase space. For instance, the classical harmonic oscillator, which holds a special place as perhaps the most fundamental of all dynamical systems, has beautiful symmetry in phase space, with momentum and position trading off in eternal sympathy. The action of Maupertuis is simply the area enclosed by the phase point tracing out its orbit. On the other hand, the quantum harmonic oscillator in phase space behaves strangely. An eigenstate of the quantum harmonic oscillator is represented by a stationary probability distribution in position and momentum, neither moving nor changing, all dynamics seemingly gone. Dynamics can return when the quantum state is in a coherent superposition of different eigenstates, the time dependences beating against each other in time-varying interference patterns. But even in this picture, there are strange extra interference effects in phase space that are troublesome, discovered by the Hungarian-American Eugene Wigner.

Eugene Wigner (1902–1995) and John von Neumann were longtime friends, meeting as boys in school in Budapest, and both were invited to Princeton University in 1930 as part of the second wave of quantum theorists who had studied under the great names of the Berlin and Göttingen schools. Shortly after arriving at Princeton, Wigner began working on the problem of how to define the phase-space trajectories of quantum systems. He discovered that the representation of quantum systems in phase space cannot be done with strictly positive probabilities, which is one of the side effects when transforming from Hilbert space into phase space. The best one can do is to define a pseudo-probability distribution function in phase space, known today as a Wigner distribution, in which negative pseudo-probability values are a fundamental signature of non-classical behavior—of Schrödinger's cat.

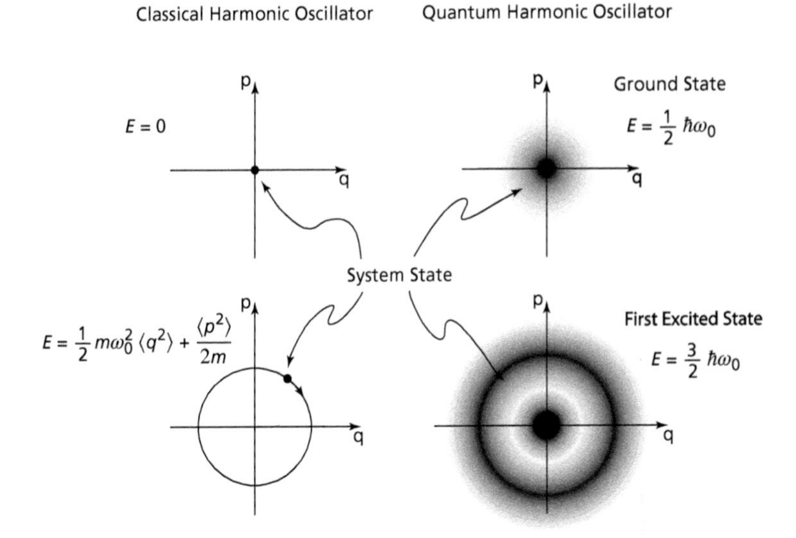

Fig. 8.4 Comparison of the phase space of the classical harmonic oscillator with the quantum harmonic oscillator. The classical system has a distinct phase point that executes a trajectory in phase-space. The quantum system in the same phase-space is described by a stationary pseudo-probability distribution that extends across phase-space and can have regions of negative values

This is the price paid for a phase-space description of quantum dynamics (see Fig. 8.4). The clean orbits of classical physics are replaced by corrugated interference patterns with negative probabilities. Dirac (who married Wigner's sister) had found negative probabilities abhorrent when he was deriving relativistic quantum theory late in 1928,[28] preferring instead to accept negative energy states that turned out to be the prediction of antimatter. Yet quantum coherences in phase space require negative probabilities if both position and momentum are to be retained in a single description of quantum phenomena. Once again, quantum phenomena and entanglement impose uncomfortable demands on our credulity.

Entanglement today has grown into an entire industry. It lies at the heart of quantum computer architectures being pursued by major information technology corporations. Quantum entanglement provides the exponential resource that allows quantum algorithms to calculate answers that would take classical computers the age of the universe to find. Schrödinger's cat and quantum entanglement are alive and well in quantum gates that seek to utilize quantum superpositions. Meanwhile, Schrödinger's micro-macro divide, once meant to serve as a philosophical argument, now has become an engineering problem that is central to the emerging field of quantum computing.

Quantum computing had its origin in 1982 when Richard Feynman was asked to give a keynote lecture at a computer science conference. Not knowing anything about the topic, he improvised, giving an impromptu speech that introduced the computer science world to quantum mechanics. Feynman's seminal role in quantum computing was part of a pattern that persisted throughout his lifetime, harking back to his early days when he already tended to be iconoclastic, never shy about challenging the paradigms of the day.

Drums in Brazil

Richard Feynman (1918–1988) was a whiz at mathematics, winning the national Putnam Math competition his senior undergraduate year at MIT.[29] But his physical intuition was visual, and his arguments were verbal, practiced best in animated debates in hallways. When he thought of physics problems, he saw pictures in his mind's eye, pictures that were dynamic and moved. While he was taking upper-division mechanics and required to learn Lagrangians, he resisted adopting the formalism because it was too abstract and mathematical. He felt that it swept the true physical processes under the rug—processes that he could see and keep track of much better by decomposing forces into their components and identifying all the action-reaction pairs at work in the old Newtonian way. It took mental gymnastics to keep all these forces and components in one's head, like tracking multiple jugglers' balls in the air, especially for those tedious problems with multiple pulleys and masses given to every undergraduate physics major. But Feynman liked to juggle.

After receiving his bachelor's degree in physics from MIT in 1939, Feynman applied to the graduate school at Princeton expecting to work with Eugene Wigner, considered one of the top physicists of the day. But when he arrived, he was assigned instead to an unknown young assistant professor, newly arrived at Princeton, by the name of John Wheeler. At their first meeting, Wheeler told Feynman that he was a busy man and could only spare a few hours per week to work with him.

He placed a stopwatch on his desk to track the time, and stopped their discussion when the time was up. At their next meeting, when Wheeler placed his stopwatch on the desk, Feynman took out his own, started it, and placed it next to Wheeler's. They both burst out laughing, and it launched a deep mentor–apprentice-friendship that lasted a lifetime.[30]

Wheeler and Feynman were a perfect fit. Wheeler was a wild open thinker, with bold ideas that spanned the breadth of theoretical physics, but with deep physical insight like Feynman's. They discussed and argued about physics for hours, the stopwatch long since forgotten. Wheeler and Feynman began talking about a central problem of quantum electrodynamics called the self-energy of the electron. Dirac, Heisenberg, and Pauli in the 1930s had made great strides in understanding the basic interactions of quantum electrodynamics, but when they tried to calculate the next order corrections to improve the accuracy of their calculations, they stumbled upon integrals that diverged to infinity. One of the most basic of these integrals described how an electron interacted with its own emitted field. It was an infinite self-recursive loop that could not be ignored, and it stopped all progress.

As Feynman and Wheeler considered how to get around the infinities, Feynman was alerted to an obscure paper published by Dirac in 1933 that had gone virtually unnoticed.[31] In that paper, Dirac had suggested that the phase of a quantum wavefunction was related to the mechanical action—Maupertuis' action, expanded by Euler, Lagrange and Jacobi—but he had not pursued this connection. Armed with Dirac's hunch, Feynman made rapid progress to show that the phase of the wavefunction not only was related to the mechanical action, but also was directly proportional to it, with Planck's fundamental constant h being the only quantity needed in the proportionality coefficient. Feynman's thesis work proceeded without mathematical surprises, but the consequences were profound. Feynman began to construct a new approach to quantum processes that was unique and unlike the formalisms of either Heisenberg's matrix mechanics or Schrödinger's wave mechanics. Feynman's work took on the form of a third fundamental approach to quantum mechanics based on the principle of least action.

To get a clear view of this radical new approach to quantum trajectories, it is helpful to return to the classic problem of Snell's Law that was discussed in Chapter 4. This problem is explained by numerous approaches—by Fermat through his principle of least time, by Huygens through his principle of primary and secondary wave fronts (Fig. 8.5),

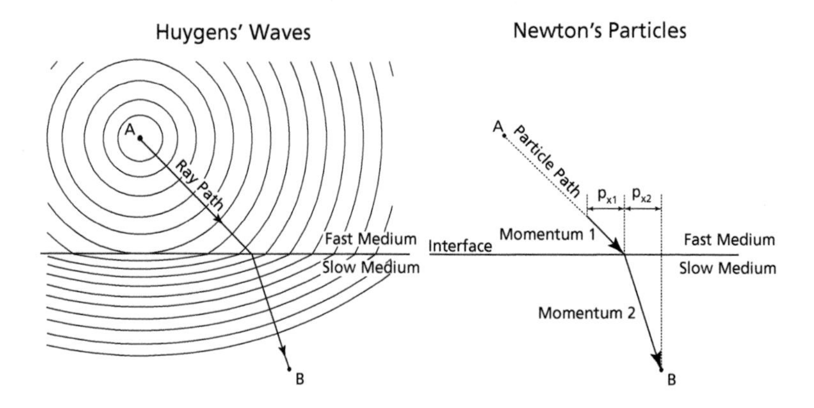

Fig. 8.5 The wave view of Snell's Law (on the left). The source resides in the medium with higher speed. As the wave fronts impinge on the interface to a medium with lower speed, the wave fronts in the slower medium flatten out, causing the ray perpendicular to the wave fronts to tilt downwards. The particle view of Snell's Law (on the right). The momentum of the particle in the second medium is larger than in the first, but the transverse components of the momentum (the x-components) are conserved, causing a tilt downwards of the particle's direction as it crosses the interface

by Maupertuis through his principle of least action (although applied incorrectly at the time), by Maxwell's equations and by momentum conservation of a photon crossing the interface. In short, there are roughly a half-dozen different derivations of Snell's law that all work to explain the change in angle of a ray at the interface.

Feynman's approach allows a quantum particle, like an electron or a photon moving from a point A to a point B, to take all possible paths, no matter how wild and unphysical the path may be. This is illustrated for Snell's Law in Fig. 8.6. A photon leaves a source at point A and travels everywhere, sampling all paths from A to B. A complex amplitude with a specific phase characterizes each possible path. The value of the quantum phase is precisely the phase that Dirac proposed in 1933 as the integral of Maupertuis' old mechanical action integrated over the path. The breakthrough that Feynman achieved in his doctoral dissertation was in developing the mathematical methods to evaluate the integral for each path, and then summing all the complex amplitudes of all the paths to yield the probability of observing the particle taking a path

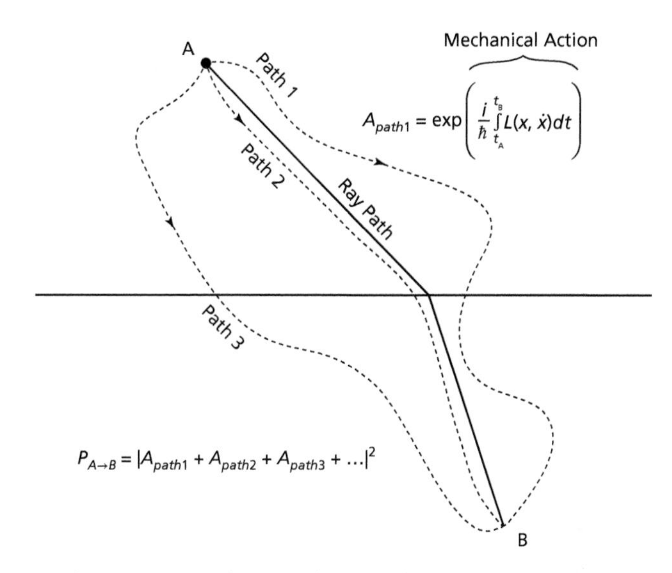

Fig. 8.6 The Feynman path integral approach to dynamical trajectories. The amplitudes of all possible paths are equal, but the phase of a given path depends on the mechanical action integrated over the entire path. The phases interfere in the sum over histories, which is squared to yield the probability. Constructive interference is strongest for the path of stationary action—the ray path

from A to B. This is the Feynman path integral approach to quantum mechanics.

In the path integral approach, illustrated in Fig. 8.6, the magnitudes of each path are all equal, but every path from A to B has a quantum phase given by the path integral. The probability for discovering the quantum particle between A and B is given by the square of the sum of all the possible amplitudes. Because each term in the sum has a quantum phase, when the series is squared, there are many interference effects among all the possible paths. On average, there is destructive interference among almost all of the paths, so the probability of finding the particle in most locations is close to zero. However, for paths that are close to the ray path, the phases build up in constructive interference, leading to larger probability for finding the particle near the physical path. The key question is what is so special about the ray path? What differentiates it from all other paths such that only the paths nearby the ray path experience constructive interference? The answer becomes clear by making small variations, small adjustments, of the paths around the ray path. These adjustments produce little to no change in the quantum phases of those paths. That is why the phases of the paths near the physical trajectory all build up constructively. The ray path is the one special path for which the variation of the mechanical action is zero—the path of least action. Feynman had recovered Maupertuis' principle of least action (actually stationary action), which began as a classical result of mechanics, but now was shown to be an underlying principle of quantum processes as well. The path of least action is the path of maximum constructive quantum interference!

Feynman's thesis made a dramatic connection between quantum phenomena and classical trajectories. Here was the correspondence principle couched in much more fundamental language, identifying the principle of mechanical action as a common thread between the strange behavior of quantum systems and the rational regularity of classical trajectories. In the limit of short wavelengths, the classical trajectory of a particle is the path of maximum quantum constructive interference. Classical physics, that is to say Galileo's trajectory, emerges from quantum interference.

Feynman's Diagrams

In the years immediately following the Japanese surrender at the end of WWII, before the horror and paranoia of global nuclear war had time to sink into the psyche of the nation, atomic scientists were the rock stars of their times. Not only had they helped end the war with a decisive stroke, they were also the geniuses who were going to lead the US and the world into a bright new future of possibilities. To help kick off the new era, the powers in Washington proposed to hold a US meeting modeled on the European Solvay Congresses. The invitees would be a select group of the leading atomic physicists, invitation only. The conference was held at the Rams Head Inn on Shelter Island, at the far end of Long Island, New York in June of 1947. The two-dozen scientists arrived in a motorcade with police escort and national press coverage. Richard Feynman was one of the select invitees, although he had done little fundamental work beyond his doctoral thesis with Wheeler. This was his first chance to expound on his path integral formulation of quantum mechanics. It was also his first conference where he was with all the big guns. Oppenheimer and Hans Bethe (1906–2005) were there as well as Wheeler and Kramers, von Neumann and Pauling. It was an august crowd and auspicious occasion.

The topic that had been selected for the conference was Foundations of Quantum Mechanics, which at that time meant quantum electrodynamics, known as QED, a theory that was at the forefront of theoretical physics, but mired in theoretical difficulties. At that time, it was knee deep in infinities that cropped up in calculations that went beyond the lowest order. The theorists could do back-of-the-envelope calculations with ease and arrive quickly at rough numbers that closely matched experiment, but as soon as they tried to be more accurate, results diverged, mainly because of the self-energy of the electron, which was the problem that Wheeler and Feynman had started on at the beginning of his doctoral studies. As long as experiments had only limited resolution, the calculations were often good enough. But at the Shelter Island conference, Willis Lamb (1913–2008), a theorist-turned-experimentalist from Columbia University, announced the highest resolution atomic spectroscopy of atomic hydrogen ever attained, and there was a surprise within the experimental results.

Hydrogen, of course, is the simplest of all atoms. This atom launched Bohr's model, inspired Heisenberg's matrix mechanics and proved Schrödinger's wave mechanics. Deviations from the classical Bohr levels, measured experimentally, had been the testing grounds for Dirac's relativistic quantum theory that had enjoyed unparalleled success until Lamb's presentation at Shelter Island. To everyone's surprise, Lamb showed there was an exceedingly small energy splitting of the ground state by about 200 parts in a billion that amounted to a wavelength of 28 cm in the microwave region of the electromagnetic spectrum. This splitting was not predicted, nor could it be described, by the formerly successful relativistic Dirac theory of the electron.

On the train ride up state after the Shelter Island Conference, Hans Bethe took out his pen and a sheaf of paper and started scribbling down ideas about how to use mass renormalization (adjusting the value of the electron mass depending on the energy of the interaction), subtracting infinity from infinity in a precise and consistent way, to get finite answers in the QED calculations. He made surprising progress, and by the time the train pulled into the station at Schenectady, he had achieved a finite calculation in reasonable agreement with Lamb's shift. When Bethe returned to Cornell towards the end of the summer, he presented his new calculations in a seminar, including the need for a relativistic theory. Feynman, his junior faculty colleague in the Physics Department, jumped up and boldly announced that he would have the answer to Bethe by the next morning. Bethe was aware that Feynman had been toying with an alternative approach to quantum electrodynamics that used path integrals over all histories. This outgrowth of Feynman's work with Wheeler had the key advantage that it used relativistic Lagrangians—exactly what was needed to provide a relativistic theory of the Lamb shift.

Feynman worked overnight, but he hit a snag. He did not know how to calculate the self-energy of the electron. The whole motivation for his work with Wheeler was to eliminate the troublesome electron self-energy, so he had never learned to calculate it in the first place. When Feynman went to Bethe's office the next day, Bethe patiently showed his junior colleague how to do the calculation. Then they got to work to see if Feynman's theory gave a finite answer to the energy difference of the Lamb shift. They worked at Bethe's blackboard, writing with chalk, and the theory failed to give the appropriate results. Feynman was shocked, but Bethe was not. The theoretical physics community had been struggling with these problems for two decades, so the idea that Feynman could solve it overnight was maybe too ambitious and definitely too arrogant.

Feynman was not so easily to be put down. He was a phenomenological thinker, and he knew in his gut that the answer was finite. He stalked away and got down to work, doing the hard slogging that is so necessary to uncovering the subtleties of physics. Somehow, he and Bethe had made a mistake on the chalk board (neither later could figure out what it was),[32] but after many months of intense work, Feynman was able to show how a clear and precise use of relativistic cutoffs allowed the self-energy of the electron to be subtracted away using mass renormalization.[33]

Feynman's work on electron mass renormalization was a breakthrough for QED. Furthermore, Feynman had also begun to condense down his complicated path integrals by encapsulating their quantitative content into visual mnemonics that looked like scribbled cartoons. These cartoons were shorthand that allowed Feynman to quickly dispense with many tedious calculations, dramatically increasing the speed with which he arrived at answers.

The first diagram that Feynman published is shown in Fig. 8.7 for the scattering of two electrons via the Coulomb interaction. The Feynman

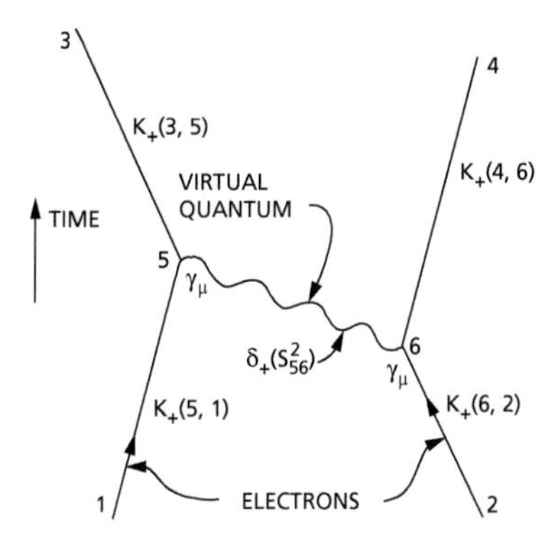

Fig. 8.7 First Feynman diagram published by Feynman, showing two electrons exchanging a virtual photon[34]

diagram is a space-time diagram with time running vertically, and position horizontally. The lines with arrows represent the electrons. The two electrons interact by exchanging a virtual photon, represented by the squiggly line. The diagram is not meant to be a physical trajectory taken literally, but is a shorthand way of visualizing the topology of the interaction. For instance, it does not matter whether electron 1 or 2 emits or receives the photon. Either process yields the same quantum amplitude that goes into the calculation of the scattering probability. The diagrams were a way of enumerating Feynman's multiple histories of a quantum process that needed to be summed as quantum amplitudes prior to calculating the probability.

In the fall semester of 1947, a brilliant young British mathematician arrived at Cornell University to begin a yearlong fellowship paid by the British Commonwealth. Freeman Dyson (1923–) had received an undergraduate degree in mathematics from Cambridge University and was considered one of their brightest graduates. With strong recommendations, he arrived to work with Hans Bethe on quantum electrodynamics. The following summer, Dyson had time to explore America before taking a temporary position at the Institute for Advanced Study in Princeton. Feynman was driving his car to New Mexico to patch things up with an old flame from his Los Alamos days, so Dyson was happy to tag along. For days, as they drove across the US, they talked about life and physics and QED. Dyson had Feynman all to himself and began to see daylight in Feynman's approach, and to understand that it might be consistent with a field theoretic approach taken by Julian Schwinger and Sin-Itiro Tomonaga.

After leaving Feynman in New Mexico, he travelled to the University of Michigan to attend summer school lectures given by Schwinger, and then to Berkeley. On the long bus ride back east, as he half dozed, and half looked out the window, he had an epiphany. He saw all at once how to draw the map from one to the other. What was more, he realized that many of Feynman's techniques were much simpler than Schwinger's. By the time he arrived in Chicago, he was ready to write it all down, and by the time he arrived in Princeton, he was ready to publish. It took him only a few weeks to do it, working with an intensity that he had never experienced before. When he was done, he sent the paper off to the Physical Review.[35]

Once he was established at the Institute for Advanced Study, and after winning over an initially skeptical Oppenheimer who was its director, Dyson was in a position to communicate the new methods to a small army of postdocs at the institute, supervising their progress on many outstanding problems in quantum electrodynamics that had resisted calculations using the complicated Schwinger–Tomonaga theory. Feynman, by this time, had finally published two substantial papers on his approach,[36] which added to the foundation that Dyson was

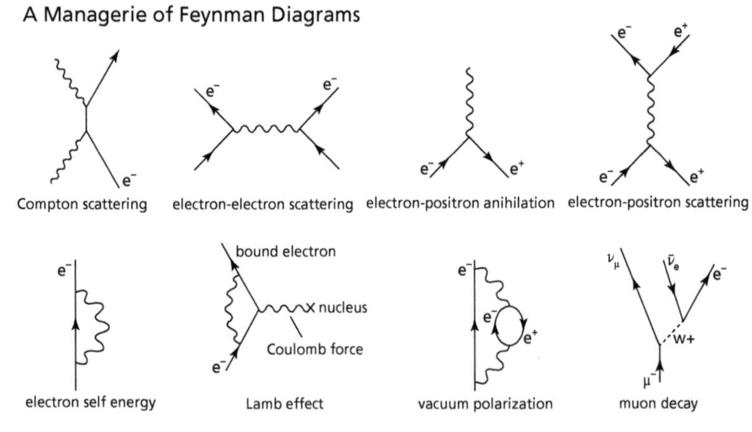

Fig. 8.8 A menagerie of Feynman diagrams showing many basic processes (upper row) as well as many of the "problem" diagrams that cause divergences in the theory (bottom row). On the far bottom right is a decay diagram for the muon that requires the electroweak interaction

building at Princeton. Although Feynman continued to work for a while on QED, the center of gravity for these problems shifted solidly to the Institute for Advanced Study and Dyson. The army of postdocs that Dyson supervised helped establish the use of Feynman diagrams in QED, calculating ever higher-order corrections to electromagnetic interactions, as in Fig. 8.8. These same postdocs were among the first batch of wartime-trained theorists to move into faculty positions across the US, bringing the method of Feynman diagrams with them, adding to the rapid dissemination of Feynman diagrams into many aspects of theoretical physics that extend far beyond QED.[37]

Galileo in Quantumland

Like Alice in Wonderland, Galileo would have found things "curiouser and curiouser" as his simple trajectory slipped into the quantum realm. It almost did not survive the trip. At one point, Heisenberg denied reality to the trajectory, either in configuration space or phase space. In his purest form of matrix mechanics, quantum reality existed merely as numbers filling the ranks of matrices—the extreme rejection of Bohr's original atomic orbits. Schrödinger and Bohr helped to resurrect quantum trajectories, although for different reasons. Schrödinger took the quantum wavefunction as his reality, evolving in time through space in response to his time-dependent equation. Bohr, on the other hand, though willing to adopt the wavefunction as a fundamental element of the emerging Copenhagen interpretation, pushed for complementarity, for wave-particle duality as waves and particles play off each other in a continuous compromise.

Heisenberg relented a little, allowing trajectories back into quantum theory as he gained a deeper understanding of his own uncertainty principle. But he set limits, insisting on a continuous trade-off in the quantum realm between pairs of phase-space observables. Position and momentum, the two canonical variables of phase space, became locked in a perfect balance—as one gained more certainty, the other became less so. With this concession by Heisenberg, the phase-space trajectory reappeared like the Cheshire Cat, fuzzy and hard to pin down, liable to disappear at any moment.

In this story of Galileo's trajectory in Quantumland, Feynman's path integral approach for summing over all quantum histories holds a nebulous place. On the one hand, it is an abstract construction that is more conceptual than useful. Few modern quantum mechanics textbooks even mention the path integral approach to the formulation of quantum mechanics. Although it stands as a third approach next to Heisenberg's matrix mechanics and Schrödinger's wave mechanics, Heisenberg and Schrödinger are mentioned on every page, while Feynman is lucky to be a footnote. On the other hand, it is fundamental and profound, providing the link that connects classical and quantum physics across a span of scales and centuries, making contact with the theory of least action by Maupertuis and Euler and explaining how a classical trajectory emerges from quantum interference. Finally, it provides the foundation for all quantum field theories, with Feynman diagrams emerging as an abstract new way to visualize Galileo's beloved trajectory, although now a trajectory with intrinsic uncertainties.

The uncertainty demanded by quantum theory is not the only uncertainty that seeps into Galileo's trajectory. In the classical dynamics of the macroscopic world, where quantum effects are not allowed to travel, fundamental uncertainties continue to haunt the trajectory, propelled by the gentle beat of a butterfly's wings.

9

From Butterflies to Hurricanes

... the swinging of a pendulum in a clock, the tumbling of a rock down a mountainside, or the breaking of waves on an ocean shore, in which ... variations are not random but look random.

EDWARD LORENZ, The Essence of Chaos (1993)

To live in our world is to live nonlinearly. Every experience, every interaction with outside reality, every beat of the heart, every nerve pulse traveling down innumerable neurons has, at its core, a nonlinear reaction in which the whole *cannot* be the sum of its parts. Linearity is a special case reserved for the flow of time, or the addition of electric fields in vacuum. But as soon as electric fields are superposed in a physical material, nonlinearities lurk beneath. Sometimes the nonlinearities of optical materials can be so strong that light can interact with light when two laser beams cross in a crystal,[1] like dueling light sabers clashing in a shower of sparks. However, the overpowering prevalence of nonlinearities in our lives is surprisingly masked by stability.

Stability is ubiquitous. Most physical objects exist in a condition of stability. A rock has stable mineral structures, and rests stably on the ground. Rolling stones are hard to find. A guiding principle of physics is the principle of least potential energy. Dissipative systems (and most real systems dissipate themselves) always find a path to a minimum in their potential energy landscape. The principle of least potential energy predates Maupertuis' principle of least action, making it more primitive and more dominant. Once a system is at its minimum potential energy, a slight deviation is met with a linear restoring force that tends to return the system to stability, and all the nonlinearities hiding beneath seem to melt away. But don't be fooled! The very linearity is a balance between nonlinear forces, and as soon as that balance is lifted, large and unexpected things may happen.

Galileo Unbound. David D. Nolte, Oxford University Press (2018).
 DOI: 10.1093/oso/9780198805847.001.0001

When the door to nonlinearity is opened, it is like Dorothy opening the door of her Aunt's ruined house and catching her first sight of Oz. Linearity is dull and drab, black and white with a lot of gray. Nonlinearity is brilliant and colorful, full of bizarre flora and wild fauna. Nonlinearity is what makes a physicist sit up and take stock. It makes life brighter and more surprising. Nonlinearities remove the ordinary sweep of the pendulum and replace it with unpredictability—deterministic unpredictability. Poincaré glimpsed such unpredictability when he was scrambling to fix his mistake in King Oscar's prize entry. Poincaré originally tried to prove the stability of the three-body problem, but when he corrected his mistake and found his unstable tangle, the ultimate fate of the Earth was called into question. Can the Earth survive the complex perturbations to its orbit caused by all its companion planets? The crucial answer to this existential question had to wait for the death of a head of state.[2]

(KAM) and the Fate of the Earth

The passing of the terrible Russian dictator Joseph Stalin provided a long-needed opening for Soviet scientists to travel again to international conferences where they could meet with their western colleagues to exchange ideas. Four Russian mathematicians were allowed to attend the 1954 International Congress of Mathematics (ICM) held in Amsterdam, the Netherlands, in that year. One of those was Andrey Nikolaevich Kolmogorov (1903–1987) who was asked to give the closing plenary speech. Despite the isolation of Russia during the Soviet years before World War II and later during the Cold War, Kolmogorov was internationally renowned as one of the greatest mathematicians of his day. As a young undergraduate student at the Moscow State University, while still a teenager, he had gained international recognition for creating a trigonometric series that diverged almost everywhere, except on a set of measure zero. This had been a major open question that pushed far beyond Weierstrass' monster (that merely had a derivative nowhere).

By 1954, Kolmogorov's interests had spread widely into topics in topology, turbulence and logic, but no one was prepared for the topic of his plenary lecture at the ICM in Amsterdam. Kolmogorov spoke on the dusty old topic of Hamiltonian mechanics. He even apologized at the start for speaking on such an old topic when everyone had expected him to speak on probability theory. Yet, in the length of only half an hour he laid out a bold and brilliant outline to a proof that the three-body problem had an infinity of stable orbits. Furthermore, these stable orbits provided impenetrable barriers to the diffusion of chaotic motion across the full phase space of the mechanical system. The crucial consequences of this short talk were lost on almost everyone who attended as they walked away after the lecture, but Kolmogorov had discovered a deep lattice structure that constrained the chaotic dynamics of the solar system.[3] His conclusion was that Hamiltonian systems were not generically ergodic and that the Earth may be saved from ejection for the remaining lifetime of the Solar System.

Kolmogorov's approach used a result from number theory that provides a measure of how close an irrational number is to a rational one. This is an important question for orbital dynamics, because whenever the ratio of two orbital periods is a ratio of integers, especially when the integers are small, then the two bodies will be in a state of resonance, which was the fundamental source of chaos in Poincaré's stability analysis of the three-body problem. Number theory has a lot to say about ratios of integers and their relationships to real numbers. Diophantus of Alexandria first studied problems like these in the third century AD. They were problems like finding when $a^2+b^2 = c^2$ could be satisfied for a, b and c all integers, like the 3:4:5 right Pythagorean triangle. A counter example is Fermat's Last Theorem that states that there are *no* integer solutions to $a^n+b^n = c^n$ for $n > 2$. These types of problems are called Diophantine problems and are addressed using Diophantine analysis. As simple and as old as these problems seem, especially the Pythagorean triplets, Diophantine analysis is a deep part of modern mathematics. For instance, Fermat's Last Theorem was proven only as recently as 1995.

One of the early developers of Diophantine analysis was the German mathematician Carl Ludwig Siegel (1896–1981). Despite being drafted into the German army in 1917, which was desperate for any semi-able-bodied man to replace the hundreds of thousands who were being slaughtered on the two fronts, Siegel was such a screw-up that the army eventually washed him out, and he never saw action. He entered Göttingen at the end of World War I, studying number theory for his doctorate, and his habilitation in 1920 was a seminal work in Diophantine analysis. He received a faculty position at the University in Frankfurt in 1920, where he remained until 1938 when he moved back to Göttingen. However, several of his close friends were Jewish, and he was increasingly distressed by their mistreatment under the Nazi regime. By 1940, he felt he could no longer live in Germany and traveled to Norway just ahead of the German invasion and then to self-imposed exile in the United States, taking a position at the Institute for Advanced Study in Princeton where Albert Einstein also had a position.

In 1942 while at the Institute for Advanced Study, seeking applications of Diophantine analysis, Siegel tackled an old problem that had been stated first in the 1870s and then slightly improved by Poincaré around the turn of the century. This problem concerned the lack of convergence of Fourier series that had small divisors, which was close to the problem of resonant denominators in the three-body problem. Siegel realized that there were Cantor-like sets (see Chapter 5) of frequencies that escaped the small divisor problem and whose Fourier series could be made to converge. These Cantor-like sets of frequencies were identified by Diophantine conditions that are used to measure how closely an irrational number can be approximated by a rational number. For instance, the most irrational number, farthest from any rational number, turns out to the golden mean $\varphi = 1.6180339887$.[4] By Siegel's analysis, if an irrational orbital frequency was sufficiently far from rational numbers, then the Fourier series would converge, and chaos would not arise for that orbit—the orbit would be stable! This is what Dirichlet, Weierstrass and Poincaré had hoped to prove half a century earlier. It was the first time that convergence could be proven if solutions were restricted to a special fractal-like domain. When Siegel published a paper on this result[5] in the midst of World War II, when attention was diverted elsewhere, few took notice, and the paper sat dormant until picked up again by Kolmogorov.

Kolmogorov realized that the dust-like nature of the fractal set of irrational orbits might protect resonant orbits from chaos and instability. Furthermore, these irrational orbits might provide barriers to prevent the chaotic orbits from spreading across the full phase space. The difficulty was finding whether irrational orbits survived if they were sufficiently far from resonant ones. Siegel's results were a starting point, but in perturbed Hamiltonian systems, frequencies shift with increasing perturbation strength, while in Siegel's analysis the frequencies were all fixed. To solve this problem, Kolmogorov used a modified version of

Newton's method (used to find the roots of equations) that worked in a general function space and which could track the shift in frequencies with increasing perturbation. What he found was that, although the frequencies did shift, the Diophantine condition still allowed solutions to converge if the original unperturbed frequencies were sufficiently irrational. With his confidence bolstered by this discovery, and convinced that he had uncovered a crucial insight into the stability of the three-body problem, Komogorov boldly presented his results at the ICM of 1954. What remained was the necessary mathematical proof of Kolmogorov's daring conjecture.

Lectures of the ICM are published in leading mathematics journals that typically appear years after the event due to delays in getting researchers to submit and review manuscripts. When Kolmogorov finally submitted the manuscript of his lecture to *Mathematical Reviews* in 1957, the journal sent the manuscript for review to a German mathematician, Jürgen Moser (1928–1999), who had taken a temporary position at MIT after receiving his doctorate at Göttingen where he had interacted closely with Siegel, who had returned from self-imposed exile to Germany in 1950. The two studied problems in celestial mechanics, especially related to the n-body problem and issues of stability. As Moser read Kolmogorov's manuscript in 1957, he struggled to understand it, and although he read Kolmogorov's other related works, he was left with the serious impression that crucial elements were missing in the proof. Subsequently, this became the focus of his own research.

Meanwhile, back in Moscow, an energetic and creative young mathematics student knocked on Kolmogorov's door looking for an advisor for his undergraduate thesis. The youth was Vladimir Igorevich Arnold (1937–2010), who showed promise, so Kolmogorov took him on as his advisee. They worked on the surprisingly complex properties of the mapping of a circle onto itself, which Arnold filed as his dissertation in 1959. The circle map holds close similarities with the periodic orbits of the planets, and this problem led Arnold down a path that drew tantalizingly close to Kolmogorov's conjecture on Hamiltonian stability. Arnold continued to his PhD with Kolmogorov, solving Hilbert's 13th problem by showing that every function of n variables can be represented by continuous functions of a single variable. Arnold was appointed as an assistant in the Faculty of Mechanics and Mathematics at Moscow State University.

Arnold's habilitation topic was Kolmogorov's conjecture, and his approach used the same circle map that had played an important role in solving Hilbert's 13th problem. Kolmogorov neither encouraged nor discouraged Arnold to tackle his conjecture. Arnold was led to it independently by the bizarre similarity of the stability problem with the problem of continuous functions. In reference to his shift to this new topic for his habilitation, Arnold stated, "The mysterious interrelations between different branches of mathematics with seemingly no connections are still an enigma for me."[6] Arnold began with the problem of attracting and repelling fixed points in the circle map and made a fundamental connection to the theory of invariant properties of action-angle variables.[7] These provided a key element in the proof of Kolmogorov's conjecture. In late 1961, Arnold submitted his results to the leading Soviet physics journal—which promptly rejected it because he used forbidden terms for the journal, such as "theorem" and "proof," and he had used obscure terminology that would confuse their usual physicist readership, terminology such as "Lesbesgue measure," "invariant tori" and "Diophantine conditions." Arnold withdrew the paper, but it appeared in print a year later in 1962 in a different journal.

By this time, Moser, with encouragement from Siegel, had made important advances on his own towards a proof of Kolmogorov's conjecture. At the ICM of 1962, held in Stockholm Sweden that year, he gave an invited lecture in which he announced a formal proof of Kolmogorov's conjecture.[8] He had to relax certain conditions of the original problem to succeed, but he felt that these were justified by the results. To his surprise, Kolmogorov and Arnold were enthusiastic about his achievement, and applauded his relaxed conditions because they made the theorem more applicable to broader classes of problems in Hamiltonian mechanics that went beyond celestial mechanics. Arnold quickly incorporated Moser's approach and published a definitive article on the problem of small divisors in 1963, followed by another in 1964.[9] The combined work of Kolmogorov, Arnold and Moser had finally established the stability of irrational orbits in the three-body problem, the most irrational and hence most stable orbit having the frequency of the golden mean. The term "KAM theory," using the first initials of the three theorists, was coined in 1968 by B. V. Chirikov, who also introduced in 1969 what has become known as the Chirikov map (also known as the Standard map[10]) that reduced the abstract circle maps of Arnold and Moser to simple iterated functions that any student

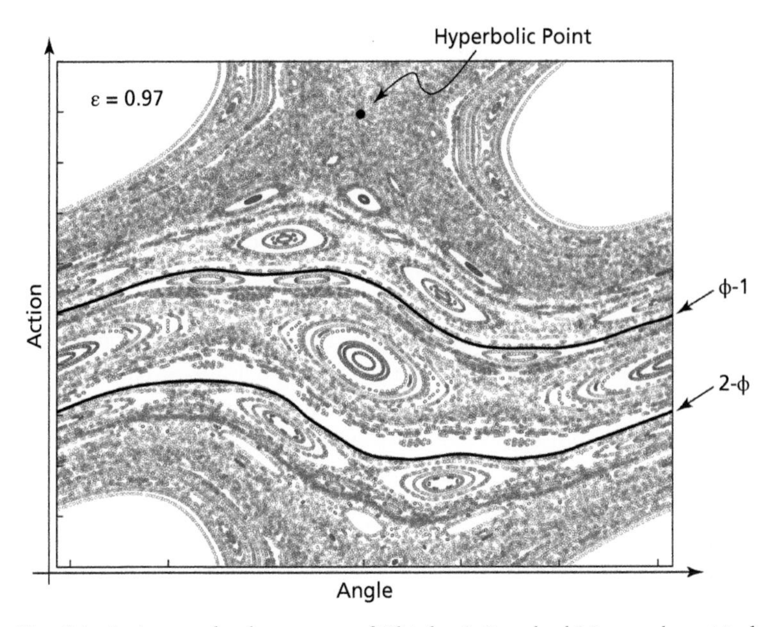

Fig. 9.1 Action-angle phase space of Chirikov's Standard Map at the critical condition $\varepsilon = 0.97$ when the last stable orbits, the golden mean orbit with orbital frequency ratios φ-1 and 2-φ, dissolve into chaos

can program easily on a computer to explore KAM invariant tori and the onset of Hamiltonian chaos, as in Fig. 9.1.

KAM theory solved a two-century-old problem in celestial mechanics concerning the stability of the solar system. The key result is that Poincaré's resonant orbits, that are unstable to perturbations, leading to chaos, are separated from each other by invariant tori that are protected from the perturbation. These invariant tori present insurmountable barriers to the chaos from the resonances, preventing the chaos from diffusing throughout the full phase space, ensuring the long-term safety of the Earth and all life on it—at least in terms of the 3-body problem. But Arnold did not stop there, considering a number n larger than 3 in the n-body problem. To his surprise, even though invariant tori continue to exist, they no longer occupy contiguous regions of phase space to act as barriers to the diffusion of trajectories across the full range of the phase space. Chirikov named this effect "Arnold diffusion" in 1969, and Arnold devised a simple mapping that mimicked the stretching and folding of phase space that occurs at Poincaré's homoclinic tangle by constructing an iterative transformation of the image of a cat, becoming known as Arnold's "cat map". Arnold's interest in stretching and folding was shared by one of the first Western mathematicians whom he had met in Moscow during a visit in 1961—Stephen Smale.

Stephen Smale (1930–) is the mathematician's equivalent of a rock star and spiritual guru. He founded a school of thought at the University of California at Berkeley in the 1960s and 1970s that influenced a generation of researchers in nonlinear dynamics. Like the mythology surrounding any good rock star, the turning point in Smale's career is said to have occurred like a divine epiphany, in his case on the beach at Rio de Janeiro, induced by sun and sand, or by heat stroke eased by a dip in the ocean.

Smale was at the end of a post-graduate fellowship from the National Science Foundation and had come to Rio to work with Mauricio Peixoto. Smale and Peixoto met in Princeton in 1960 where Peixoto was working with Solomon Lefschetz (1884–1972) who had an interest in oscillators that sustained their oscillations in the absence of a periodic force. For instance, a pendulum clock driven by the steady force of a hanging weight is a self-sustained oscillator. Lefschetz was building on work by the Russian Aleksandr A. Andronov (1901–1952) who worked in the secret science city of Gorky in the 1930s on nonlinear self-oscillations using Poincaré's first return map. The map converted the continuous trajectories of dynamical systems into discrete numbers, simplifying problems of feedback and control.

The central problem of mechanical control systems, even self-oscillating systems, was how to attain stability. By combining approaches of Poincaré and Lyapunov, as well as developing their own techniques, the Gorky school became world leaders in the theory and applications of nonlinear oscillations. Andronov published a seminal textbook in 1937 *The Theory of Oscillations* with his colleagues Vitt and Khaykin, and Lefschetz had obtained and translated the book into English in 1947, introducing the West to this important topic. When Peixoto returned to Rio his interest in nonlinear oscillations captured the imagination of Smale even though his main mathematical focus was on problems of topology.[11] Back on the beach in Rio, he had an idea that topology could help prove whether systems (known as structurally stable[12] systems) had a finite number of periodic points.

Peixoto had already proven this for two dimensions, but Smale wanted to find a more general proof for any number of dimensions.

Norman Levinson (1912–1975) at MIT became aware of Smale's interests and sent off a letter to Rio in which he suggested that Smale should look at Levinson's work on the triode self-oscillator[13] (known as a van der Pol oscillator, to be discussed in more detail in Chapter 11), as well as the work of Cartwright and Littlewood[14] who had discovered strangely quasi-periodic behavior hidden within the equations. Smale was puzzled but intrigued by Levinson's paper that had no drawings or visualization aids, so he started scribbling curves on paper that bent back upon themselves in ways suggested by the van der Pol dynamics. During a visit to Berkeley later that year, he presented his preliminary work, and a colleague suggested that the curves looked like strips that were being stretched and bent into a horseshoe.

Smale latched onto this idea, realizing that the strips were being successively stretched and folded under the repeated transformation of the dynamical equations. Furthermore, because dynamics can move forward in time as well as backwards, there was a sister set of horseshoes that were crossing the original set at right angles. As the dynamics proceeded, these two sets of horseshoes were repeatedly stretched and folded across each other, creating an infinite latticework of intersections that had the properties of the Cantor set (Chapter 5). Here was solid proof that Smale's original conjecture was wrong—the dynamics had an infinite number of periodicities, and they were nested in self-similar patterns in a latticework of points that map out a Cantor-like set of points. In the two-dimensional case, shown in Fig. 9.2, the fractal dimension of this lattice is $D = \ln 4/\ln 3, = 1.26$ somewhere in dimensionality between a line and a plane. Smale's infinitely nested set of periodic points was the same tangle of points that Poincaré had observed while he was correcting his King Otto Prize manuscript but was unable to draw. Smale, using modern principles of topology was finally able to put rigorous mathematical structure to Poincaré's homoclinic tangle. Ironically, Poincaré launched the modern field of topology, so in a sense he sowed the seeds to the solution to his own problem.

Smale's horseshoe is important for several reasons. It explains, using topological methods, Poincaré's homoclinic tangle, and it was the first demonstration of an attractor with an infinite nesting of periodic points with repeating patterns at all scales. In this way, it helped launch a new branch in the field of geometric mechanics. It also explains the

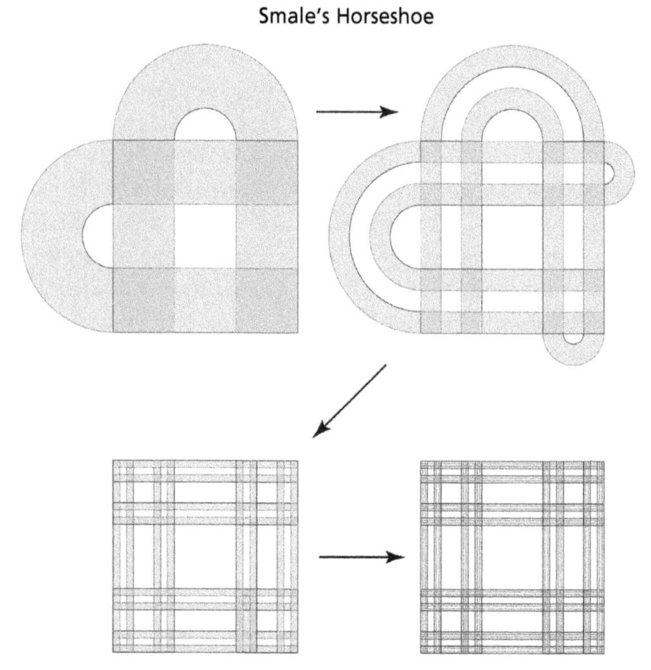

Fig. 9.2 The Smale Horseshoe construction of the nested lattice of points around a hyperbolic fixed point. The fractal dimension of the final set, after an infinite number of iterations, is $D = \ln 4/\ln 3 = 1.26$—the same as the Cantor Set

origins of sensitivity to initial conditions, because each nesting of the transformation guarantees eventual full mixing within phase space, a feature that was pursued by Arnold nearly a decade later with his cat. Perhaps most important was the effect the discovery had on Smale himself. He was so thrilled with the results and the potential power of geometric and topological approaches to dynamics, that he established a school of dynamical system studies at the University of California at Berkeley, a school that spanned several decades and launched a new community devoted to understanding the complexities of dynamical systems of all kinds, from nonlinear oscillators to economic dynamics, from evolutionary processes to mathematical mappings. The school at Berkeley was the vanguard of a new wave that was to wash over the entire field of physics. But Smale was not alone, and others were poised to help launch what would come to be called chaos theory, most notably Edward Lorenz at MIT, studying atmospheric stability using the newest computing machines.

Lorenz's Butterfly

John von Neumann, at the Institute for Advanced Study (IAS) in Princeton in the 1950s, was an intellectual force of nature like a hurricane. He had so many interests and so many ideas that it was nearly impossible for his students and colleagues to keep up with him. The breadth of his activities, although rooted in mathematics, ranged widely across fields as diverse as number theory and geometry, economics and social games, quantum logic and computers. His favorite approach was to axiomatize a field, stripping it down to bare fundamentals that captured the simple essence of behavior. In this manner, he sought to establish a theoretical foundation for automatic computers. He had participated in the heavy computational work of designing and building atomic bombs during the war and had worked with Eckert and Mauchly, who were the inventors of the ENIAC computer at the University of Pennsylvania. As an improvement over the ENIAC, that was physically rewired to execute different programs, Eckert and Mauchly proposed a stored-program approach, which von Neumann formalized into a general structure that is now called the von Neumann architecture. Following the war, von Neumann coordinated an effort to develop such a stored-program computer at the Institute for Advanced Study. One of the problems that he thought could benefit from such a computational machine was the complex physics of the atmosphere and of long-range weather prediction.

In 1948, von Neumann invited Jule Charney (1917–1981), an American meteorologist then working in Olso, Norway, to head the meteorology group of his Electronic Computer Project that would use the IAS computer as its key resource. Charney was an expert in computational approaches to the nonlinear dynamics of geostrophic flow (the jet stream) and of global weather patterns. However, progress on von Neumann's IAS computer was slower than planned, so early computational runs were performed on the ENIAC computer at the Aberdeen Proving Ground of the US Army in Maryland. The weather runs took more than 24 hours of round-the-clock attention by the meteorology staff, mainly because of persistent breakdowns of vacuum tubes, but the first one-day nonlinear predictions were successful in

April 1950. Two years later, in 1952, the IAS computer was ready for use in Princeton, and Charney became the driving force behind the meteorological applications of von Neumann's IAS computer, generating spectacular results. However, von Neumann was increasingly absent from Princeton because of his involvement in the H-bomb project, and Charney was unable to convert his appointment at the Institute for Advanced Study into a permanent position. By 1956, it was time to find a permanent position elsewhere. Charney received several offers for a faculty position as a senior hire, but he was most interested in the offer from MIT. A young meteorologist there by the name of Edward Lorenz had made important advances in nonlinear theoretical models of the weather, but despite being the same age as Charney, was still in a non-tenure-track position as a university researcher. As one of the conditions for accepting the MIT offer, Charney requested that Lorenz be promoted to assistant professor.

Edward Lorenz (1917–2008) had always been a wiz at numbers.[15] From his earliest childhood, he was fascinated by how numbers combined or factored, memorizing all perfect squares up to ten thousand. This interest led him through a bachelor's degree in mathematics at Dartmouth, followed by a master's degree in mathematics at Harvard in 1940 under the supervision of Georg Birkhoff. The mentorship under Birkhoff was a portent for the important role Lorenz later would play in the modern discovery of chaos. Birkhoff had proven Poincaré's Last Geometric Theorem[16] that underpinned the behavior of homoclinic points where Poincaré first observed chaos in 1890. However, Lorenz's master's thesis was on an aspect of Riemannian geometry rather than dynamics. His foray into nonlinear dynamics was triggered by the intervention of World War II. Only a few months before receiving his doctorate in mathematics from Harvard, the Japanese bombed Pearl Harbor and his life changed abruptly, as it did for many young men across the country.

Lorenz left the PhD program at Harvard to join the United States Army Air Force to train as a weather forecaster in early 1942. This allowed him to stay in Boston as he took courses on forecasting and meteorology at MIT. After receiving a second master's degree, this time in meteorology, Lorenz was posted to Hawaii, then to Saipan and finally to Guam. His area of expertise was in high-level winds, which were important for high-altitude bombing missions during the final months of the war in the Pacific. After the Japanese surrender, Lorenz returned to MIT, where he continued his studies in meteorology, receiving his doctoral degree in 1948 with a thesis on the application of fluid dynamical equations to predict the motion of storms. Victor Starr, newly arrived at the meteorology department at MIT, offered Lorenz a position as a researcher on a project concerning the global circulation of the atmosphere. Lorenz distinguished himself in the field as an expert in nonlinear dynamics of the atmosphere, producing a significant theoretical analysis of the available potential energy that drives weather patterns. During this time, he arranged a visit to see von Neumann's computer at the Institute for Advanced Study, where he met Charney. Lorenz was not initially inclined to believe that the computer would be a helpful tool for meteorology, but he left having made a favorable impression of Charney. By the time Charney was offered the position at MIT, Lorenz had been in his research position with Starr for eight years and was approaching his fortieth birthday. With Charney's stipulation that Lorenz should receive an appointment as a faculty member in the meteorology department, Lorenz finally was on a tenure track.

One of Lorenz' colleagues at MIT was Norbert Wiener (1894–1964), with whom he sometimes played chess during lunch at the faculty club. Wiener had published his landmark book *Cybernetics: Control and Communication in the Animal and Machine* in 1949. This was the result of one of the most formidable mathematicians of the age focusing his talents on the apparently mundane problem of gunnery control during the Second World War. As an abstract mathematician, Wiener attempted to apply his cybernetic theory to the complexities of weather, but he developed a theorem concerning nonlinear fluid dynamics which appeared to show that linear interpolation, of sufficient resolution, would suffice for weather forecasting, possibly even long-range forecasting. Many on the meteorology faculty embraced this theorem because it fell in line with common practices of the day in which tomorrow's weather was predicted using linear regression on measurements taken today. However, Lorenz was skeptical, having acquired a detailed understanding of atmospheric energy cascades as larger vortices induced smaller vortices all the way down to the molecular level, dissipating as heat, and then all the way back up again as heat drove large-scale convection. This was clearly not a system that would yield to linearization. Therefore, Lorenz determined to solve nonlinear fluid dynamics models to test this conjecture, but integrating the equations was too complex to do by hand. It required a computer.

With the help of a friend at MIT, Lorenz purchased a Royal McBee LGP-30 tabletop computer. It was, in its day, to von Neumann's IAS machine what a laptop is, today, to a cluster of servers. Where the IAS computer filled an entire room, the Royal McBee could sit on a large table. It used 113 of the latest miniature vacuum tubes, and also had 1450 of the new solid-state diodes made of semiconductors rather than tubes, which helped reduce the size further, as well as reducing heat generation. The McBee had a clock rate of 120 kHz and operated on 31-bit numbers with a 15 kb memory.[17] Under full load it used 1500 Watts of power to run. But even with a computer in hand, the atmospheric equations needed to be simplified to make the calculations tractable. Lorenz was more a scientist than an engineer, and more of a meteorologist than a forecaster. He did not hesitate to make simplifying assumptions if they retained the correct phenomenological behavior, even if they no longer allowed for accurate weather predictions. He was more interested in gaining insight into the physics. He had simplified the number of atmospheric equations down to twelve, and he began programming his Royal McBee.

Progress was good, and by 1961, he had completed a large initial numerical study. He focused on nonperiodic solutions, which he suspected would deviate significantly from the predictions made by linear regression, and this hunch was vindicated by his numerical output. One day, as he was testing his results, he decided to save time by starting the computations midway by using mid-point results from a previous run as initial conditions. He typed in the three-digit numbers from a paper printout and went down the hall for a cup of coffee. When he returned, he looked at the printout of the twelve variables and was disappointed to find that they were not related to the previous full-time run. He immediately suspected a faulty vacuum tube, as often happened. But as he looked closer at the numbers, he realized that, at first, they tracked very well with the original run, but then began to diverge more and more rapidly until they lost all connection with the first-run numbers. The internal numbers of the McBee had a precision of six decimal points, but the printer only printed three to save time and paper. His initial conditions were correct to a part in a thousand, but this small error was magnified exponentially as the solution progressed. When he printed out the full six digits (the resolution limit for the machine), and used these as initial conditions, the original trajectory returned. There was no mistake. The McBee was working perfectly.

At this point, Lorenz recalled that he "became rather excited."[18] While he had set out to discredit linear regression, he had done much more. He was looking at a complete breakdown of predictability in atmospheric science. If radically different behavior arose from the smallest errors, then no measurements would ever be accurate enough to be useful for long-range forecasting. At a more fundamental level, this was a break with a long-standing tradition in science and engineering that clung to the belief that small differences produced small effects. What Lorenz had discovered, instead, was that the deterministic solution to his twelve equations was exponentially sensitive to initial conditions (known today as SIC). Over the following months, he was able to show that SIC was a result of the nonperiodic solutions. The more Lorenz became familiar with the behavior of his equations, the more he felt that the twelve-dimensional trajectories had a repeatable shape. He tried to visualize this shape, to get a sense of its character, but it is difficult to visualize things in twelve dimensions, and progress was slow.

Fortunately, science is never done in a vacuum, and Lorenz was not alone in devising nonlinear models of the atmosphere. While visiting Barry Saltzmann, a fellow theoretical meteorologist, he learned that Saltzmann had reduced the number of nonlinear equations to seven. Lorenz tried these out in his own computations and discovered that when the solution was nonperiodic (the necessary condition for SIC), four of the variables settled down to zero, leaving all the dynamics to the remaining three variables. Lorenz immediately rewrote the equations to isolate the relevant behavior to only those three variables: the stream function, the change in temperature, and the deviation from linear temperature. Here he finally had a three-variable dynamical system that displayed chaos (see Fig. 9.3). Moreover, it had a three-dimensional state space that could be visualized directly. He ran his simulations, exploring the shape of the trajectories in three-dimensional state space for a wide range of initial conditions, and the trajectories did indeed always settle down to restricted regions of state space. They relaxed in all cases to a sort of surface that was elegantly warped, with wing-like patterns, as the state point of the system followed its dynamics through time. This surface behaved like an attractor, the thing to which all trajectories relax, but it was not a true surface, and the individual trajectories comprising this surface were never the same. Two close trajectories, even though they lived on this subspace of the full three-dimensional state space, separated exponentially. In a word,

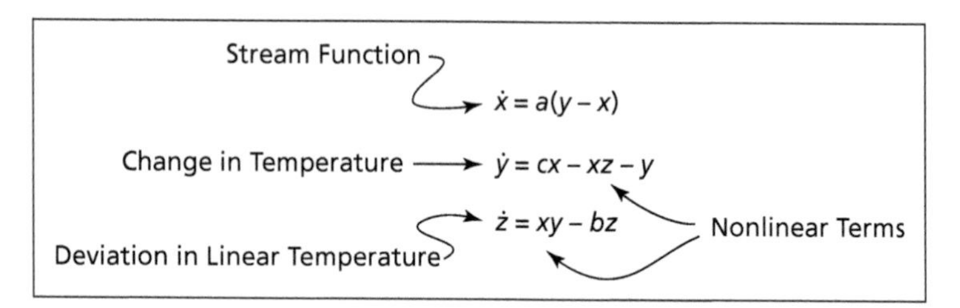

Fig. 9.3 The Lorenz equations for three-dimensional chaos

the attractor of the Lorenz equations was *strange*. Later, in 1971, David Ruelle (1935–), a Belgian-French mathematical physicist named this a "strange attractor," and this name has become a standard part of the language of the theory of chaos.

Lorenz published his three-dimensional equations and solutions in 1963 in what has become one of the most celebrated and highly cited papers in chaos theory,[19] although the initial community response at publication was overwhelming silence. The world was not primed for the deep consequences of Lorenz' work, and in a sense, neither was he. Some of his fellow theoretical meteorologists showed interest, but many others in the field were less than impressed, because the simplifications required to reduce the dimensions of the dynamics down to three also made the model useless as a tool for accurate modeling of the atmosphere. Fortunately, Lorenz was not concerned with accurate modeling and continued to explore the beautiful dynamics.

In 1972, Lorenz was invited to speak at the annual meeting of the American Association for the Advancement of Science (AAAS). He accepted the invitation, but missed the deadline to supply a title, so the conference organizer picked a title for him. He called it "Does a Butterfly Flapping its Wings in Florida cause a Tornado in xx?" This was an appropriate choice, because the Lorenz attractor looks like a butterfly's wings (see Fig. 9.4), and the phrase has become a self-propagating meme that has mutated in many allied forms, some turning Florida into China and others turning the tornado into a hurricane. Collectively, these are all called the "Butterfly Effect," capturing the sensitivity to initial conditions that Lorenz discovered after his coffee break in 1961. By 1975, the news of chaos theory had spread, and it was going viral. In that year, James Yorke of the University of Maryland coined the phrase "chaos

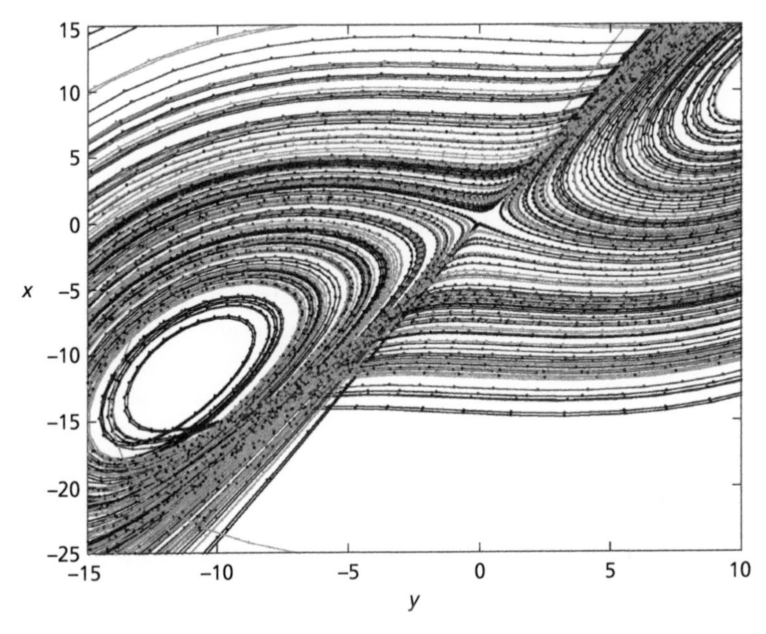

Fig. 9.4 A projection of the Lorenz Butterfly[20] onto two dimensions. This shows a single trajectory arising from a single initial condition that relaxes onto the strange attractor

theory" which stuck. Chaos theory, as a field, can be said to have coalesced at a conference in New York on bifurcation theory held in 1977 in honor of Eberhard Hopf (1902–1983) on his seventy-fifth birthday, marking the official beginning of an important and vibrant new field of science. All the leading figures who had launched the field were there: Hopf, Ruelle, Smale, Mandelbrot, Yorke, Haken and Prigogine, among many others. Notably, Lorenz was given a triumphant ovation, because, by now, his 1963 paper had become famous and he himself had become a celebrity.

Feigenbaum's Ratio

The study of physics tends to attract quiescent geniuses that, despite their obvious gifts, lie dormant like a chrysalis waiting for the right time and the right place to emerge. Most of them lead unremarkable careers, never finding the right problem on which to spend their talents, never emerging from obscurity. But a few persist, sometimes working on seemingly arcane problems, unnoticed by their colleagues, until they suddenly issue forth with a great work, like an emerging butterfly taking wing and soaring above the crowd. This was the situation for two Cornell physicists in the early 1970's. One was Kenneth Wilson, and the other was Mitchell Feigenbaum. Although they never worked together, a common thread known as the renormalization group strangely linked their careers.

Kenneth Wilson (1936–2013) was a graduate in high-energy physics theory from Cal Tech, receiving his doctorate degree under the Nobel Laureate Murray Gell-Mann and joining the faculty at Cornell University in 1963 as an assistant professor. No one who met him doubted that he was brilliant. He had deep insights into physics, but he didn't produce much concrete work (i.e., publications), and his promotion several years later was hard fought within the faculty. Not long after receiving tenure, he began to attend a weekly theory discussion group led by Michael Fisher (1931–) and Benjamin Widom (1945–) of the chemistry department. Fisher had been working on problems of critical phenomena, thinking about how physical systems appear at different scales, when Widom made an important discovery in the behavior of phase transitions very close to the transition temperature.[21] The temperature dependence of the phases did not obey classical predictions, but displayed power-law dependence. More importantly, different systems, composed of different materials with different transition temperatures, obeyed the same power law with the same critical exponent, displaying unexpectedly universal behavior. Leo Kadanoff (1937–2015), then at the University of Illinois, using similar arguments as Fisher, was able to explain aspects of the universality by considering how fluctuations in the system behaved when viewed at different scales.[22] It was around this time, in the mid-60s, that Wilson joined the weekly discussion group at Cornell.

Why a high-energy particle theorist, working in the realm of quarks and high-energy collisions, would take an interest in critical phenomena that occur in condensed materials was not immediately clear. These two fields of science have diverged so thoroughly during the era of the specialization of physics through the twentieth century that they are almost never mentioned together. However, Wilson was focused on a deep problem in theoretical particle physics known as renormalization. When Richard Feynman and Freeman Dyson, working with Hans Bethe at Cornell some years before, had developed quantum electrodynamics (QED), the success of the theory had come with a cost—in many of the calculations there were infinite sums that diverged to infinity. This had been a major obstacle to progress, until it was realized that calculating quantities over defined scales could eliminate the infinities. This was a mathematical trick that kept things finite, and it worked, but it also meant that fundamental constants of nature, like the electric charge, took on the appearance of being scale dependent, having different values when observed at different scales—this was renormalization. The mathematical trick worked for QED, and most theorists stopped worrying about the infinities. However, when quantum field theory was applied to the more difficult problem of interacting quarks, the infinities got much worse, and renormalization, the original fix, itself needed fixing.

This was the situation when Wilson recognized that the renormalization procedure of high-energy particle physics shared much in common with the scaling theories of Fisher and Kadanoff for phase transitions. The high-energy problems were not phase transitions, but renormalized quantities in particle interactions did have power-law scaling behavior. It was a subtle connection that, on the surface, seemed tenuous, but Wilson was tuned in to a deeper correspondence that Fisher and Widom did not initially appreciate. Quickly, Wilson learned the scaling techniques of Fisher and Kadanoff, and then generalized them by defining a renormalization procedure that used a scaling transformation to successively rescale the physical process. At the critical threshold, this scaling transformation became invariant—it was a fixed point of the transformation. Every time you applied the transformation, you got the same thing back. Wilson called this process, and the analysis it enabled, the renormalization group, and he quickly provided it with a firm theoretical foundation. The physics faculty at Cornell realized that Wilson had produced something extraordinary and promoted him to full professor the year before he published two landmark papers in 1971[23] in the Physical Review B. Building upon this foundation, in which he began by making important inroads into phase transitions in solid-state physics, Wilson went on to apply the renormalization group to a theoretical construct called lattice gauge theory that yielded new insights into the interactions of quarks. Wilson had taken a roundabout detour from high-energy physics, through solid-state theory, to arrive back again in high-energy physics. For his work on the renormalization group, Wilson received the Nobel prize in 1982, receiving it alone, without Fisher and Kadanoff, but whose work and influences he readily acknowledged.

Just as Wilson was being promoted to full professor in 1970, Mitchel Feigenbaum arrived at Cornell as a fresh PhD in particle physics from MIT. Feigenbaum had a penchant for numbers, just like Lorenz, and spent hours at a time as a youth thinking about and playing with numbers, learning to calculate logarithms in his head. At Cornell he got access to an early HP desktop calculator, whose only other user was Wilson, and he used it to explore numerical solutions to nonlinear equations. But Feigenbaum's postdoctoral position was only part time, and Feigenbaum was also a lecturer, leaving him little time to do serious work that could lead to published papers. After two years, the postdoc position was up, and he moved to Virginia Tech, where he ended up in the same situation and soon was out looking for a job again with virtually nothing to show for four years of postdoctoral work. Part of his problem was that as soon as he had understood a problem, he would leave it behind and start thinking about something else, without writing it up. Without publications, physicists are generally not employable, so his prospects after Virginia Tech were bleak.

Fortunately, there are people who recognize genius, even if they have no genius of their own. The character of Salieri in the stage play *Amadeus* springs to mind. He was the jealous rival of Mozart at the Austrian court. Mozart easily outshone him and made Salieri's music seem trite in comparison to his own, earning him his enmity. Yet, Salieri alone had a deep and thorough recognition of Mozart's genius. He was enthralled by it even as he sought to destroy it. Others, better than Salieri, can recognize genius and nurture it. At the Los Alamos National Laboratory, Peter Carruthers, the new director of the Theoretical Division, recognized Feigenbaum's talent and offered him a job. Feigenbaum arrived in his new position in late 1974, and Carruthers asked him to look into the problem of the onset of turbulence.

The onset of turbulence was an iconic problem in nonlinear physics with a long history and a long list of famous researchers studying it. As far back as the Renaissance, Leonardo da Vinci made detailed studies of water cascades, sketching whorls upon whorls in charcoal in his famous notebooks. Heisenberg, oddly, wrote his PhD dissertation on the topic of turbulence even while he was inventing quantum mechanics on the side. Kolmogorov in the 1940s applied his probabilistic theories to turbulence, and this statistical approach dominated most studies up to the time when David Ruelle and Floris Takens published a paper in 1971 that took a decidedly nonlinear dynamics approach to the problem rather than statistical, identifying strange attractors in the nonlinear dynamical Navier–Stokes equations[24]. This paper coined the phrase "strange attractor." One of the distinct characteristics of their approach was the identification of a bifurcation cascade. A single bifurcation means a sudden splitting of an orbit when a parameter is changed slightly. In contrast, a bifurcation cascade was not just a single Hopf bifurcation, as seen in earlier nonlinear models, but was a succession of Hopf bifurcations that doubled the period each time, so that period-two attractors became period-four attractors, then period-eight and so on, coming fast and faster, until full chaos emerged. A few years later Gollub and Swinney experimentally verified the cascade route to turbulence,[25] publishing their results in 1975.

When a new field of science is "hot," there is an onslaught of new ideas and new results tumbling one after another. Furthermore, when the new science is cross disciplinary, there can be a crush of advances as fast or faster than everyone can keep track of. This was the case for the new science of chaos in the mid-70s. A confluence of traditions came from many directions—from mathematical dynamical systems theory (Poincaré, Birkhoff, Kolmogorov, Smale), from hydrodynamics (Ruelle), from meteorology (Lorenz), from control systems (Andronov, van der Pol, Wiener), and from population ecology (Volterra, Lotka, May, see Chapter 11). The signatures of chaos were being observed in all of these fields and reported in the literature.

Robert May (1936–) is an Australian population biologist at Merton College, Oxford, who is also a life peer of the House of Lords with the title Baron of Oxford. In 1976 he published a review of the logistic equation (shown in Fig. 9.5), first developed and named by Pierre-François Verhulst in 1838, to describe the population-carrying capacity of an environment, as a simple discrete model of how populations change in successive generations.[26] The equation is deceptively simple, just the equation for an inverted parabola as in Fig. 9.6, but it is iterated repeatedly so that this year's population is a simple parabolic function of last year's population. The only adjustable parameter is the yearly growth rate. For low values of the growth rate, the population is stable, so that this year's population is a bit more than last year's. But as the growth rate increases above a critical value, the yearly population

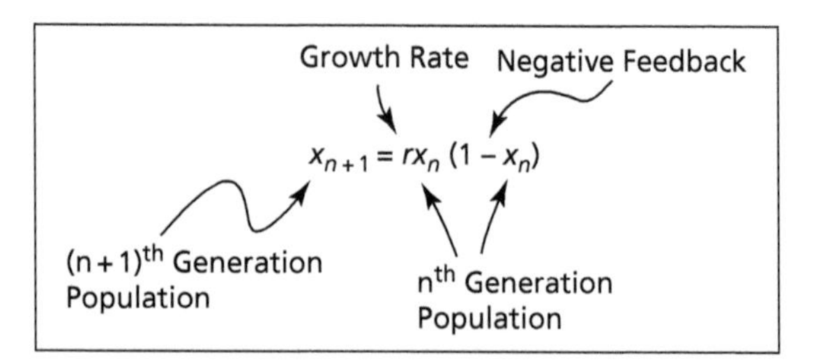

Fig. 9.5 Logistic map equation. The n+1th generation depends on the nth generation with a growth rate r. The nonlinearity of the model is contained in the negative feedback that limits the population if it grows to large

bifurcates into a period-two limit cycle. As the growth rate continues to increase, there are more and more bifurcations, coming faster and faster as the growth rate increases, until full-blown chaos emerges. This simple model, possibly the simplest of all nonlinear models, showed a bifurcation cascade (see Fig. 9.6) remarkably similar to the onset of turbulence proposed by Ruelle and Takens. Feigenbaum at Los Alamos was intrigued by the similarity, abandoning his study of turbulence to pursue this model that had no physics—just pure numbers. What he discovered was that pure numbers *were* the physics.

In 1976, computers were not common research tools, although handheld calculators now were. One of the most famous of this era was the Hewlett-Packard HP-65, and Feigenbaum pushed it to its limits. He was particularly interested in the bifurcation cascade of the logistic map—the way that bifurcations piled on top of bifurcations in a forking structure that showed increasing detail at increasingly fine scales. Feigenbaum was, after all, a high-energy theorist and had overlapped at Cornell with Kenneth Wilson when he was completing his seminal work on the renormalization group approach to scaling phenomena. Feigenbaum recognized a strong similarity between the bifurcation cascade and the ideas of real-space renormalization where smaller and smaller boxes were used to divide up space. One of the key steps in the renormalization procedure was the need to identify a ratio of the sizes of smaller structures to larger structures. Feigenbaum began by studying how the bifurcations depended on the increasing growth rate. He calculated the threshold values r_m for each of the bifurcations, and then took the ratios of the intervals, comparing the previous interval $(r_{m-1} - r_{m-2})$ to the next interval $(r_m - r_{m-1})$. This procedure is just like the

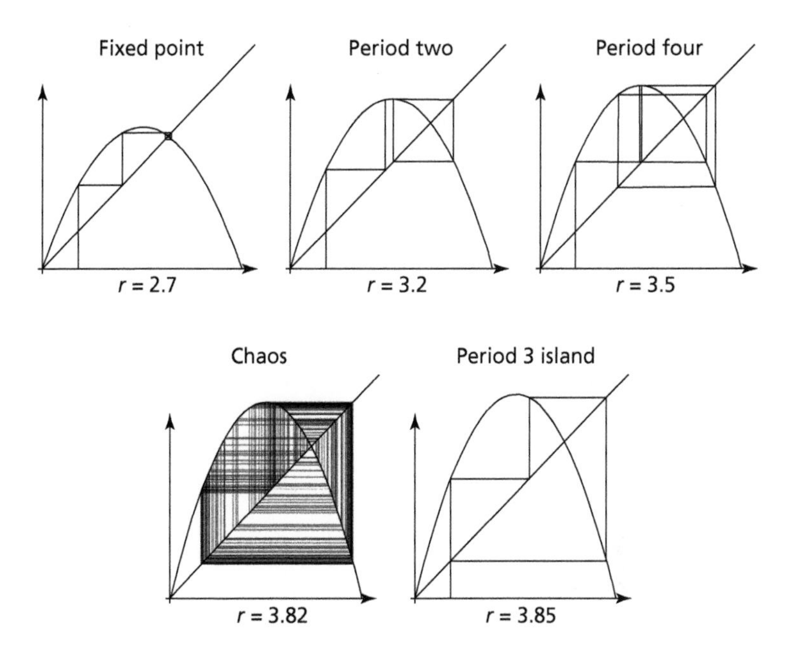

Fig. 9.6 Graphical representation of the discrete dynamics of the logistic map, plotting the n+1th value against the n-th value. The inverted parabola is the logistic function. A starting x value is evaluated on the parabola, this point is then translated horizontally to the diagonal line, where it becomes the next x-value to be evaluated, and so on. At $r = 2.7$, the successive values settle down to a single point. For $r = 3.2$ the attractor is a period-2 limit cycle. For $r = 3.5$ it is a period-four limit cycle. The transition to chaos occurs at $r = 3.57$. However, for some values above the chaos limit it is possible to find "islands of stability," in the case of $r = 3.85$ it is a period-three limit cycle[27]

well-known method to calculate the golden ratio $\phi = 1.61803$ from the Fibonacci series, and Feigenbaum might have expected the golden ratio to emerge from his analysis of the logistic map. After all, the golden ratio has a scary habit of showing up in physics, just like in the KAM theory. However, as the bifurcation index m increased in Feigenbaum's study, this ratio settled down to a limiting value of 4.66920. Then he did what anyone would do with an unfamiliar number that emerges from a physical calculation—he tried to see if it was a combination of other fundamental numbers, like pi and Euler's constant e, and even the golden ratio ϕ. But none of these worked. He seemed to have found a new number (see Fig. 9.7).

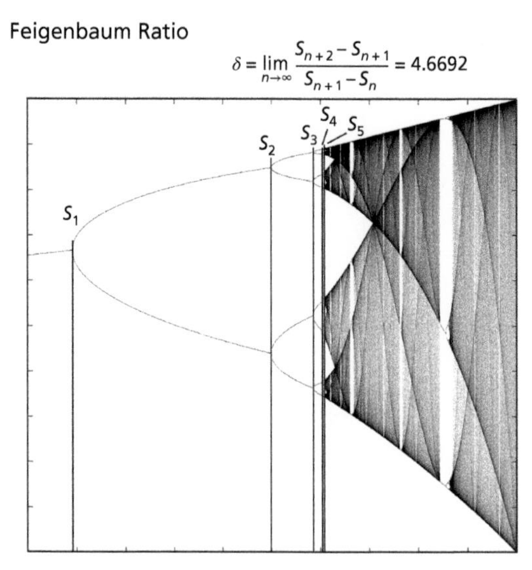

Fig. 9.7 The stable population values of the logistic map as a function of the growth rate r showing the bifurcation cascade to chaos. The bifurcation thresholds have values whose relative ratios approach the Feigenbaum number[28]

At this point, he got serious, and put in a requisition order for a desktop computer. He taught himself Fortran (the ancient computer language for engineering applications) and pushed the accuracy of his calculations. He began to wonder if other mathematical models that showed similar kinds of bifurcation cascade would have their own characteristic number, so he began to explore sinusoidal maps that shared some properties in common with the logistic map, but whose iterated maps looked very different. To his surprise, when he calculated the ratio of successive intervals in the sine map, he obtained the same value of 4.66920 to within numerical precision. Then he tried other discrete maps known to have bifurcation cascades, and this same number emerged from these too. Here was something fundamental. Here was a commonality, universality, among different mathematical models used to describe very different physical systems.

In the excitement of discovery, Feigenbaum forgot to eat and sleep, pushing himself for nearly two months, until doctors had to intercede for fear of his health.[29] What Feigenbaum had hit upon was a potential new use of the renormalization group that he had seen Wilson develop at Cornell. The bifurcation cascade of many of the maps he studied had clear scaling properties through their branching structure. In fact, the cascades were scale invariant, just like a phase transition at the critical temperature. By applying Wilson's renormalization group in real space (known as real-space renormalization group), Feigenbaum was able to show that the number 4.66920 was a fundamental constant of nature, independent of the microscopic details of whatever map was being considered. All it depended upon was the lowest-order polynomial expansion of the mapping function. As long as the lowest-order expansion was a quadratic function, the simplest and most common nonlinearity, then the bifurcation cascade would be governed by this number—his number—known today as the Feigenbaum ratio.

There are many fundamental numbers, like $\pi = 3.14159$ and the golden mean $\phi = 1.61803$ and Euler's number $e = 2.71828$, but not so many have been discovered recently that govern a whole class of phenomena. Furthermore, in the early development of chaos theory, the field had the appearance of being filled with lots of special cases, each different than the other, without common principles. With the discovery of universality in the bifurcation route to chaos, Feigenbaum united many far-flung parts of the young field. He was right to be excited. One evening in 1976, while in the midst of his discovery, Feigenbaum called home to his parents to tell them he had discovered something that would make him famous. But fame is fickle, and he struggled to get his discovery published. It finally appeared in 1978 in a lesser journal after a difficult review process. By 1984, scaling in complex systems was all the rage. Wilson had received the Nobel prize in 1982 for his renormalization group in the same year that Mandelbrot had gained high visibility with the publication of his popular book on fractals. In 1984, Feigenbaum and Mandelbrot were invited to a symposium of the Nobel Committee in Stockholm, Sweden. Both were unofficial "candidates" for a possible future prize, either jointly or exclusively. Feigenbaum's lecture was performed with aplomb and excitement—a young scientist soaring high. Mandelbrot's lecture, on the other hand, was dismissive and has come to be known as his "Anti-Feigenbaum lecture" during which he ruthlessly belittled Feigenbaum's contributions.[30] Neither ever received the Nobel prize.

Galileo Poised

The physics of two interacting bodies is trivial and barely deserves our attention, while three bodies can explode into chaos. The transition is abrupt and astonishing and opens a brave new world of dynamics. The early discoverers of dynamic uncertainty, Poincaré,Lyapunov and Birkhoff,[31] took steps that were as fundamental as they were minimalist. The two traditions that led to chaos, celestial mechanics on the one hand and control theory on the other, were each concerned with small deviations from stability. In celestial mechanics, the deviations were perturbations, for instance the effect of Jupiter upon the Earth-Sun system, while in control theory, the deviations were error signals and feedbacks. These systems of interest had phase-space dimensionalities that tended to be low. For instance, Lorenz reduced his twelve dimensions to three so that he could study the barest minimum that retained the complex behavior he was interested in. Furthermore, there was the difficulty of hyper visualization—visualizing twelve dimensions is beyond challenging, while visualizing three dimensions is second nature to us.

Yet, mathematics knows no limits to degrees of freedom. The hyper-dimensional manifolds upon which dynamical systems live in phase space are as easy to define and calculate as a three-space geometric figure, even if they cannot be pictured. Furthermore, the physical world cares nothing for limits, and systems having an Avogadro's number N_A of degrees of freedom are the rule rather than the exception. Boltzmann understood this, as did Maxwell and Gibbs, which is why, as soon as they defined the system dynamics in phase space, they abandoned the system trajectory in favor of probability distributions and ensembles. For gases and liquids in thermodynamic equilibrium, ensembles and probabilities are a simplification of great utility, leading to powerful conclusions about the complex behavior of such large systems.

There is a lot of room between the few-body problems and the N_A-body problem, even on a logarithmic scale. Yet in this wide middle kingdom reside problems of the greatest importance to our lives—life and consciousness. Life and consciousness are possibly the greatest complex systems for which we would like answers to deceptively simple questions. How did life begin? What is consciousness? No three-dimensional phase-space portrait, no statistical ensemble, will answer these questions. These systems are neither in equilibrium, nor reducible to low-dimensional manifolds. New tools and approaches are needed that push the concept of Galileo's trajectory to extremes as it threads its way into existential problems of fine filigree, touching our souls. The descendants of Galileo's trajectory are now poised to measure the dimensions of our existence. To begin exploring this final frontier, we turn next to Darwin and the dynamics of evolution.

10

Darwin in the Clockworks

. . . I use the term Struggle for Existence in a large and metaphorical sense, including dependence of one being on another . . .

Charles Darwin, *On the Origin of Species* (1859)

In the sciences, three great questions stand above all others. One is the origin of the Universe and of all the matter and forces within it. Another is the origin and evolution of life. The third is the origin and nature of consciousness. Of these, the emergence of life takes the middle ground. The distant past (four billion years ago when life first appeared on Earth) may be out of reach, but much of the subsequent fossil record is not, tracing the progress of life through stages of increasing complexity. Life progresses—that is a fundamental observation of the fossil record. Complexity rises—that is a fundamental principle of self-organizing systems. Information increases—that is a fundamental law of competition and selection. Hence, complex self-organizing systems, composed of interacting subparts, competing with other systems, evolve.

Complex self-organizing systems are studied in *evolutionary dynamics*, a field that tackles problems of changing complexity. Evolutionary dynamics draws heavily from principles of nonlinear dynamics and chaos, but extends those principles to systems having a large number of degrees of freedom. Chaotic systems that are studied classically usually have only a few dimensions (three or a few more) that are fixed. By contrast, the systems treated by evolutionary dynamics can have hundreds or thousands of dimensions in an expanding hyper-dimensional hyperspace. In this dynamically changing context, Galileo's trajectory slips ever farther from his simple parabola, twisting through abstract high-dimensional spaces with new dimensions popping casually into existence. But the principles stay the same—the state of the system is a single point in an expanding hyperspace, evolving according to fixed laws of dynamics.

Galileo Unbound. David D. Nolte, Oxford University Press (2018).
 DOI: 10.1093/oso/9780198805847.001.0001

The Origin of Species

The wind was blowing briskly from the lee as the *HMS Beagle*, driven back twice before by southwestern gales, sailed from the harbor at Devonport, England on 27 December 1831, outfitted for a five-year mission of exploration.[1] On board, looking over the rails at the receding coastline was twenty-two-year-old Charles Robert Darwin (1809–1882), recently graduated without distinction from Cambridge University, without professional prospects, without a paid salary as the expedition's volunteer naturalist, but armed with a great optimism. Darwin was young, eager and thrilled at the prospects of adventures in exotic lands. Most importantly, his voracious mind was empty, ready to devour observations of the finest detail that would fuel a growing conviction that the world was dynamic, that the lands and species contained within it experience steady transmutation rather than remaining the unchanging God-given species of Scripture.

The five-year voyage provided Darwin with a wealth of data as well as several nascent ideas. After returning to England in October 1836, he began categorizing, analyzing, and making sense of all he had seen. Within a year, Darwin began writing down his ideas in a series of journals that have come to be called the "Transmutation Notebooks." These fascinating documents show the evolution of his thought, revealing a process of gradual crystallization. Reading them is like reading Sherlock Holmes, as each piece of evidence is cleverly connected to an emerging comprehensible theory of the crime—for instance, the finches of the Galapagos Islands.

Upon his return to England, Darwin sent off his collection of birds to John Gould, an ornithologist and artist (art and ornithology tended to be dual careers, viz. Audubon). During the voyage, Darwin had assumed that the many specimens of collected finches, displaying a remarkable variety of adaptations suited for their specific behaviors, had all been terrestrial finches of South America transplanted to the young volcanic islands. In March of 1837, Gould informed Darwin that these finches represented eleven different species of a single group that was not represented anywhere on the mainland.[2] Darwin was faced with a sudden decision. According to the thinking of his day, all species appeared suddenly in an act of creation. Yet given the young nature of the Galapagos Islands, this creation event of the finches must have been relatively recent, within the range of human memory. Therefore, common sense dictated an alternative: these species had originated from the mainland, but had changed in a short time, transformed to adapt to their new and varied environments among the islands. Darwin was exceedingly careful not to use such direct language as we use today, as accustomed as we are to natural selection. His thoughts were of consistency. The presence of the eleven species of Galapagos finches was consistent with transmutation of species, not proof. After all, proof required the transmutation process to be observed directly, which was not possible with the finches.

On the other hand, the fossil record did provide snapshots of such transmutation processes. In Darwin's time, the fossil record was not nearly as complete as it is today, but during Darwin's travels in Patagonia he uncovered amazing fossils of horses and armadillo's that were surprisingly similar to modern species, and yet not the same, with differences that made it clear that these were other species. The gradations were paramount to watching transmutation at work. Once again, this was circumstantial evidence, consistent with the transmutation of species, but not proof.

The underlying half-seen pattern of so many puzzling pieces nagged at him, begging him to draw it forth. For instance, he understood the basics of artificial selection, easily seen in the cultivation of hybrid varieties by farmers who selected the best features for transmission to the next generation. These were active intentional choices made by agents acting with forethought. The central question troubling Darwin was what force caused selection in nature without forethought. What caused *natural* selection? He could feel meaning behind the details—he just needed the right way of looking at it, of describing it, and most importantly, of explaining it.

A book he happened to open in September of 1838 was titled *An Essay on the Principle of Population* by Robert Malthus (1766–1834), first published in 1798, but extensively expanded in many later editions, such as the 6th Edition from 1826 that Darwin held in his hands.[3] Malthus was an Anglican cleric and scholar who taught at the college founded by the East India Company to train its employees. As a young curate, he had worked with the urban poor in his parish and had come to reflect on the causes of their miserable lives. Around the same time, the French Revolution had introduced philosophical ideals of egalitarianism and the perfectibility of man, which Malthus found contrary to his own experience. The central thesis of his *Essay* was that population, if unchecked, would grow geometrically—a result proven fifty years earlier by Euler—while food supplies could grow only arithmetically (linearly) with time. Malthus was no mathematician, and even the later editions of his *Essay* contain no equations, but he made clear that finite resources would curb population growth, imposing hunger, misery and vice—the struggle for existence, the burden of which would fall mainly to the poor.

Evidenced by entries in his Transmutation Notebooks, as Darwin read Malthus he had an epiphany. The efficient cause, the mechanism, of natural selection was the geometric growth of population, limited by the resources available in the environment. To explain the origins of species, Darwin adopted the metaphor of a "wedge."[4] Geometric growth was like a wedge being driven into wood, sinking ever deeper while pushing the halves apart, opening gaps—what we today call an ecological niche. Therefore, in Darwin's new theory, transmutation was the trigger, geometric growth was the amplification (or wedge), and the struggle for life was the judge and jury of selection favoring adaptations that would propagate.

Darwin's book *On the Origin of Species by Means of Natural Selection, or the Preservation of Favored Races in the Struggle for Life* had an extremely slow gestation. It was already written out in draft form in 1844, and Darwin spent the next fifteen years assembling evidence to support it. He was concerned that his theory of natural selection, based on the transmutation of species, would be rejected, or worse, ridiculed, by the scientific establishment that he worked so hard to cultivate and yearned so much to be a part of. To ease reception of his radical ideas, he spent much of those fifteen years introducing his ideas to a broadening circle of friends and supporters, mounting a political campaign for the acceptance of natural selection.

By June 1858, the groundwork was laid, and his voluminous evidence assembled, ready to be put into final manuscript form, when he received a battered packet from Indonesia, mailed several months before by a distant acquaintance, Alfred Russel Wallace (1823–1913). The packet contained a manuscript that Wallace respectfully asked Darwin to read for him before the Linnaean Society.[5] No eyewitness account has survived of how Darwin reacted when he opened and read the manuscript that Wallace had written on his own theory of natural selection, and Darwin's autobiography of many years later was less than forthcoming about his intense personal ambition and jealous ownership of his pet idea, but it is clear that Wallace's nearly simultaneous discovery of natural selection was one of the great shocks of Darwin's life. He moved immediately to rally his formidable scientific friends to his support, accepting (or tacitly eliciting) their advice to write up a short manuscript to be read, side by side with Wallace's, at the meeting of the Linnaean Society in London on 1 July 1858. Despite the rush, and in spite of his worries, both papers were received with general acceptance—the Darwin–Wallace theory of natural selection had been successfully proposed without ridicule or immediate negative reaction, perhaps due to the years Darwin spent preparing the ground. Darwin also took the opportunity, at the meeting, to present copies of letters he had written to colleagues, years earlier, staking his own claim of scientific priority. In the end, Wallace apparently was content to yield to Darwin the lion's share of the credit, as well as the lion's share of the storm of criticism that was coming.

Fortunately, Darwin did not have to fend off all the criticism by himself. His assiduous cultivation of supporters now paid off as others of like mind took up the zealous battle. This was all the more important for the protection of the fragile new idea because Darwin by this time was withdrawing from public life due to troubled health, retreating to the confines of his family home at Down House in Kent outside London. Nonetheless, he was like a general behind the lines, in constant contact through letters and reports. It was with no little satisfaction that he read the report of the meeting of the British Society in Oxford on 30 June 1860 where Samuel Wilberforce, a bishop of the Church of England, arguing against Darwin's supporter T. H. Huxley (1825–1895), asked whether Huxley was descended from monkeys on his grandmother's or his grandfather's side. Huxley's immediate rejoinder put Wilberforce in his place before their assembled colleagues, by saying that he would rather have a monkey for a grandparent than to be associated with a man who used God's gifts to obscure the truth.

Huxley was the leading expert in the comparative anatomy of hominids and was making noise that humans were related, through evolution, to our simian brethren. Huxley had come from modest origins, but worked his way up through the ranks as a self-taught naturalist. He had his own voyage of discovery, like Darwin, but on the HMS *Rattlesnake* from 1846 to 1850 sailing to New Guinea and Australia. He was only twenty years old when he shipped as assistant surgeon, despite no formal training, but he had impressed the Admiralty with his deep knowledge of anatomy, which apparently was all it took to be a surgeon on one of Her Majesty's ships. Three years after the famous Oxford meeting, Huxley published a landmark book *Man's Place in Nature* on comparative anatomy, showing the striking resemblances among the hominid species. Darwin relied heavily on Huxley for his own book *The Descent of Man*, published eight years later. Huxley became known as Darwin's bulldog because he was a formidable combatant in the early struggles to establish evolution as the origin of species. He carried the weight willingly, as he carried the weight of Darwin's coffin in 1882 as pallbearer at his funeral.

The origin of species and the descent of man represent special types of dynamic trajectories. In the years before genetics, these trajectories were envisioned as trees of descent, with upper branches splitting off from lower branches and roots. In Darwin's *Origin*, the only illustration in the book is a conceptual diagram of the tree of descent, shown in Fig. 10.1. Darwin states "Thus the small differences distinguishing varieties of the same species, will steadily tend to increase till they come to equal the greater differences between species of the same genus, or even of distinct genera." He goes on to say:

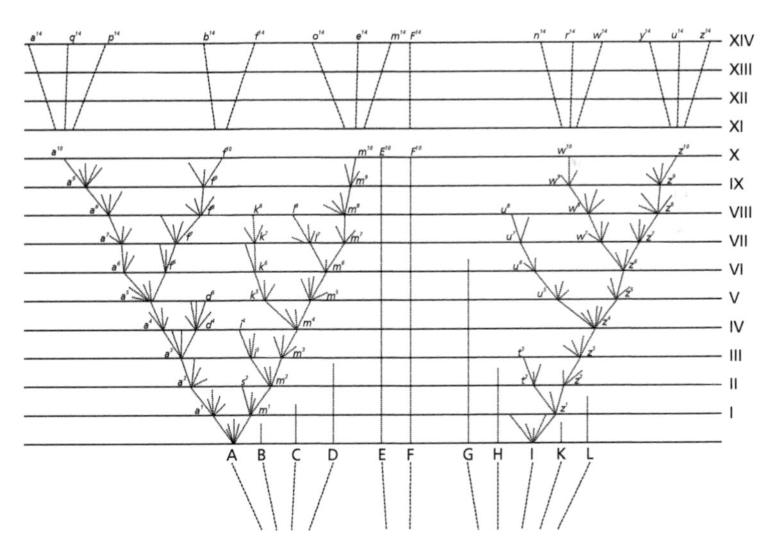

Fig. 10.1 This conceptual illustration of the tree of descent is the only figure in Darwin's *Origin*. The roots are A thru L. Each time period on the vertical labeled I thru XIV is a thousand generations

> The affinities of all the beings of the same class have sometimes been represented by a great tree. I believe this simile largely speaks the truth . . . . As buds give rise by growth to fresh buds, and these, if vigorous, branch out and overtop on all sides many a feebler branch, so by generation I believe it has been with the great Tree of Life, which fills with its dead and broken branches the crust of the earth, and covers the surface with its ever branching and beautiful ramifications.[6]

This sole illustration captured his abstract ideas on evolution and made them concrete, connecting allegorically to the biblical Tree of Life, but transcending God's creation with the mechanics of the struggle for life.

The Limits of Growth

The Moorish beekeepers near the port city Bejaia on the southern coast of the Mediterranean were busy with their bees when young Leonardo, visiting with his father from Pisa, may have been drawn to the problem of their multiplication. Female bees, the queens, come from fertilized eggs, while the male bees, the drones, come from unfertilized eggs. Therefore, queens have two parents, but drones have only one. As queens mature and lay eggs, a new generation of bees emerges whose ancestry depends on the numbers of the previous two generations. Leonardo was a mathematical prodigy, and he must have been intrigued by the intricate sequence of ancestry that emerged from this process. As he spent his time immersed in the rich cultural life of the Berber coast, Leonardo learned how the Arabic scholars used Hindu numerals, a counting system far superior to the awkward Roman system and the anachronistic abacus still in use at the turn of the thirteenth century in Europe. Using the Hindu numerals, the number of ancestors for successive generations of bees is 1, 1, 2, 3, 5, 8, 13, 21, 34, 55, 89 ad infinitum (see Fig. 10.2).

After traveling widely around the Mediterranean, visiting Arabic ports with his father who was an emissary of the Republic of Pisa, at that time a powerful seafaring state, Leonardo returned home with his mind full of the possibilities for improved calculation. In 1202, he published a comprehensive work called the *Liber Abaci* that demonstrated in detail how to do calculations without an abacus using the new Hindu–Arabic numerals. His *Liber Abaci* is credited with introducing Hindu–Arabic numerals into Europe, and Leonardo, later known as Fibonacci, became

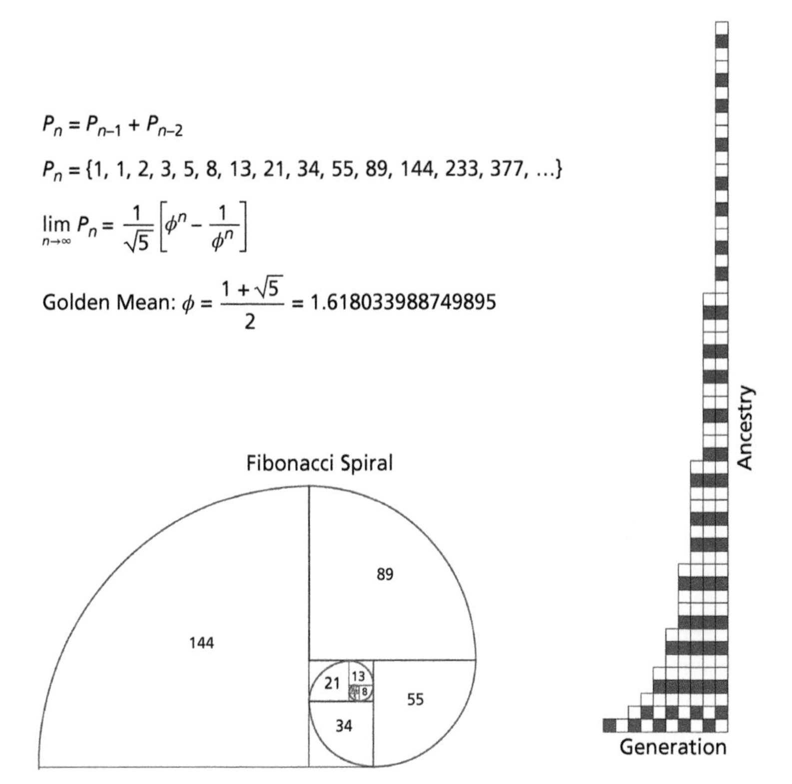

Fig. 10.2 Fibonacci numbers and the ancestry of bees. All starting values of the series converge on the golden mean as the ratio between successive numbers. The golden mean appears often in mathematical evolutionary dynamics. It is the "most irrational" number[7]

famous in his own time for his prowess in arithmetic and his practical application of mathematics to trade and business.

As one of his examples, he might have remembered the problem of the bees of Bejaia. But he was also the son of a merchant and recognized the importance of trade secrets. The beekeepers of Bejaia had a lock on the production of beeswax across the Mediterranean, and their secrets were dear. So, no bees appeared in *Liber Abaci*, rather, they morphed into rabbits.[8] Of course, the ancestry of rabbits has nothing in common with bees, but they were convenient to illustrate the infinite sequence generated by the recursive formula $P_n = P_{n-1} + P_{n-2}$. This simple example of the sequence of numbers generated by a recursive equation is called the Fibonacci sequence, and it may represent the first demonstration of mathematical population dynamics, although this part of *Liber Abaci* was not well known.

Half a millennium later, Leonhard Euler took the next decisive step in the mathematical understanding of populations. In 1748 Euler was in Berlin at Frederick the Great's Academy of Sciences under the direction of Maupertuis, newly arrived from Paris, when he published *Introductio in analysin infinitorum* (Introduction to Analysis of the Infinite). This book established the modern form of calculus by employing concepts of infinite series and limits. The work was monumental in size and scope, and among its many problems and examples were solutions to basic questions of population growth. Euler constructed a recursive model in which numbers of successive populations depended on the numbers of previous populations. The simple mathematical expression of this population growth is $P_n = (1+r)P_{n-1} = (1+r)^n P_0$, which represents the geometric growth of the population with the compounded yearly rate r.

Euler, a devout protestant, used this formula to refute critics of the Bible who argued that the world population could not have grown so large since the Flood of Noah, which was only some four thousand years in the past. Using logarithms to deal with the large numbers involved, he showed that the growth rate since the Flood, assuming the world population started with the six survivors on the Arc, needed to be only half a percent per year, which was well below the rate of growth in his day of 15 percent per year. Furthermore, if one did assume a growth rate of 15 percent, then in only 400 years after the Flood, the world population would have grown to one billion people, and Euler doubted that the Earth could support such a number. Ten years later, when Euler was back in St. Petersburg, he published a refinement of his growth model that included more realistic parameters that captured the rates of births and deaths, generating the kinds of actuarial tables that were beginning to be standard tools for a budding insurance industry. As Euler had remarked, there would be limits to growth when the growing population exhausted existing resources, which Malthus later amplified to famous effect.

In 1838, Pierre-Francois Verhulst (1804–1849), a Belgian mathematician at the Free University in Brussels, put this principle to a mathematical equation. He considered a population with a natural growth rate denoted by r within an ecosystem with a carrying capacity K. The

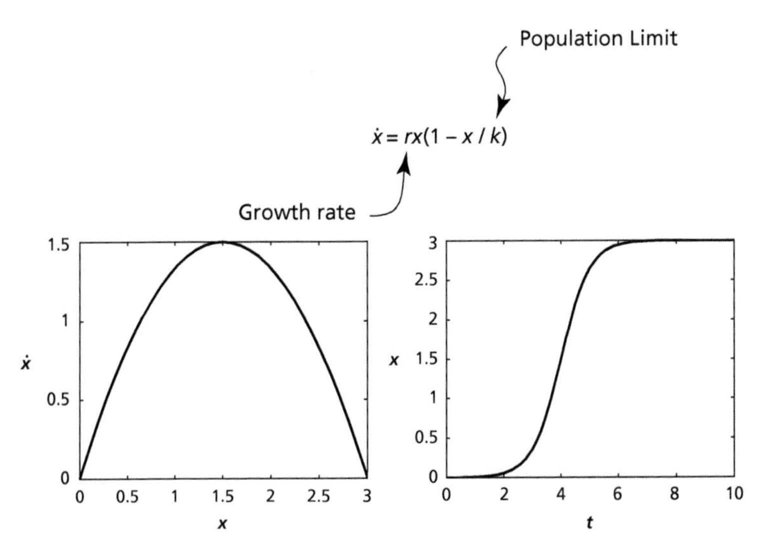

Fig. 10.3 The logistic equation (on the left) is an inverted parabola. The logistic curve (on the right) is a "sigmoid" function as the population x grows exponentially at short times with rate r, but limits to k at long times

resulting growth equation was an ordinary differential equation with a solution that grew exponentially for short times, but later saturated to the carrying capacity, shown in Fig. 10.3. He called this solution a "logistic" curve, and he used it to analyze the populations of the United States, France and Belgium.[9] Of special notice were the estimates for the carrying capacity of France at 40 million and Belgium at 5.6 million. In 1900, 60 years after Verhulst's projections, France had a population of 38 million and Belgium had a population of 6.1 million, both very close to Verhulst's estimates. However, by the year 2000 the populations were 59 million for France and 10.5 million for Belgium. Clearly, the carrying capacity of a country changes in time as production efficiency improves. In addition, Verhulst's model for steady growth and steady saturation failed to capture more complex dynamics that were observed in ecosystems where natural animal populations went through times of boom and bust. Extensions were needed to improve the descriptive power of population dynamics.

After the end of the First World War, the American demographer Alfred James Lotka (1880–1949) constructed a pair of ordinary differential equations that modeled the ecological interaction between the vegetation in a local habitat and a species of herbivores.[10] If the herbivores overgrazed, then both herbivore and vegetation became extinct, which was one steady state. However, other solutions underwent steady oscillations in which the herbivore population expanded and contracted, following an oscillating density of plants. The solution represented a dynamic steady state, like one of Poincaré's limit cycles. Lotka published the work in 1920 shortly before he became head of the research department at the Metropolitan Insurance Company in New York.

Several years later, the Italian mathematician and physicist Vito Volterra (1860–1940) used the same equations to model the proportion of sharks and rays caught in the Adriatic Sea.[11] The interruption of the fishing industry during the war had somehow caused a large increase in predator populations. Volterra showed that the densities of predators and prey went through oscillations of boom and bust, much as the herbivore-vegitation system had, although Volterra had been unaware of Lotka's work. After a dispute on issues of priority between Lotka and Volterra, the equations eventually became known as the Lotka–Volterra model for the oscillating dynamics of two components of an interacting ecosystem.

The phase-space portrait of the Lotka–Volterra system dynamics is shown in Fig. 10.4. The two populations go through a steady cycle of boom and bust. Both the period and the amplitude of the oscillation depend on the initial condition. A special fixed point exists where the populations are static, but for any other initial condition, the populations oscillate. The dependence of the period of oscillation on amplitude is a characteristic of nonlinear systems, known as amplitude-frequency coupling that is much different than a harmonic oscillator that has the same frequency independent of amplitude of oscillation. Furthermore, the predator-prey model is self-sustaining, while a harmonic oscillator, for any finite amount of dissipation, rings down to zero.

Lotka and Volterra's population dynamics in ecosystems have an intriguing similarity to monetary dynamics in economies. Some of the driving forces are the same, such as competition and predation as well as survival of the fittest. Even reproduction has a strong analog in the economic growth of individuals and companies. Therefore, the simple equations of Verhulst, and Lotka and Volterra, intended to model herbivores and fisheries, can just as easily describe the wealth of nations.

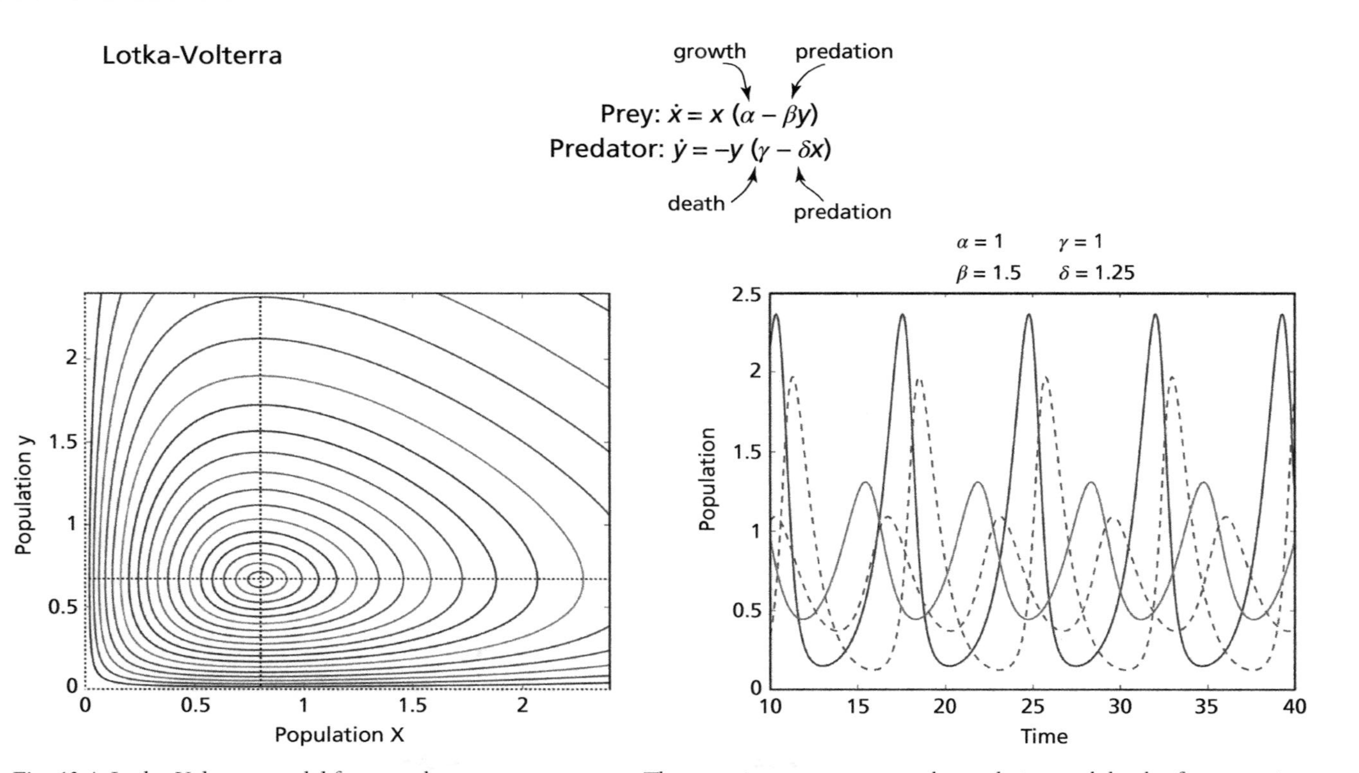

Fig. 10.4 Lotka-Volterra model for a predator-prey ecosystem. The equations capture growth, predation and death of two species, one of which subsists on the other. The solutions are oscillatory steady states that go through regular boom-bust cycles

The Wealth of Nations

Coins within economies are like individuals within populations. Money replicates like rabbits, increasing geometrically through compounded interest until limited by land, labor, or capital. The dynamics of money are governed by hidden laws, buried deep within the structure of a society like a genetic code. People too, as individuals, are economic agents in a web of monetary interactions. To survive in a monetary society, one must earn wages for food and for shelter. The private business owner is an economic agent dispensing those wages and competing against other small companies, struggling for greater market share through better products or better strategies. Market segments in a national economy expand or contract in response to the contraction or expansion of other market segments, modulated by governmental forces like taxes and interest rates, while nations compete and collude with one another in trade. Therefore, economies are complex systems, and problems of economic dynamics have consistently challenged scientists and mathematicians.

There has long been a close association between economics and the sciences, whether by analogy with biology and evolution, or by drawing upon mathematical theorems or laws of physics. Jean Buridan and Nicole Oresme, the French theologians and philosophers who grappled with the latitude of forms and the arc of trajectories, were among the first to study the changing roles of money in society. Nicolaus Copernicus turned his eyes away from the heavens for a moment to publish a small treatise on the value of money supply in 1519, about twenty years before he shocked the world with his heliocentric Universe. The first Nobel Prize in Economics was shared by Ragnar Frisch, a Norwegian, and by Jan Tinbergen, a Dutch economist who received his undergraduate degree in physics at the University of Leiden, studying under Paul Ehrenfest, and mingling with Lorentz and Zeeman and Einstein.

Economic dynamics taps into the same mathematical tools used to describe nonlinear dynamics and the stability of complex systems. The phase-spaces of economic dynamics have dimensions defined by economic degrees of freedom like cost and income, unemployment and inflation, interest rates and savings. In microeconomics, the wealth of individuals and the wealth of companies are represented as trajectories through the phase–space of economic dynamics whose coordinate dimensions may consist of supply and demand of price and labor, among others. Macroeconomics, on the other hand, captures the trajectories of national or world economies that are modeled in stable or unstable configurations. Spirals model the response of the economy to sudden changes, called shocks, when stability (or instability) is determined by the same criteria identified by Poincaré and Lyapunov for planets in orbit. Saddle points are of particular concern as economies overheat with excessive interest rates, or go into deep depressions. Because economics affects all of us every day, the dynamics of economic trajectories—the descendants of Galileo's trajectory—are a topic of personal interest and concern, amenable to mathematical analysis and powerful prediction that can help us live our lives better.

Antoine Augustin Cournot (1801–1877) was an aspiring French mathematician in 1829 when his doctoral dissertation on mathematical physics came to the attention of Siméon-Denise Poisson, who was the acknowledged leader of the French mathematics community. Poisson helped to secure posts for Cournot at the Faculty of Sciences in Lyon and then Grenoble, holding great hopes for Cournot's future in mathematics. But his trusted disciple took an unexpected turn when he published in 1838 a treatise called *Recherches sur les Principes Mathematiques de la Theorie des Richesses* (Researches on the Mathematical Principles of the Theory of Wealth). There was little new to mathematics in *Recherches*, but Cournot's application of mathematics to economic problems *was* entirely new, ushering in the field of mathematical economics. Cournot introduced ideas of functions and partial differentiation as well as probabilities into economic analysis. Partial differentiation became an essential tool in mathematical economics through the principle of *ceteris paribus* (other things held constant) that sought to isolate individual cause and effect, when so many forces are in play.

One of Cournot's concerns was the tendency of monopolies to take over key markets with deleterious effects on the common good. However, when two companies shared a market, a situation known as a duopoly, a healthy give-and-take could emerge with benefits for both consumers and suppliers. Cournot constructed a simple mathematical model of a duopoly in which each company attempts to maximize its profits. If one company raises the price on its own product, profits increase, but customers may decide to switch to the competitor's product instead (if the competitor has a price advantage) so profits might actually decrease. Without knowing what the other competitor will do with pricing, each company must act independently (without collusion) to maximize one's own profits based on what price its competitor has set. The mathematical derivation of this maximum uses a partial derivative, thereby assuming *ceteris paribus*. As each firm acts to maximize profits, market equilibrium is established, possibly to the benefit of all. This solution became known as Cournot's Duopoly Equilibrium, and it anticipated the Nash Equilibrium of John Nash[12] and game theory by more than a century. Cournot also drew the first graphical diagrams of supply and demand based on price, anticipating Alfred Marshall's Supply-and-Demand curves by half a century.

One of Cournot's economics students, August Walras, had a son Léon who did not know what to do with himself. Marie-Esprit-Léon Walras (1834–1910) cast about as a young man trying his hand at writing romantic novels between jobs as a railway clerk and bank teller. But try as he might to be unconventional, Léon kept gravitating back to his father's profession of economics. He finally gave in and became an economist himself, obtaining a faculty position at Lausanne, Switzerland. In 1874, Walrus published *Éléments d'économie politique pure*, (Elements of Pure Economics) in which he generalized Cournot's equilibrium solution to multiple companies competing within a market, and multiple markets competing within an economy. Walras' central principle assumed that a general equilibrium existed across all the markets. If excess demand increased in one market, it would be balanced by excess supply in another. The equilibrium operated mechanically through adaptive changes in prices. Walras' treatise launched the major branch of economics known as general equilibrium theory, which has reemerged recently as a major trend in economic thought.

As Walras extended his ideas of market equilibrium to include dynamic changes in prices, he adopted a mechanistic process that he called *tâtonnement*, from the French expression for "groping towards" something. Tâtonnement is a trial-and-error process in which a price is set, and the market responds by saying how much it is willing to demand or supply at that price. This groping towards the equilibrium point represents an economic trajectory, and Walras was one of the first to define dynamic economic systems evolving within a high-dimensional phase space of price values. Walras' work predates Boltzmann's and Poincaré's work on dynamical systems in phase space, but there are parallels between Walras' Law and Poincaré's phase space trajectories for the three-body problem. In the case of the three-body problem, energy is conserved, supplying a constraint that reduces the dimensionality of the trajectory by one dimension, simplifying the analysis of the trajectory properties. Similarly, Walras' Law conserves fluxes of prices or goods among markets. Therefore, one market price can be chosen as a *numeraire* against which all others are compared. This has the same effect of reducing the dimensionality of the problem by one dimension. Poincaré's integral invariants arose from fundamental laws of physics. Walras' numeraires arose from laws of general market equilibrium. The underlying mechanisms are as different as can be imagined, but the mathematical descriptions are surprisingly convergent. There is a unity in visualization.

The wealth of nations rests on the success or failure of business strategies through competitive balances. Adopting winning strategies leads to survival and proliferation, possibly at the expense of others, but also possibly symbiotically. Strategies in economics share much in common with survival strategies among species in an ecosystem. However, there is a fundamental difference, because underlying the behavioral strategies of a species is the genetic wiring that defines what the species is and how it acts.

Genetics in the Mix

The exponential nature of geometric growth represents a breathtaking power of increase that slips beyond simple human intuition. People typically overestimate what can be accomplished in five years, but underestimate what can be gained in ten. It was while Malthus was considering Euler's theory of geometric population growth that he had his epiphany about the unsustainability of such growth and the natural limits that would be imposed upon it. Subsequently it was while Darwin was reading Malthus that he had *his* epiphany about the power of geometric growth as a metaphorical wedge to create ecological niches through natural selection. The power of geometric growth provided Darwin with a principle, and natural selection provided the process, but neither provided a mechanism. Around the same time that Darwin was rocketing to fame with his theory of natural selection, unbeknownst to him or to anyone in the midst of the evolution revolution, an unassuming monk in the Austrian empire was providing that mechanism by watching pea pods grow in his garden.

Johann Mendel (1822–1884) was born on a farm in Moravia but aspired to a life of books that could be afforded only by a clerical life. He became a priest and joined an abbey in the town of Brünn that supported his studies at the University of Vienna where he took classes in physics (one from Johann Christian Doppler), as well as mathematics and the natural sciences. After returning to Brünn, he taught physics at the local technical school. In his spare time he began cultivating a garden of peas at his abbey. Peas reproduce by self-fertilization and display a wide range of easily recognizable inheritable traits, such as their sizes and shapes and textures. Some traits might be expressed in the next generation, while others would stay hidden until the generation after that. With keen insight, Mendel guessed that the hereditary traits were somehow particulate in nature. These particles would be paired in all possible combinations from generation to generation, with some traits dominant and others recessive. To test this hypothesis, he systematically grew and hand-fertilized thousands of pea plants, sorting and counting tens of thousands of seeds. The emerging statistical data indeed supported his hypothesis, and with some eagerness, he prepared a paper and read it at a meeting of the Natural History Society of Brünn in 1865.

In his paper, he considered two hidden hereditary particles for a given trait that he called A and a. Each pea would have a pair of particles, with the possible combinations being (AA), (Aa), (aA) or (aa). Whenever A was present, its trait would be expressed in the appearance of the pea seed. The particle A was dominant and would be expressed in three out of four seeds. The fourth type of seed with only the pair aa would express the recessive trait. Beginning with a seed with the pair Aa, the next generation would have the combinations given by $(A+a)^2 = AA + 2Aa + aa$. The generation after that would have the possible combinations $(A+A)^2+2(A-a)^2+(a+a) = 6AA+4Aa+6aa$ and so forth (see Fig. 10.5). Therefore, a recursive relation can represent heredity of a single trait as,

$$\begin{aligned}(AA)_{n+1} &= (Aa)_n + 4(AA)_n \\ (AA)_{n+1} &= 2(Aa)_n \\ (aa)_{n+1} &= (Aa)_n + 4(aa)_n.\end{aligned}$$

For the single dominant-recessive trait, there are three possible configurations (aa, AA and Aa). However, Mendel could describe many

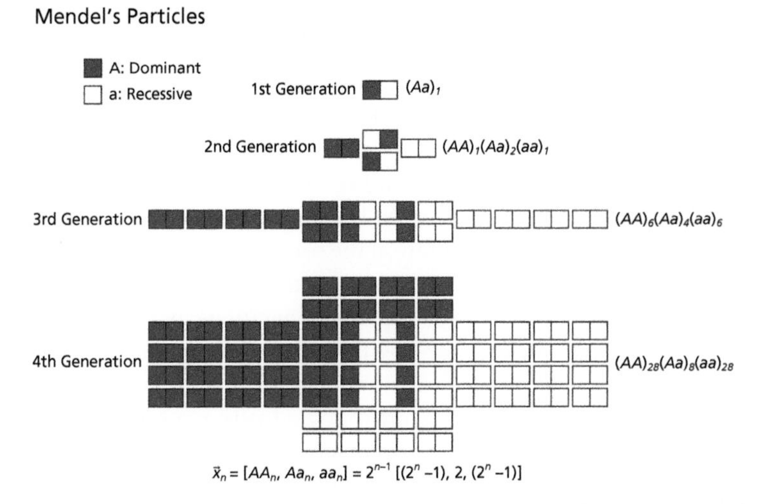

Fig. 10.5 Successive generations of pairs of hereditary particles beginning with one pair of dominant-recessive genes *Aa*

pairs of traits for the seeds. For instance, two pairs of traits would have the possible combinations (*AABB*), (*AABb*), (*AAbb*), (*AaBB*), (*AaBb*), (*Aabb*), (*aaBB*), (*aaBb*) and (aabb) for $3^2 = 9$ configurations. For N pairs of traits, the number of configurations becomes 3N. The successive generations represent a type of trajectory in a hyper-dimensional configuration space, although Mendel certainly was not thinking in terms of trajectories or spaces.

Mendel was destined never to see the fruits of his labor. His work received little attention at the time, and the scientific journal in which he had published was less than obscure. Forty-five years after he presented his paper at the Natural History Society of Brünn, and fifteen years after his death, the work was repeated independently by De Vries in Amsterdam, by Correns in Tübingen and by von Tschermak in Vienna. As these three were publishing their results, they stumbled upon the prior work by Mendel and admitted his priority by citing him. This launched Mendel to a level of fame and recognition that probably would have overwhelmed his delicate nature. Mendelian genetics rapidly became a vibrant new field of biological science at the turn of the century, opening an avenue to solve many outstanding problems in the variation of species. William Bateson coined the word "genetics" in 1905, and Mendel's "particles" of inheritance were renamed *allelomorph* "other form," later shortened to *allele*.

The power of the concept of discrete alleles lies in the power of combinatoric mathematics enlisted to solve puzzling aspects of inheritable characteristics. For instance, one immediately notices that the frequency of the *Aa* combination in Mendel's example starting with a single *Aa* variety becomes swamped by the much more frequent *AA* and *aa* combinations. Furthermore, if the original population consisted of a mixture of *AA*, *Aa* and *aa*, it seemed that the dominant trait *A* would become more frequent over the generations, edging out the recessive gene and making the population more homogeneous. Yet, genetic variance is the substrate upon which natural selection operates. Therefore, under this process known as gradualism, a species would become so genetically homogeneous that natural selection would cease to act, and the species would become static. At the turn of the century, this was a serious challenge to Darwinism, which was under assault from many corners and was tottering precariously despite its merits.

This problem was mentioned one day in 1907 on the cricket field at Cambridge University by the biologist Reginald Punnett to Godfrey Harald Hardy (1877–1947) who was a pure mathematician with little interest in practical problems. In spite of himself, Hardy was intrigued and tackled the problem by assigning probabilities to the discrete combinations of alleles to see what patterns might emerge upon successive generations. He discovered, using his combinatoric analysis, that there were mixtures of alleles that maintained steady proportions from generation to generation—a fixed point in the dynamics, keeping the genetic diversity constant through the years. In a stroke, Hardy's solution, published in 1908, showed that genetic diversity of a species could be conserved through the generations. After Hardy returned to his usual diet of pure mathematics, the question of genetic diversity was taken up by a brilliant yet volatile statistician working at the Rothamsted Experimental Station north of London where half a century of data had been amassed on the application of manure and the yield of crops from the farms of England.

Ronald Alymer Fisher (1890–1962) was born in London, received his degree in mathematics from Cambridge, went to work on a farm in Canada for a year, and returned to England in 1919. He took a position at Rothamsted Station as the perfect place where he could combine his interest in farming with his interest in mathematics. Faced with an early type of "massive data" in the form of manure reports, he became the world's leading expert in multivariate statistical analysis, inventing many of the key analysis procedures used today, such as analysis of variance (ANOVA), maximum likelihood, the null hypothesis and the Design of Experiments (DOE). He also popularized the use of Student's t-test. The name Student had been the pseudonym of William Gossett, a statistician working for the Guinness Brewing Company in Dublin Ireland. Gossett was tasked with finding the best way to measure the quality of the raw ingredients for Guinness beer by using the minimum possible number of samplings. Because Guinness did not want their competitors to know what they were up to, Gossett published his findings under the pseudonym *Student*. Fisher expanded on Gossett's work and forever immortalized him, as Student, to future generations of biostatisticians.

Despite Hardy's demonstration of constant genetic variance, there was still disturbing uncertainty about whether Mendel's laws were compatible with Darwinian selection. Fisher revisited Hardy's problem, which had not included any measure of fitness or reproduction rates, and introduced differential fitnesses for the different combinations of genetic alleles. Under the pressure of selective reproduction, the dynamics of allele frequency became dependent in an essential way on the fitness that each allele conferred. Fisher showed that if the combination *AA* has a better chance of survival over the other two genotypes *Aa* and *aa*, then the recessive allele *a* slowly disappears from the population, eventually becoming extinct. However, if the *Aa* combination has a better chance of survival than either *AA* or *aa*, Fisher found that there is a steady-state solution in which all three genotypes can persist through the generations as a fixed point of the dynamics with unchanging proportions. Therefore, by combining Mendel's laws with Darwinian selection, Fisher had shown not only the compatibility of these two theories, but also how they worked in concert to cause and sustain genetic diversity. Nonetheless, a key element was still missing in the theory, an element needed to move beyond simple combinations of alleles. This was the emergence of new alleles through the process of mutation.

The understanding of mutations in the early twentieth century was not based on molecular theory (coming only later after the discovery of the role of DNA in heredity in 1943), but relied on the observation of new phenotypes that could appear in offspring and be inherited as a distinct allele. In 1922, Fisher tackled the problem of how mutations, once they appeared, would disappear from the genome of a population. For instance, in the first generation, only a single individual has the mutation, but if the individual is fertile and mates, several offspring in the second generation would carry the mutation. Fisher realized that a mathematical description of this process could be constructed from permutation probabilities, and there was a finite chance, at each successive generation, that the mutation would disappear. Fisher's calculations showed that the rate of removal depended on the relative fitness the mutation conferred, even if the rate of removal was exceptionally slow. After a hundred generations, a single mutation with an enhanced fitness of only a percent could still be present with several percent probability. Despite Fisher's melding of Mendelian genetics with Darwinian selection into what is today called the modern synthesis, or Neo-Darwinism, the new field still lacked a comprehensive foundation. This was provided by a contemporary of Fisher's, one of those larger-than-life individuals who sweep in to a field of study like a whirlwind, change it in fundamental ways, and just as suddenly leave it to pursue other interests.

J. B. S. Haldane (1892–1964) was born a Scottish patrician with a sense of purpose and an intrinsic self-confidence that convinced him that nothing was beyond his powers. He was educated in mathematics and biology at the best schools and moved in the highest circles. In the First World War he fought with personal ferocity in the bloody trenches of the Western Front as well as in the deserts of Mesopotamia, wounded twice, and witnessed the first use of weapons of mass destruction—chemical gas. He helped develop the first gas masks with his father, a famous physiologist at Oxford who was rushed to France by the War Office to counter this appalling new type of warfare. Through this experience, Haldane became interested in respiration and immersed himself in a series of experiments in which he himself played Guinea pig. After the war, he began lecturing at Oxford in physiology, like his father, but his interests ranged widely, soon turning to genetic experiments.

Haldane's special talent was the combination of experimental biology with applied mathematics, and in the ten years between 1924 and 1934 he published ten journal papers that established the mathematical foundation of population genetics. Each paper was a cornerstone, providing the firm base for others in the expanding field to build upon. For instance, the fifth paper in the series took up the problem of mutations where Fisher had left off. Haldane flipped the problem, asking *not* how rapidly the mutation would disappear from the genome, but how probable it was that the mutation would become widespread, invading the genome entirely and becoming fixed in the genetic code for all future generations. Building on Fisher's mathematical analysis using permutation probabilities, Haldane showed that the probability of fixation was inversely proportional to the size of the population, while the time to fixation was inversely proportional to the relative fitness that the mutation conferred. Based on his experiments, he also estimated the rate at which mutations become fixed in the human genome, at approximately one significant mutation per gene per million years. While the steady accumulation of mutations provided a means for the rise of new species, Haldane was conservative among the neo-Darwinians, believing that natural selection played a dominant role over mutations. The cornerstones of Neo-Darwinism were now in place: the genetic theory of traits, and the payoff concept of fitness. What remained was to fuse these ideas into a coherent field—the new field of evolutionary dynamics.

Landscapes of Life

In 1925, the American biologist, Sewall Wright (1889–1988), took up a faculty position at the University of Chicago, poised to contribute a central concept to evolutionary theory. Wright started out in mathematics as an undergraduate student but was drawn to biology when he had the chance to work summers on biological research projects that fascinated him. For his PhD, completed at Harvard in 1915, he combined his mathematical skills with biological experiments, a connection that he continued to strengthen during the ten years he spent at the USDA. Much of his research focused on genetics, evolution, and the application of multivariate statistics to these problems. By the time he moved to Chicago, he had become accustomed to thinking about how complex systems with a very large number of variables progressed—literally evolved—in time. It was a natural step for Wright to begin thinking about the pressing questions of evolutionary genetics and how he could apply his new tools and techniques to answer them. Wright's secret for understanding such exceedingly difficult processes was a simple metaphor—the idea of fitness landscapes.[13]

Landscapes are common metaphors in the sciences. For instance, a potential energy landscape is one of the most common metaphors in physics, visualized as a terrain of peaks and valleys expressing a potential energy function. Energy landscapes are useful because one can imagine placing a ball on a hillside and watching it roll downhill, subject to forces caused by the local gradient in the potential energy, like a gravitational potential energy of a hillside. In chemistry, chemical potential is typically visualized as a landscape, and in biology protein folding is often visualized as a funnel-shaped landscape in an attempt to understand how a linear chain of amino acids fold into their eventual three-dimensional structures. In each of these examples, the idea of a landscape is only a metaphor, usually drawn in three dimensions—two dimensions of coordinates and the third dimension given to the dependent variable—even though the underlying dimensions of the system can be very large. The image gives a flavor of how a system is influenced by its environment, providing some intuition about how it responds.

Wright's landscapes are metaphors that help express the properties of evolving species in much the same way that configuration space and phase space capture the global properties of evolving dynamical systems. Instead of energy or chemical potential, the dependent variable in Wright's landscapes is fitness—the likelihood for passing on inheritable traits. For instance, in Chapter 1, we reviewed two different types of trajectories through dynamical spaces for the swallows of Wildcat Creek. In the first kind of space, each bird executes its own trajectory within its individual space of three dimensions. At a given moment in time, each bird is a point in this space, and the population of birds is a cloud of points. As the wind blows or as insects dart away, this cloud moves, changes shape and can split into multiple groups, filling space with a spaghetti of trajectories. In the second kind of space, the entire population trajectory can be described as a single point moving through a configuration space defined by all the degrees of freedom. These same two types of space also lead to two types of fitness landscapes.

For instance, the individual space of a species is defined by the number of gene locations and the number of possible gene alternatives at each location. For instance, the coordinates of a multidimensional space can be connected explicitly to the genome. Consider a genome of 4 loci (4 locations on the DNA) and 2 alleles (two possible genetic traits). All 16 possible genomic configurations are represented by a binary string of

4 bits: (0000), (0001), (0010), (0011), (0100), (0101), (0110), (0111), (1000), (1001), (1010), (1011), (1100), (1101), (1110), and (1111). Many of these are related by the flip of a single bit. For instance, (1010) is one bit-flip away from (0010), (1110), (1000) and (1011). Therefore, it is possible to say these are a "distance" of one bit-flip away. Measuring distances by bit flips is called the Hamming distance. Continuing this line of thinking (1010) is a distance of two bit-flips away from (1111). Therefore, there exists a configuration space of four dimensions that has a metric structure of distances assigned on it. It can be visualized as a four-dimensional cube, the hypercube shown in Fig. 10.6, with 16 vertices and 32 nearest-neighbor links representing the edges.

Visualizing binary hypercubes of increasing dimension is easy. An example of a genome space with seven loci and two alleles is shown in Fig. 10.7 as a seven-dimensional hypercube. Each circle is a genotype, and the fitness is represented by shades of gray, black having high fitness

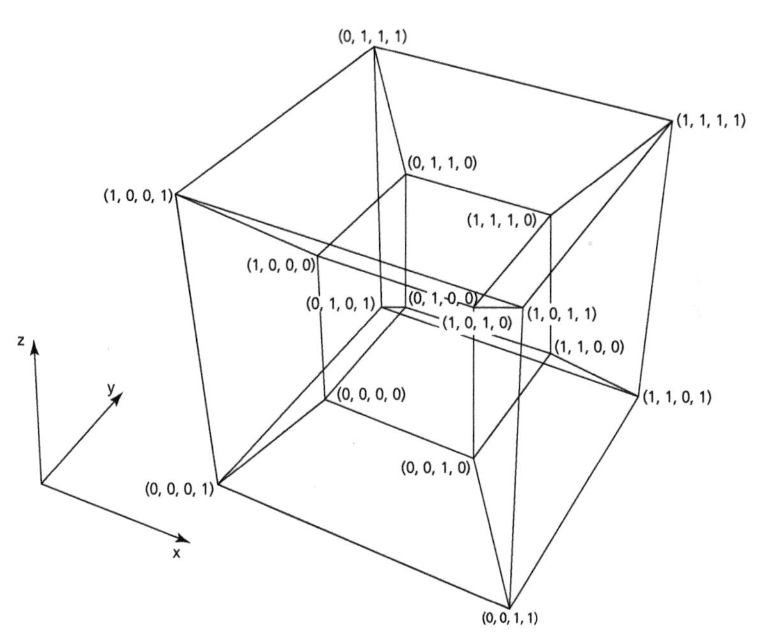

Fig. 10.6 A four-dimensional hypercube with 4-loci binary genomes at its vertices. The links are the nearest neighbors related by a single bit flip[14]

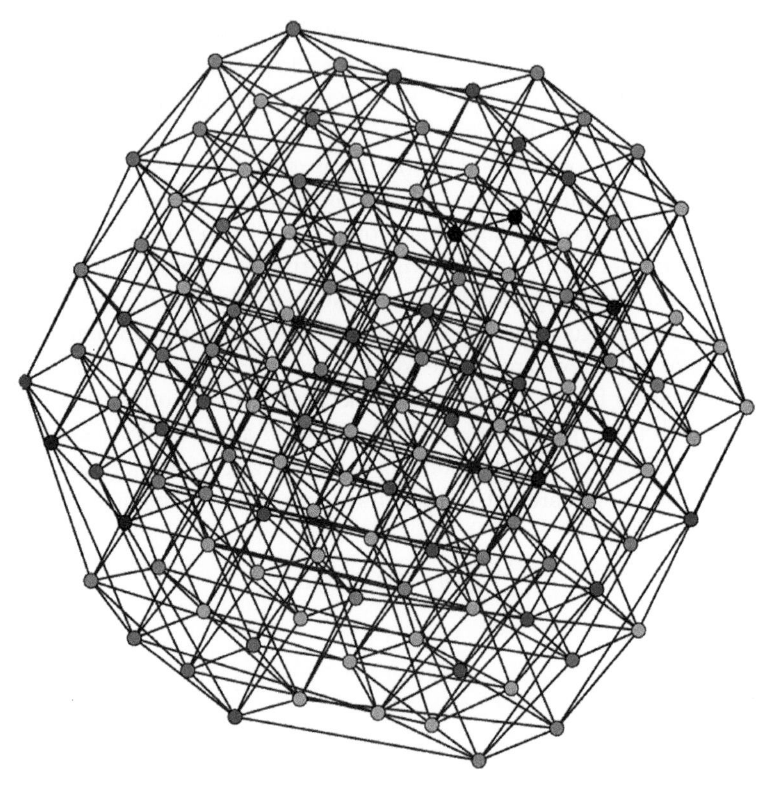

Fig. 10.7 A seven-dimensional hypercube containing 128 different binary genomes. Each link connects two genomes that are related by a single bit flip

and white having low fitness. The links are the single-mutation steps to nearest neighbors. In this figure, the fitness landscape is relatively smooth, with nodes having high values of fitness located near other nodes of high fitness, and similarly for nodes with low fitness, capturing the topography of the fitness landscape.[15] In this two-dimensional representational figure, you are visualizing an eight-dimensional fitness landscape: seven dimensions for the coordinates plus one more dimension (the shade of gray) for the value of the fitness. In Wright's original 1932 paper, long before the properties of DNA were understood, he estimated that species might have 1000 gene loci with ten alleles per locus. The dimensionality of this space would be 9000, with each locus one Hamming step away from 9000 other genetic patterns. This is already a large dimensionality, impossible to visualize directly, but it can still be approached from the point of view of a landscape.

The second kind of Wright's landscapes is known as an adaptive landscape in which each coordinate represents the frequency of a genetic trait occurring within a population. With 1000 loci and ten alleles, the dimensionality of such a space was immense, equal to 10^{1000}, which is a larger number than all the particles in the Universe. A species, consisting of a large population of individuals, is located in this space as a single point executing a single trajectory analogous to a point in state space occupied by a dynamical system. The landscape is formed by taking the multidimensional state space one dimension farther by assigning a fitness value to every point in it. Fitness in this adaptive context represents how well that population, with its frequencies of traits, flourishes. For large genomes and large populations, a population with one mixture of traits might have a similar fitness to other nearby mixtures. In this sense, the fitness values distributed over the coordinates of the space vary more or less smoothly, creating a hyperdimensional surface that can be envisioned as a landscape with peaks and valleys. Regions of high fitness are peaks in the landscape, and regions of low fitness are valleys.

The fitness/adaptive landscape metaphor helps to visualize what happens as a species evolves, as in Fig. 10.8. A population may occupy a region of the space located on the lower slope of a fitness peak. Over time, the population slowly drifts upwards as it "climbs" the fitness peak, eventually arriving at the top of the peak and stabilizing. Within the fitness landscape metaphor, one says that populations climb fitness peaks, adapting their average genetic traits in the process, i.e., evolution. Because fitness depends on competition and environmental factors, ecological niches in the environment can contribute local peaks to the fitness landscape, providing a means for a new species to arise as it evolves to fill that niche by climbing that local peak. If the environment changes, the peaks themselves may shift. If they shift slowly enough, then the species can adapt and move with the peaks. But if the peaks shift too swiftly, adaptive selection may not be fast enough, and the species become extinct.

Despite its heuristic appeal, the evolutionary landscape metaphor is only that—a metaphor. It provides a simple way to visualize very complex high-dimensional spaces, and to visualize the dynamic evolution of

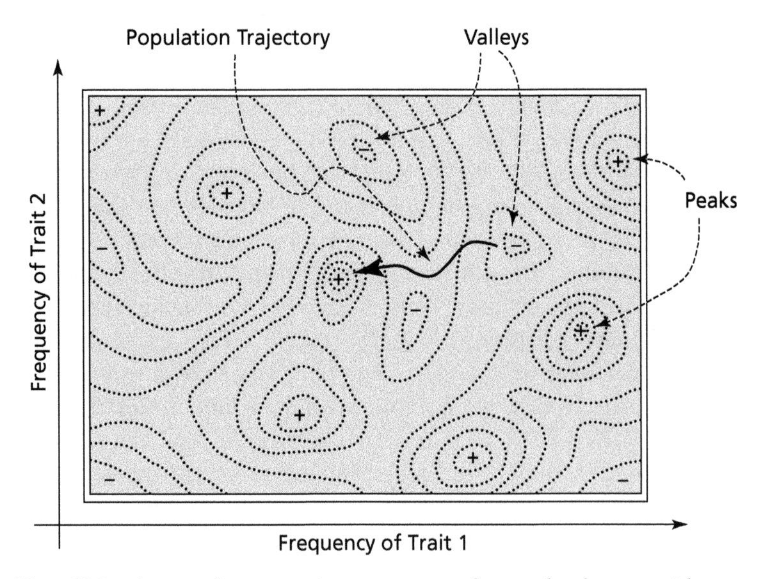

Fig. 10.8 A population trajectory on a fitness landscape with two dimensions[16]

populations within it. The mental pictures of peaks and valleys resonate with us. However, metaphors go only so far in capturing fine details. For instance, our intuition about peaks and valleys comes from living in our three-dimensional space. To cross from one peak to another higher one it is necessary to descend into the valley, which in this metaphor is a valley of low fitness, like a desert. How does a species survive as it amasses mutations needed to cross the valley of death? In this picture, ecological niches would seem to be inaccessible, and speciation, the rise of new species, would not be likely, yet life is living proof that evolution does occur across all scales, from molecular evolution to ecosystem evolution. This apparent paradox was solved as an unlikely spin-off from the anti-nuclear weapon activities of one of the most influential scientists of the twentieth century.

Pauling's Molecular Clock

In the late 1950s genetic mutations were an increasing concern in an era of open-air nuclear weapons tests as they salted the Earth with radioactive fallout. The adverse effects were widespread enough that radioactive elements were showing up in baby's teeth. Linus Pauling (1901–1994), a physical chemist at Cal Tech, became alarmed at the connection between radiation and mutations and started to push the scientific community to support an open-air test ban. At the same time, he began studying mutation rates in protein molecules. Protein chemistry was, for Pauling, just his latest interest. He was a peripatetic scientist whose interests ranged widely, making fundamental contributions to older fields while sometimes launching entirely new fields, such as the field of molecular evolution.

Pauling was born in Oregon and educated at Oregon State University before attending Cal Tech to obtain his PhD in 1925, working with Richard Tolman for his thesis on X-ray diffraction from crystals. For postgraduate work, he received a Guggenheim Fellowship that took him to Europe in 1926 to work with Sommerfeld, Bohr and Schrödinger. Pauling fortuitously found himself deposited right in the center of the most exciting moment in the history of quantum mechanics. Heisenberg, Born and Jordan had recently published their quantum mechanics, and Schrödinger had quickly replied with his own wave mechanics, launching the famous wave-particle duality debate. Pauling soaked up the new quantum theories and, with his concentration on physical chemistry, began to explore how quantum principles governed the formation of molecular bonds. After returning to Cal Tech in 1927 to take up a faculty position in theoretical chemistry, he developed his ideas and published, beginning in 1932, a succession of papers on the quantum nature of the chemical bond. This work established the field of quantum chemistry for which Pauling received the 1954 Nobel Prize in chemistry.

The Nobel Prize of 1954 was a welcome salve to cover Pauling's wounds received the previous year over his mistaken announcement that the structure of DNA was a triple helix, quickly overturned by Watson and Crick when they published their definitive results on the double-helix structure. Fortunately, Pauling was an irrepressible optimist and a grand thinker who was already moving on to other frontiers. His work on DNA gave him a taste for biological molecules, which captured his imagination around the same time that he was becoming concerned about the radiation effects of nuclear fallout. He became a vocal anti-testing activist in 1957, using his recently gained status as a Nobel Laureate to give him political clout. In 1958, he published a polemical book *No More War* that detailed, among other compelling arguments, the effects of radiation damage on biological molecules. In 1963, he received his second Nobel Prize, this time the Peace Prize, on the same day that the open-air test ban treaty went into effect, adopted by the major nuclear powers of the world.

In the midst of his anti-war stance and his focus on mutations, Pauling met an Austrian biochemist, Emile Zuckerkandl, during a visit to Paris. They had a mutual interest in measuring mutation rates in proteins using recent advances in protein separation. Pauling arranged for Zuckerkandl to come to Cal Tech as a visiting scientist in 1959. They concentrated on mutations in hemoglobin, measuring the number of alterations in the molecules among humans, gorillas and horses. In late 1960, Pauling was invited to present a talk at a symposium in honor of Albert Szent-Györgyi, who discovered vitamin C (a topic that later became one of Pauling's obsessions). Such symposia always had an associated paper submitted to a journal as a special issue, so Pauling suggested to Zuckerkandl "it is for Szent-Györgyi, so we should say something outrageous."[17] As Zuckerkandl began writing the paper, his thinking about mutations coalesced on the emerging principle that amino acid mutations accumulate regularly over time, providing something like an internal clock. By comparing human hemoglobin to horse, he calibrated the time since the species diverged from a common ancestor, which was known from the fossil record to be between 100 million and 160 million years. Using this calibration, the time to the divergence of human and gorilla was estimated to be about 11 million years. The 1962 paper[18] by Zuckerkandl and Pauling caused a sensation, and over the following years, Pauling promulgated the idea of steady mutation in a series of talks and associated papers written with Zuckerkandl. The most famous paper in this series was published in 1965 that gave the name "molecular evolutionary clock" to the process, including a mathematical derivation of the clock rate by Pauling.

Zuckerkandl and Pauling's molecular clock launched the new field of molecular evolution, growing quickly to include mutations in the information molecules DNA and RNA. A few years later in 1968 the Japanese biologist Motoo Kimura, working with Tomoko Ohta at the National Institute for Genetics in Mishima near Mt. Fuji, refined the concepts of the molecular clock by proposing a theory of neutral drift.[19] Kimura had worked with Sewall Wright (then at the University of Wisconsin) during his PhD studies in the United States, concentrating on the statistical mechanics of random walks and their relation to mutation rates within high-dimensional genetic configuration spaces. After returning to Japan in 1956, he refined his theories by distinguishing molecular evolution (genes) from phenotypic evolution (behavior), proposing that molecular changes could be mostly neutral because they had no effect on the phenotype of the individual. In other words, random walks in genetic configuration space were different than phenotypic trajectories in fitness landscapes. This launched a selectionist-neutralist controversy that continues today as opposing sides debate to what degree natural selection can influence molecular changes. However, the debate over the details takes little away from Kimura's powerful new idea of neutral genetic drift in extremely high-dimensional configuration spaces.

Our three-dimensional intuition fails us in high dimensions because it downplays the existence of mountain ridges. In three dimensions, ridges are not a prevalent geographical feature. They obviously occur, but not nearly as frequently as peaks and valleys. However, with increasing dimensions, the density of ridges in the landscape relative to peaks and valleys increases, and as the number of dimensions grows into the thousands and beyond, ridges dominate.[20] The importance of ridges in a fitness landscape is that most steps taken in almost every direction are neither uphill nor downhill. In other words, most mutations are neither advantageous nor disadvantageous—they are neutral. A species can diffuse along neutral networks of genotypes, easily moving through the hyper-dimensional landscape until the side of a fitness peak is probed, and then the species can begin climbing. Therefore, moving from one peak to another does not require a deep descent into a valley of death. It only requires a short descent to a neutral network of mutations, where it can diffuse to the side of a distant fitness peak. This is the concept of a *hyper-dimensional bypass*. It is like Abbott's famous Flatland allegory, where three-dimensional spheres avoid walls (lines) by moving into the third dimension to get over them. The hyper-dimensional bypass provides a compelling explanation for the evolution of molecular complexes that seemingly could never arise from random mutations.

Dawkin's Meme

In 1976 Richard Dawkins (1941–), before he became the famous zoologist and noted author, invented a concept that took on a life of its own, without his intention, and he subsequently disowned it. This concept is called the *meme*. At the time, he was writing his book *The Selfish Gene* about the smallest replicating unit of selection, and he wanted to create a lively metaphor to illustrate the behavior and fate of a self-replicating unit. The meme is a social construct that, using today's language, goes viral. It might be a word, or a phrase, that simply catches on and suddenly starts appearing everywhere, used by everyone. It survives and expands as it is copied and transmitted from person to person. Naturally, Dawkins' meme became a meme, propagating and expanding its range of usage so that today virtually everyone knows and uses meme as if it had always been a part of the English language. Dawkins, as father of the meme, had no control on the life choices made by his child. As the idea of the meme expanded beyond his intent and began to be misused in inappropriate ways, he distanced himself from it.

That is the problem with children—hard to control and impossible to anticipate. But it is the problem with any replicator, as Dawkins was well aware. Evolution is the process by which replicators replicate, growing with rates proportional to their fitness within a competitive environment. In the competition among individuals in a population, or among species within an ecosystem, or among economic agents within an economy, each has a rate of growth (or decay) dependent on the allocation of finite resources among the individuals—a zero-sum game. The frequency (or fraction of the population) of other members impacts the survivability of each individual, because someone, or some genome, gains resources as others lose.

Incorporating frequency-dependent fitness into population dynamics makes a strong connection to the field of game theory. Game theory is at once trivial yet inscrutable. Its rules can be stated with the greatest simplicity, yet its consequences can be as opaque and convoluted as the most difficult laws of physics. These consequences interested John von Neumann. In 1928, at the same time he was laying the foundations of quantum theory and tying together the warring views of Heisenberg and Schrödinger,[21] he proved a little theorem called *minimax* in a paper on *Parlour Games*.[22] Later, during World War II, taking snatches of time between designing explosive lenses for atomic bombs, von Neumann finished a manuscript, working with the economist Oskar Morgenstern, that extended the simple parlour game results to the much larger field of economics. Their book *Theory of Games and Economic Behavior* was published in 1944, establishing the new field of mathematical game theory. The book had more impact initially on political science than on economics or evolutionary theory. In his capacity as one of the first game theorists, Von Neumann, having helped build both the A-bomb and the H-bomb, was hired as a government strategist at the height of the Cold War and is partially credited for developing the principle of mutually assured destruction (MAD). He was supposedly the inspiration for Dr. Strangelove in the Stanley Kubrick and Peter Sellers 1964 film *Dr. Strangelove, Or How I Learned to Love the Bomb*. Game theory had become deadly serious, but it also was about to expand into the arena of evolutionary dynamics.

In 1972 the science magazine *Nature* received an odd manuscript from an odd American scientist, John Price (1922–1974), who had a wandering resume, including a year spent on the Manhattan project helping to build the atomic bomb and a period spent as a journalist, before settling down in England where he worked at the Galton Laboratory of Genetics at University College of London. The manuscript posed the ecological question of why animals with horns or claws so seldom are injured in their battles over territory or mates. His manuscript was reviewed by the evolutionary biologist John Maynard Smith (1920–2004) who found the paper unsuitable for publication in *Nature* because it was far beyond the length and scope of acceptable *Nature* publications. However, Smith saw an intriguing feature in Price's paper that touched on aspects of game theory as they applied to animal behavior and evolution. Smith contacted Price directly, and the two worked together to submit a joint paper titled *The Logic of Animal Conflict* to Nature in 1973. This paper was one of the first to apply game theory to evolutionary biology.[23] Later, as Price's interests wandered elsewhere, Smith continued to apply game theory to evolutionary biology and published his classic book *Evolution and the Theory of Games* in 1982.

A key element in game theory is a mathematical device known as a payoff matrix. It states how one player wins or loses relative to others as each player adopts strategies. The application of the payoff matrix to populations with a large number of components was first put into mathematical form[24] in 1978 and immortalized in the Replicator Equation[25] shown in Fig. 10.9. The Replicator Equation uses the payoff

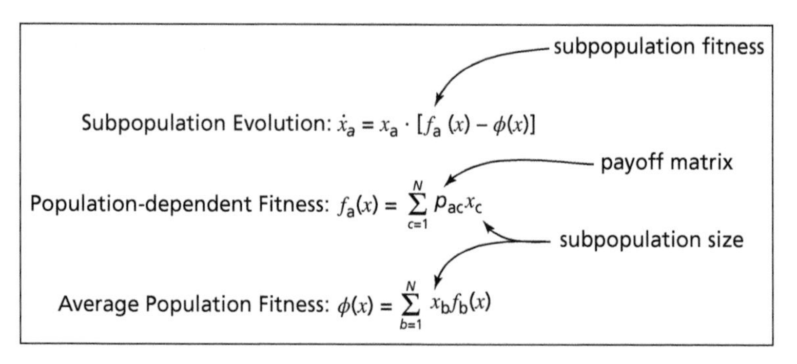

Fig. 10.9 Replicator dynamics among N populations

matrix to construct an average fitness that depends on the numbers of individuals in all the different subpopulations. The growth rate of each subpopulation depends on the fitness of the subpopulation relative to the average fitness of the entire population. Despite the simple form of the growth dynamics, the zero-sum payoff matrix introduces interesting and sometimes unintuitive behavior into the growth, decay and oscillations of the subpopulations. The trajectories of such replicators are difficult to visualize in many dimensions, but intuition can be built by starting with simple triangles.

A three-subpopulation replicator system is plotted on a triangular domain in Fig. 10.10. In two dimensions, triangles have the convenient property that they have as many vertices as edges, and triangles can be used to represent a three-component mixture, such as a three-metal alloy or a population with three possible genotypes. This space is known as a simplex, and can be extended to arbitrarily high dimension. Within the simplex, replicator dynamics determine how the different subpopulations compete and gain or lose. In the simplex in Fig. 10.10, each vertex represents a pure population, and points inside the triangle represent mixtures. The individual trajectories arise from initial conditions distributed over the face of the triangle. If a subpopulation has a fitness higher than the average fitness, then more will reproduce, and that subpopulation fraction will expand. Through the payoff matrix, an expansion in one subpopulation influences the relative fitnesses and reproduction rates of the others. Many types of behavior are possible, including the ability to strike a dynamic equilibrium when all the subpopulations are stable, or the population oscillations of the Lotka–Volterra equations. When the payoff matrix is non-zero-sum, as in this

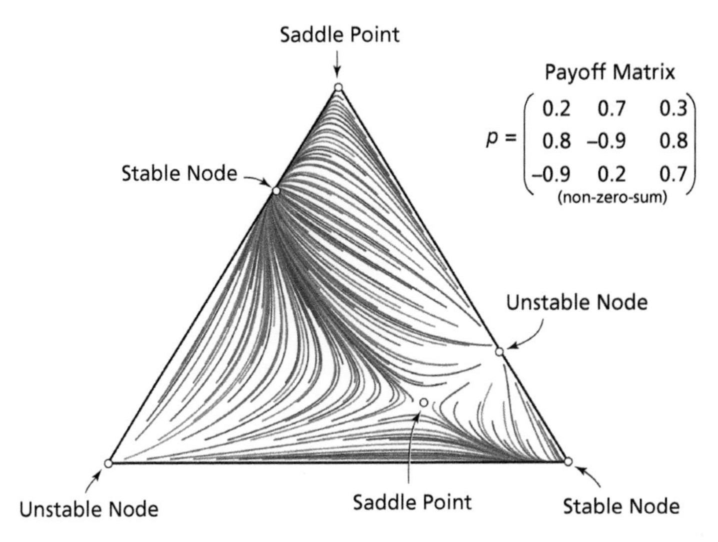

Fig. 10.10 Non-zero-sum dynamics on a simplex. Each vertex represents a pure population, each edge represents a two-component population, and all interior points represent a three-component population. For the given payoff matrix, there are two stable points, two unstable points and two saddle points

example, symbiosis becomes possible, and groups of subpopulations can be mutually beneficial to each other, as in the stable node on the upper left of the triangle that supports two of the three subpopulations in a stable equilibrium.

Variations on replicator models abound.[26] A famous version developed by Manfred Eigen and Peter Schuster in 1977 is called the quasi-species equation,[27] including mutations within a genome so that multiple genotypes become quasi-species that can mutate back and forth. The fitness of a quasi-species might be defined within a fitness landscape, with peaks and valleys and neutral networks in high dimensions, allowing for hyper-dimensional bypasses. A popular recent variant on the quasi-species equation incorporates frequency dependence into the fitness landscape itself as competition with other individuals of the same population alters the overall probability of reproduction. This extension elicited the colorful metaphor of a fitness *seascape*[28]—hyper-dimensional ocean swells with many neutral ridges between the crests and troughs, changing constantly, rising and falling.

The trajectories of evolutionary dynamics are abstract, tracing the state of multi-component systems through spaces of ultra-high dimension with generalized coordinates that can describe almost anything—a language group, a species in an ecosystem, or a base pair at a site on a DNA molecule. The forces of the environment on these elements in the context of competition and selection drive the evolution of the state of the system, establishing transformational fields and driving the flow of all possible outcomes. Yet as Galileo's trajectory expands into abstract hyper dimensions, the complex interplay among all the parts can take on a life of its own, forming complex networks of dynamical systems that support the most complex system of all. In this way, life itself becomes the final frontier for Galileo's trajectory.

11

The Measure of Life

The impressive variety of biological rhythms leaves no doubt that autonomously periodic processes contribute to the coordination of life-processes.

Art Winfree (1966)[1]

The trajectory of life is perhaps the epitome in the abstraction of Galileo's trajectory. The life of the cell and of the organism trace a dynamic arc through a hyperspace of unimaginable dimension. The coordinates of this hyperspace represent the grand ensemble of components and interactions and motions that collectively create a living thing. To apprehend this arc would be to unlock the secrets of life. A recurring motif in the trajectory of life is the presence of cycles upon cycles—life built upon epicycles, like Ptolemy's ancient devices that he used to patch up the sacred heavens, except that these epicycles keep us alive, a truly sacred task. The epicycles of our lives are sustained oscillations, self-oscillations within us, that drive other oscillations that feed-back on the originals, generating rhythms that are like clockworks.

Self-oscillating systems eagerly interact with each other, creating networks of influence that can modify the behavior of each other or crystallize into synchronization. For instance, the beating heart represents the uniform synchronization of a million self-oscillating heart muscle cells, kept in lock step by only a few thousand pacemaker cells. This ability of self-oscillators to accept feedback from their environment makes them important as components of larger systems, supporting emergent behavior that transcends all the little inputs.

The measure of life is the goal of doctors and the hope of patients, held within Galileo's trajectory, transformed into life's trajectory. The efforts to trace that path constitute one of the final frontiers of science. With the stakes so high, we begin simply, with beautiful yet rare waves of light.

Galileo Unbound. David D. Nolte, Oxford University Press (2018).
 DOI: 10.1093/oso/9780198805847.001.0001

Firefly Waves

Some years ago, before the new Wal-Mart lit up our night sky with its parking lot lights, my wife Laura and I experienced one of the great displays of nature. It was late dusk on a warm summer evening, and the fireflies were out in numbers above the hay field behind our house on the banks of Wildcat Creek. As we watched, we became aware of waves of light washing back and forth across the field at great speed, swirling and undulating. We watched in awe and wonder until the darkness deepened and the fireflies extinguished their lights. The synchronization of the flashes of multitudes of fireflies is very rare. A species in Southeast Asia does it, but in North America only a small enclave of fireflies in the Smokey Mountains is known to display perfect synchrony—an entire population of fireflies flashing on and off in unison.[2] What we witnessed in our hay field in Indiana was far more ephemeral and chaotic, not a perfect synchrony, but clearly a dynamic process as multitudes of fireflies formed an intricate network of light.

Every firefly has roughly equivalent rates of firing—natural oscillators with a natural period of oscillation. When they are isolated, they flash on and off randomly—they are out of synch. But in groups, they each watch and wait for a competitor to light up and then try to beat them to it, their timing coalescing to match one another, a phenomenon known as entrainment. In our hayfield in Indiana, the entrainment was not complete, although among the fireflies of the Smokey Mountains, the entrainment becomes perfect. In 1990, Steven Strogatz (1959–) at Cornell and Renato Mirollo at Boston College showed how pulse-coupled oscillators, like the fireflies, become entrained into perfect synchrony. This launched them to minor celebrity status because the results were universal, applying to a broad range of important biological systems like the operation of the heart's pacemaker and neural dynamics. At the heart of synchronization, as its elementary unit, is the self-sustained oscillator, like Galileo's pendulum clock.

Galileo lived his entire adult life under the cloud of money problems. He was an aristocrat without means, and many of his career choices were dictated by his need. Even his publication of the *Two Chief World Systems*, that got him into such trouble, was partially motivated by money. Finally, while under house arrest, Galileo had one last scheme to make money, not so much for himself, but for his son Vincenzo.

He had long envisioned building a new type of clock, one that would be far more accurate than the existing clocks that relied on a horizontal beam rocking back and forth called a verge escapement. Since his student days at Pisa, when he noted the isochrony of the swinging lamp in the Duomo, he had thought to replace the horizontal rocking beam with a vertical pendulum. Now, with time on his hands, he hoped to leave one final lucrative legacy to his son by constructing a pendulum clock.

Because he had become completely blind, Galileo enlisted his son to help with the designs and drawings. The central engineering problem was the escapement, the mechanism that repeatedly reverses the motion of a periodic element. The verge escapement reverses the motion of the horizontal beam when it reaches the limits of its swing, sending it back the other way until it hits the other limit, reverses, and so on. This is called an integrate-and-fire oscillator. The verge beam has no intrinsic natural frequency other than the one set by the time it takes to swing from one limit to the other. The challenge for Galileo was to design a new type of escapement that would work on a vertical pendulum that *did* have a natural period. The escapement would need to inject a tiny amount of energy, a tiny push on the pendulum taken from a falling weight, on each cycle to keep the pendulum swinging. Working with his son Vincenzo, Galileo devised a new escapement that he thought would do the trick. This intimate father-son project was interrupted by Galileo's death on 8 January 1642. Vincenzo was a skilled musician, like his grandfather, and was adept at building fine instruments. After his father's death, he attempted to build the pendulum clock, but declining health and his eventual death only seven years after Galileo's once again delayed the project. Galileo's disciple Vincenzo Viviani, who was obsessed with Galileo's legacy, continued the work on Galileo's final invention, overseeing the construction of a great clock in the tower of the central plaza of Florence. It is claimed that the clockworks, still working to this day, carry elements of Galileo's final design.[3]

By the mid seventeenth century, the isochrony of the natural pendulum had become an established fact, and others saw the same potential in the pendulum for accurate time keeping that had occurred to Galileo. Christiaan Huygens in the Netherlands became engrossed in the physics and engineering of the pendulum, seeking a design more practical and efficient than Galileo's. Huygens had the advantages of full eyesight, leisure time, sufficient money and sound engineering that Galileo had lacked. Huygens also was a theoretician (as well as a practitioner) and so was guided by rigorous principles and mathematics where Galileo worked mainly by instinct. Huygens devised a working design and model for the pendulum clock in 1656, patenting the invention the next year.

For Huygens, the invention of the pendulum clock was merely the beginning of his interest in the physics of pendula as he studied their theoretical and practical aspects over more than a decade. As practical problems went, the determination of longitude at sea was one of the highest priorities, and accurate sea clocks, if they could be constructed, were the most direct means to accomplish this. The Netherlands in the seventeenth century had become one of the greatest seafaring nations, despite its small stature, deriving its wealth and power from its ships and their ability to find safe passage from distant ports. About ten years after inventing the pendulum clock, Huygens was perfecting designs of sea clocks and had constructed two prototypes in preparation for a sea trial. These hung on the wall within a few feet of each other in his new workshop.

In a letter to his father dated 26 February 1665, Huygens recounted an amazing discovery.

> While I was forced to stay in bed for a few days and made observations on my two clocks of the new workshop, I noticed a wonderful effect that nobody could have thought of before. The two clocks, while hanging side by side with a distance of one or two feet between, kept in pace relative to each other with a precision so high that the pendulums always swung together, and never varied. While I admired this for some time, I finally found that this happened due to a sort of sympathy: when I made the pendulums swing at differing paces, I found that after half an hour they always returned to synchronism.

Huygens concluded this part of the letter by expressing his amazement that ". . . here we have found two clocks that never come to disagree, which seems unbelievable and yet it is very true. Never before have other clocks been able to do the same things as those of this new invention."[4] Huygens experimented with the clocks, shielding the air between then, or hanging them farther apart, and discovered that the strange "sympathy" was transmitted through barely perceptible vibrations in the wall or floor or even through the air.

This famous letter describes Huygens discovery of synchronization. Two autonomous oscillators of slightly different frequency, if coupled even very weakly, will each adopt an exact compromise frequency and subsequently stay in synch indefinitely. Synchronization of separate oscillators is a ubiquitous phenomenon, arising not only in wall clocks, but also in electronic circuits, in lasers, in social networks, in cellular metabolism, in brain waves and the beating of the heart. Many of the concerted actions of your body are a form of synchronization, supporting your life. Despite Huygens' early discovery of synchronization, the effects are subtle and difficult to describe, requiring mathematical techniques not available to Huygens, nor even available through the eighteenth and most of the nineteenth centuries, until an English nobleman, with a privately funded laboratory, uncovered its mathematical form.

Autonomous Oscillations

The scene at Terling Place, an elegant country house in Essex about an hour's train ride outside London, could have been straight from the popular BBC TV series Downton Abbey. The family would be assembled before dinner in the drawing room, dressed in fine attire, the Baron and his wife Lady Clara surrounded by their children waiting for the butler to announce. But one of them is late—the eldest son and heir to the title and estate—John William Strutt. Clara, his younger sister, might roll her eyes with scorn, and Richard, the second son, would scowl, as Charles and Edward, the two youngest, would squirm with impatience. John was the eccentric of the family. The others reveled in balls and hunts, while John buried himself in the makeshift science laboratory he had set up after he returned home from Cambridge. John had done well in applied mathematics and physics at Cambridge, receiving the highest honors and deeply impressing teachers like Routh and Stokes. Although an education was just one of the stepping stones for an heir to a peerage (in other words a good way to make connections), John took it very seriously, attracted to it like a vocation, and he decided that he would become a scientist of the first rate, regardless of what his family thought of him—and he did just that. John Strutt became the leading British physicist of the late nineteenth century, even as he became also the Third Baron Rayleigh.

Lord Rayleigh's (1842–1919) interests ranged across the full spectrum of physics of his time, and he was a gifted theorist as well as a skilled experimentalist. He was unusually prolific, publishing over 400 papers, chapters, encyclopedia articles and books during his lifetime, plus three completed yet unpublished journal articles that lay on his desk when he died at the age of seventy-six. His talent lay in identifying the critical gaps and inconsistencies in existing physical theories and either devising and performing decisive experiments to solve the controversies, or else supplying the missing interpretations. He was an intuitive thinker, working as much by analogies as by rigor, relying on simple physical models that captured the essential parts of a phenomenon. He became famous while still young for explaining why the sky is blue and why optical systems had finite resolution. His name is associated with numerous physical phenomena, such as the Rayleigh waves of earthquakes, the Rayleigh criterion of optical resolution, Rayleigh scattering of light by particles, the Rayleigh–Ritz method to solve differential equations, and Rayleigh–Benard convection, among others. His greatest fame came from extremely precise experiments in which he discovered the new element Argon, for which he received the fourth Nobel Prize in Physics in 1904. Perhaps his favorite topic was acoustics. He wrote the definitive treatise *The Theory of Sound*, beginning the work as he boated up the Nile on his honeymoon and publishing it in two volumes in 1877 and 1878 (still in print today as a classic). Throughout his career, he returned many times to the study of acoustics and musical instruments.

One not-so-musical instrument that caught his attention was an ingenious self-oscillator invented by Helmholtz in which a tuning fork was placed near an electromagnet that was powered by a mercury switch attached to the fork. As the tuning fork vibrated, the mercury came in and out of contact with it, turning on and off the magnet, which fed back on the tuning fork, and so on, enabling the device, once started, to continue oscillating without interruption. This device is called a tuning-fork resonator, and it is the basis of modern doorbell buzzers. Rayleigh constructed several of his own to study, and in a paper published in the Philosophical Magazine issue of 1883 with the title *On Maintained Vibrations*,[5] he introduced an equation to describe the self-oscillation by adding an extra term to a simple harmonic oscillator. The extra term depended on the cube of the velocity, representing a balance between the gain of energy from a steady force and natural dissipation by friction. Rayleigh suggested that this equation applied to a wide range of self-oscillating systems, such as violin strings, clarinet reeds, finger glasses, flutes, organ pipes, among others.[6]

Rayleigh explained how the mathematical solution to this equation describes a system that displays spontaneous self-sustained oscillations in response to a steady applied force. For instance, the pendulum clock is the quintessential autonomous oscillator, transforming the steady force of the weight into the regular periodic swing of the pendulum. If started with moderate amplitude, the oscillation increases in energy until it reaches steady oscillation. Or, if started with large amplitude, it decreases in energy until it reaches the same steady oscillation state. In his laboratory, Rayleigh focused on Helmholtz' tuning-fork oscillator because of the ease of experimentation. In one of his signature papers on acoustics and vibrations, he described how two oscillators, if placed on the same table, became entrained to a common frequency,[7] just as Huygens had discovered the sympathetic interaction of two pendulum clocks. Rayleigh's interest in self-oscillating systems, like so many of his scientific contributions, was insightful and ahead of its time. His mathematical description of self-oscillation, a central operating principle in the control of autonomous systems, helped to lay a paving stone on the path of a scientific revolution that was well underway.

The Electric Age was born in 1867 with the invention of the electromagnetic dynamo by Sir Charles Wheatstone (1802–1875) in England and Werner von Siemens (1816–1892) in Germany. Based on Faraday's Law of Induction, these were industrial-scale machines that could power factories and light entire cities. Thomas Edison (1847–1931) and Nikola Tesla (1856–1943) were early practitioners of the electrical arts, and the invention of the light bulb in 1878 by Edison stands out as the iconic event of that new technology. The lighting of houses and cities quickly became an industry, launching major electrical companies in many countries.

After the discovery of Hertzian waves by Heinrich Hertz (1857–1894) in 1886, and the demonstration of intercontinental radio transmission by Guglielmo Marconi (1874–1937) between 1901 and 1902, there was a race to find better radio detectors and amplifiers. In 1904, the British engineer John Fleming at University College in London invented the first vacuum tube diode for radio receivers, and in 1906, the American inventor and businessman Lee De Forest (1873–1961) invented the vacuum tube triode for amplification and control. Both of these devices worked on the common principle of the "Edison Effect," as thermionic emission was called, leading to bitter patent feuds waged with ferocity among many parties. The De Forest tube was called an audion, and when driven by a steady external current, it showed strong self-oscillations that were useful for radio applications, for which it was developed during the First World War by the French military. However, the physics underlying the triode oscillator was an enigma.

In 1920, the Dutch physicist Balthasar van der Pol (1889–1959) was completing his PhD thesis at the University of Utrecht on the topic of radio transmission through ionized gases because the ionosphere plays a central role in long-distance radio reception. Van der Pol's thesis combined insights he had gained during his student days visiting Fleming at University College and J. J. Thompson at Cambridge. Thompson was a staunch advocate for combining theory and experiment, and van der Pol had built a short-wave triode oscillator to perform experiments on radio diffraction to compare with his theoretical calculations of radio transmission. Van der Pol's triode oscillator was an engineering feat that produced the shortest wavelengths of the day, making van der Pol intimately familiar with the operation of the oscillator. After filing his doctoral thesis at Utrecht, van der Pol published a paper in Dutch and then in English in the British journal *Radio Review* that proposed a general form of differential equation for the triode oscillator. Two years later, in 1922, van der Pol was hired as the Chief Physicist of the Phillips Corporation in Eindhoven, The Netherlands, and was placed in charge of Phillips' activities in radio engineering. He continued working on the triode oscillator and developed a simple differential equation in 1926 that he published in the Philosophical Magazine of the Royal Society of London.[8] He had stripped away all but the most essential terms, while the equation still retained persistent oscillatory behavior.[9] A wide variety of self-sustained oscillators, from "singing arc" lamps to coupled dynamos, were guided by the same physical principles. Hence, this equation, in its simple dimensionless form that captured the physics of so many different systems, came to be called by others the "van der Pol Equation."

The van der Pol equation is a nonlinear differential equation (see Fig. 11.1). It contains an "intrinsic nonlinearity" that cannot be approximated by a linear system. Any attempt to replace the nonlinear terms in the equations with linear approximations makes the oscillations disappear. The differential equation has a gain term, in which energy is steadily delivered to the oscillator, and in the absence of loss, the

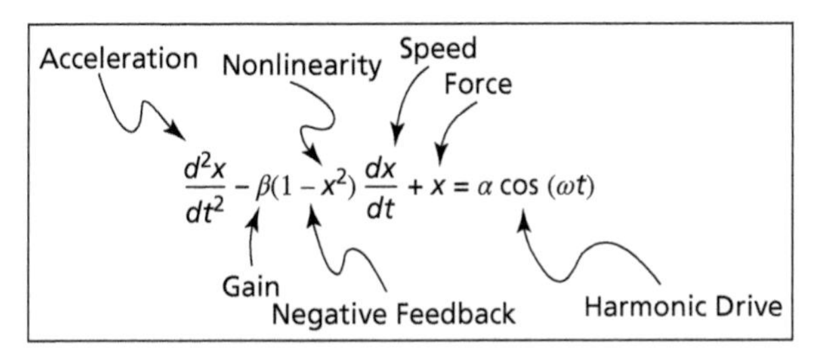

Fig. 11.1 The van der Pol equation. Gain is balanced by the negative feedback of the self-limiting nonlinearity. The oscillator can be driven by a harmonic function that can lead to synchronization as well as to chaotic behavior

amplitude of the oscillator would grow exponentially. However, a loss term gets disproportionately bigger as the amplitude grows, until the amplitude gets big enough that the gain is balanced by the loss, and the oscillator oscillates in steady state with a steady amplitude and frequency. This is a perfect example of Poincaré's limit cycle, which executes a closed loop on the phase-space plane. All transient behavior of the oscillator converges on this limit cycle (see Fig. 11.2).

The solution to the van der Pol equation has acquired a host of names and aliases through the years, each name emphasizing a different physical aspect.[10] Van der Pol called it a *relaxation oscillator*, because all perturbations relax back to steady state as a damped oscillation. Sometimes it is called an *autonomous oscillator*, because it oscillates on its own without a frequency input. At other times it is called a *limit-cycle oscillator*, emphasizing its phase-space signature. When it is simplified to its essential behavior as a mere circle on the phase plane, it is called a *Poincaré oscillator*, or simply a *phase oscillator*, which emphasizes the central importance of the phase of the oscillator, downplaying or ignoring the role of amplitude.

Radio engineers were always tinkering and were interested in how oscillators behaved when they were driven by a harmonic drive. When linear oscillators are driven, they display resonance behavior when the drive frequency is near the natural frequency of the oscillator. Nonlinear oscillators, on the other hand, behave very differentially. In 1927, van der Pol showed that as the drive frequency is swept through the natural frequency of his relaxation oscillator, there was perfect frequency locking between the oscillator and the external drive.[11] This was

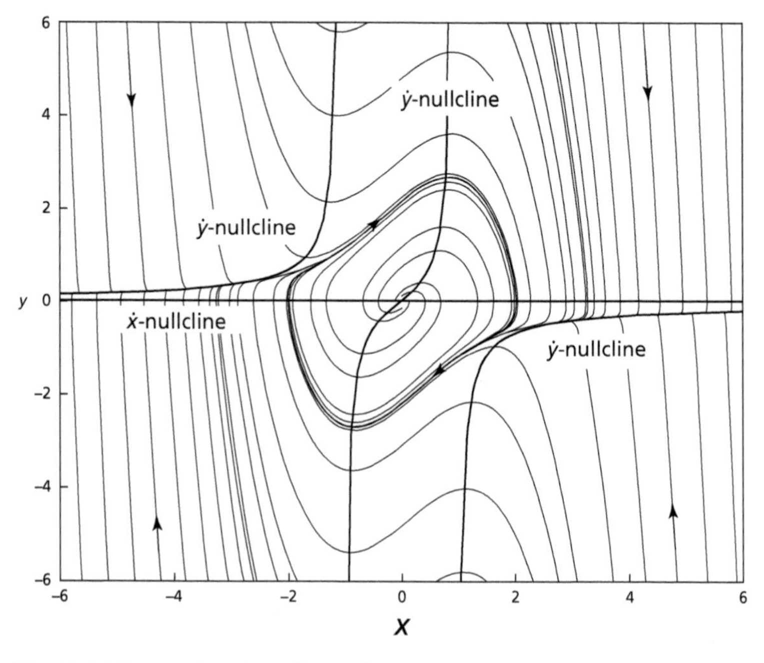

Fig. 11.2 The van de Pol oscillator phase portrait. The flow lines converge on the limit cycle where gain is balanced by nonlinear self-limiting[12]

a direct demonstration of oscillation entrainment. The synchronization of Huygens' two clocks and Rayleigh's two doorbell buzzers had been mutual self-synchronization between oscillators. In van der Pol's experiment, a single limit-cycle oscillator is entrained to the external control frequency, within limits. If the frequency difference is too large, then the limit-cycle oscillator cannot match the external frequency, creating a beat frequency between the two. This beat frequency is altered by interaction with the oscillator, as the oscillator frequency is "pulled" towards the drive frequency.[13]

During World War II, self-oscillations and nonlinear dynamics became strategic topics for the war effort. High-power magnetrons had been invented for long-range radar, and the dynamics of these oscillators could be represented by the van der Pol equation. There was an intense period of experimental and theoretical work on driven nonlinear oscillators, including the discovery that a driven van der Pol oscillator could break up into wild and intermittent patterns. This "bad" behavior of the oscillator circuit (bad for radio applications) was the first discovery of chaotic behavior in man-made circuits. These irregular properties of the driven van der Pol equation were studied by Mary-Lucy Cartwright (1990–1998) (the first woman to be elected a fellow of the Royal Society) and John Littlewood (1885–1977) at Cambridge who showed that the coexistence of two periodic solutions implied that discontinuously recurrent motion—in today's parlance, chaos—could result,[14] which was clearly undesirable for radar applications.[15] The work of Cartwright and Littlewood later inspired the work by Levinson and Smale as they introduced the field of nonlinear dynamics (see Chapter 9).

By the middle of the twentieth century, autonomous oscillations and frequency entrainment had become the standard workhorses of control theory. In a nonlinear world, van der Pol's nonlinear oscillator replaces the simple harmonic oscillator as a fundamental element of dynamical systems. Magnetrons for radar, wheel shimmy of landing gear and propeller vibrations on aircraft, even violin strings and clarinet reeds all are nonlinear limit-cycle oscillators. Likewise, if we should happen to look inward, inside ourselves, we would find an astounding array of nonlinear oscillations in nearly every aspect of our biological selves.

Unruly Heart

The primordial soup would seem the last place on Earth where order and structure should emerge. Yet life, the ultimate informational edifice, full of meaning and content and form and function, might have emerged around volcanic pools or deep-water smokers. How ordered structure and information can arise spontaneously out of formless entropy is one of the most interesting questions in science. Crystals arise out of the melt, so crystallization is one way that order emerges from disorder. There is also a strange and highly visible form of order that can emerge from a liquid soup of chemicals.

In 1950, B. P. Belousov (1893–1970), the head of the Laboratory of Biophysics at the Ministry of Health in the USSR, was trying to replicate in a test tube some energetic reactions that occur in mitochondria. He created a mixture of citric acid with cerium bromate. To his surprise, the mixture began to oscillate, changing between a colorless solution and a yellow one with a period of several minutes, maintaining the oscillations for nearly an hour before dissipating. He studied the reaction with painstaking care and wrote up a paper for publication about this highly unusual discovery, only to be rejected by the journal editors because such behavior was not possible in a homogeneous mixture.[16] He tried several times to publish the results, each time being turned down, until in exasperation he threw the manuscript into a drawer of his desk and moved on to more "publishable" work.

Several of Belousov's colleagues were aware of the reaction and continued to study it because it was so remarkable, although eventually everyone forgot where the original recipe had come from. In 1961, a fresh graduate student at Moscow State University, A. M. Zhabotinsky (1938–2008), was given the project to find out what time-dependent reactions were taking place. Zhabotinsky made important inroads into understanding the reactions. When it was time to publish a manuscript, he sent a draft to Belousov for comments, only to find that it was Belousov himself who had originated the oscillating recipe. Zhabotinsky cited Belousov and went on to publish several more papers, followed by rapidly expanding interest inside Russia in this natural oscillator. In the summer of 1968, a Symposium on Biological and Biochemical Oscillators was held in Prague, Czechoslovakia at the height of the Cold War. Scientists from East and West attended, and the western researchers first learned about the Belousov–Zhabotinsky reaction (BZR). This caused a burst of interest with a rapidly growing literature on the topic.

As interest in the BZR was about to ramp up, Art Winfree (1942–2002) was finishing his bachelor's degree in the Department of Applied Physics at Cornell University. He was a highly creative thinker, never quite in step with anyone else, and always coming at problems from unusual directions. As a requirement for the undergraduate degree, he needed to select a senior thesis topic. His choice was the venerable problem of synchronization pioneered by Huygens and Rayleigh and van der Pol. But while these towering figures had worked with one or two oscillators that became entrained, Winfree wondered what would happen if there were an entire population of oscillators that interacted with each other, as in the pacemaker region of the heart and neural oscillations in the brain. Winfree became captivated by this project and continued to work on it even after graduation, publishing his first paper on this topic in 1967 as *Biological Rhythms and Behavior of Populations of Coupled Oscillators.*[17] What he discovered was a way that the phases of individual oscillators could become locked by each other. His arguments were qualitative, based on geometric reasoning, but the mechanism for phase locking was clear.

Winfree continued working on populations of biological oscillators for his PhD in biology at Princeton, exploring circadian rhythms among populations of flies, using a geometric argument about phase singularities on a torus to propose a way to use a small pulse to quench natural oscillation. To test his ideas, he built a small fly machine where he tortured fruit flies with short pulses of light. Around this time Winfree learned about the Belousov–Zhabintsky (BZ) reaction and realized that in lower dimensions it could be represented as a collection of coupled oscillators. By compressing the three-dimensional reaction into two dimensions by soaking filter paper with the reagents and placing the filter paper in a Petri dish, Winfree demonstrated that the reaction had the same singular phase point on the torus as his fruit flies. Seeing a connection between the circadian rhythms of fruit flies and the oscillations of a chemical reaction might have been a stretch for most, but to Winfree it was obvious. Unknown to Winfree at the time, the connection between chemical oscillations and biological oscillators was also coming into focus for a young Japanese researcher.

Yoshiki Kuramoto (1940–) at Kyoto University saw Winfree's first paper on the synchronization of a population of oscillators around the same time that he learned of the BZ reaction, and he too realized that its behavior was like a population of coupled oscillators. As a theorist in statistical mechanics, he had a mathematical toolbox that allowed him to make mathematically rigorous what Winfree was doing through geometric argument. Kuramoto recognized that synchronization among multiple oscillators was a phase transition like spontaneous magnetization or the transition of a metal to a superconductor at low temperature. Phase transitions, like magnetization, are singular phenomena in the sense that above a critical temperature there is no magnetization, but as a magnetic material is cooled below the critical temperature, a spontaneous magnetization occurs suddenly. Pierre Curie and Pierre Weiss had tackled the problem of the magnetic phase transition in the first decade of the 1900s using a mean-field theory.

Mean-field theory is like a chicken-and-egg approach to solving a physics problem. In the case of magnetization, magnetization can only arise from the alignment of quantum spins, and the alignment of quantum spins can only arise through interaction with the magnetization, but which comes first? The answer is that they both arise self-consistently together. Mathematically, a self-consistent equation describes how aligned spins create a magnetization M and how that same magnetization M aligns spins. This transcendental equation is solved for the magnetization M. To a theorist in statistical mechanics, this type of mean-field solution is child's play. The problem confronting Kuramoto was how to construct a mean field theory of a large population of oscillators.

Kuramoto's key insight was to create highly idealized oscillators that varied their phases in complete regularity while not varying their amplitudes at all—Poincaré phase oscillators. Furthermore, every oscillator had to interact equally with all others, allowing them to create a mean field of oscillating phase. By using this simplification, Kuramoto succeeded in finding a unique solution to the synchronization problem. Below a critical coupling strength, all the oscillators run through their phases independently. Yet, as the coupling strength increases, there is a critical value above which the entire network of oscillators locks frequencies. Kuramoto had succeeded in generating an exact theoretical model for the synchronization of a large number of autonomous oscillators, publishing his results in a conference proceeding in 1975. As often happens with singular discoveries or accomplishments in science—no one noticed. A decade or two passed before the importance of the Kuramoto model began to be appreciated as synchronization emerged as a new paradigm for the dynamic interplay among the biological systems that support us and keep us alive, until things go wrong, as they sometimes do, for instance with our heart.

During the latter half of the 1920s, Van der Pol had become a sort of evangelist, promoting the universal nature of his relaxation oscillators. In one of his published conference papers van der Pol noted that

> [M]any other instances of relaxation oscillations can be cited, such as: a pneumatic hammer, the scratching noise of a knife on a plate, the waving of a flag in the wind, the humming noise sometimes made by a water tap, the squeaking of a door, a steam engine with much too small flywheel . . . the periodic reoccurrence of epidemics and economical crises, the periodic density of an even number of species of animals, living together and the one species serving as food to the other, the sleeping of flowers, the periodic reoccurrence of showers behind a depression, the shivering from cold, the menstruation and finally the beating of the heart.[18]

Van der Pol was convinced that the heart was a relaxation oscillator and wrote papers about its natural oscillation,[19] devising an electronic model with three relaxation oscillators, one for the sinus, and another for the auriculum and the third for the ventriculum. The heart, the seat of human passion, was becoming a mechanical phase-space portrait.

Meanwhile at Purdue University in 1977, Art Winfree was deep into his study of phase resetting of coupled oscillators when he turned his attention to the neural model of Hodgkin and Huxley. As was his habit, he was already thinking broadly about the many ways that networks of neurons could interact, and about the important role they played in the human body, especially the pacemaker region of the heart.[20] Around the same time at McGill University in Montreal, Leon Glass (1943–) and Michael Mackey were studying the effects of time delays in physiology, developing what is today called the Mackey–Glass equation to describe oscillatory physiological effects such as respiration[21]. Glass and his colleagues gathered a collection of heart cells in a Petri dish. The cells tended to beat erratically when isolated, but when grown together, they began to beat in unison.[22] When they stimulated a single cell of the group with periodic electrical pulses, the entire assembly became frequency locked on the drive frequency, beating in synchrony. In addition, they could induce irregular beating behavior in the collection of heart cells, mimicking what might happen during arrhythmia. Winfree also was gaining insights into irregular heartbeats, studying the dynamics of three-dimensional nonlinear waves, that he named scroll waves, which propagated across the surface of the heart as it contracted.

Prior to these experiments, Glass had begun to formulate the idea of a dynamical disease, identifying pathologies that had resisted deep comprehension, but which could be ascribed to healthy regular periodicities that became irregular, or to healthy irregular periodicities that became regular, governed by dynamical consequences of nonlinear oscillations with feedbacks and time delays. Examples of dynamical health and disease abounded. The pacemaker of the healthy heart yields to fibrillation and tachycardia. The waves of action potentials that wash back and forth across the brain in healthy thought yield to migraines and epilepsy and Parkinson's disease. The steady rate of breathing yields to Cheyne–Stokes syndrome. The healthy replenishment of cells through the natural cell cycle yields to uncontrolled growth in cancer.

In much of the work by Winfree, Glass, Strogatz and others, the toolbox used to tackle the complexities of dynamical disease was the familiar phase-space portraits of Poincaré describing the system trajectory. When the complexities of a subsystem are averaged together to define dynamical dimensions in phase space, it becomes a powerful way to visualize and analyze the dynamical behavior of the whole system—deriving conditions for stability or instability, and for health and disease. It is also an ideal tool to tackle the tricky problem of thought and intelligence and maybe even consciousness in the neural mesh of our minds.

Ghost in the Shell

The squid is a ghost of the deep. Nearly invisible in its natural habitat, it haunts the pelagic zones of all oceans at virtually all depths. It is an elusive prey, possessing a hydraulic jet propulsion system that enables bursts of incredible speed to flee danger. To drive large volumes of water suddenly requires a rapid, nearly simultaneous, contraction along the length its body. A solution to this mechano-hydraulic problem was provided by evolution, endowing the squid with an extremely large neural axon that supports a large and hence fast conduction of nerve impulses down its length. The squid giant axon can grow as large as a millimeter in diameter, making it one of the largest cells in the animal kingdom, also making it one of the easiest to study using simple electrodes.

In the summer of 1939, Alan Loyd Hodgkin (1914–1998) was looking out of a window of the Marine Biological Association in Plymouth, England, waiting for the daily catch from the trawlers to come in. The MBA, as it is called, is a classic stone building three stories tall situated at the base of the Royal Citadel, an ideal location to look straight out the mouth of Plymouth Harbor to the Sound beyond the breakwater. Hodgkin had learned how to dissect squid giant axons the previous summer at Woods Hole on Long Island, New York, and was hoping to continue the work with his assistant Andrew Fielding Huxley (1917–2012) newly arrived from Cambridge. The squid in Plymouth Sound, usually so plentiful, had other plans, eluding capture for several weeks. Finally, the catch came in with several healthy North Atlantic squid, and Hodgkin and Huxley got to work.[23]

The squid giant axon is so large that they found they could insert a fine capillary along its length without piercing the cell membrane. The capillary was filled with an ionic solution that conducted current and transmitted voltage to an external amplifier where they recorded the time-dependent potential across the cell membrane. They were

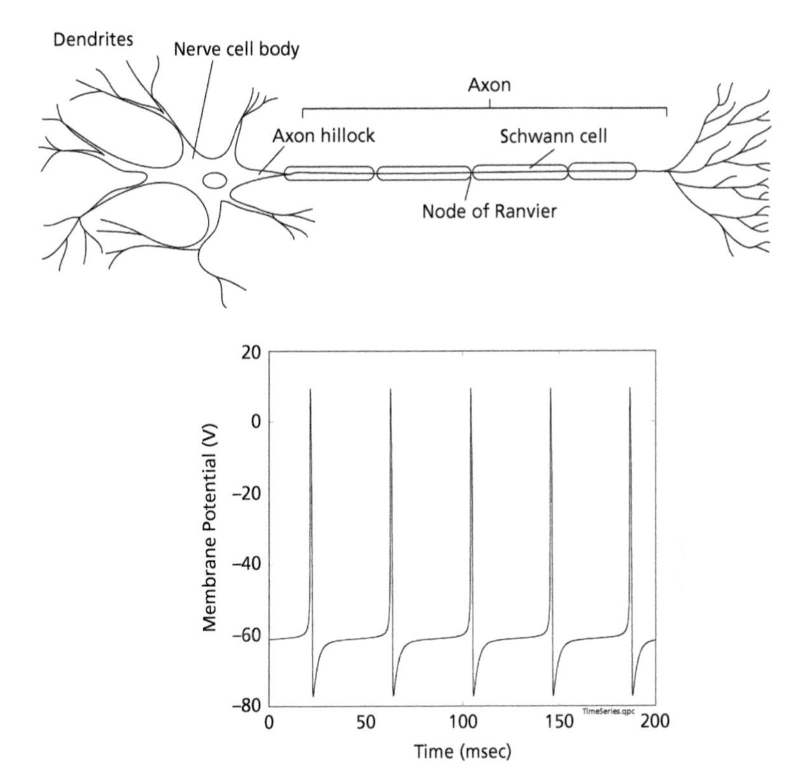

Fig. 11.3 Example of a biological neuron and the series of voltage spikes, known as action potentials, traveling down the axon

immediately rewarded with the first observation of the pulsing voltages traveling down a single nerve fiber—pulses known as *action potentials*. The pulses consisted of nearly periodic voltage spikes of 100 mV amplitude, spiking with frequencies in the range of a few to many hundreds of spikes per second (see Fig. 11.3). The work was exciting and epic, but it was interrupted abruptly by the invasion of Poland by Hitler's storm troops. Hodgkin and Huxley published their observations in *Nature* in 1939, then both turned to the war effort, Hodgkin helping with the development of the first airborne radar, and Huxley working on gunnery systems. Seven years passed before they returned to the Marine Biology Association building on Plymouth Sound to take up the work on the squid giant axon, but they returned with new skills that proved essential to their next breakthrough.

Hodgkin had gained deep experience in feedback and control systems, and Huxley had become adept at mathematical manipulations and computations. When they resumed their experiments in 1946, these skills allowed them to construct a mathematical model, a differential equation, for the spiking action potential. The biggest conceptual challenge was the presence of time delays in the voltage signals that were not explained by linear models of the neural conductance. They began exploring nonlinear models, using their experiments to guide the choice of parameters, and they settled on a dynamical model in a four-dimensional phase space. One dimension was voltage, while another was inhibitory current. The two remaining dimensions were sodium and potassium conductances, which they had determined were the major ions participating in the generation and propagation of the action potential. The nonlinear conductances of their model aptly described the observed time delays, and captured the essential neural behavior of the fast spike followed by a slow recovery. The first electronic computers had been developed for the war, and Huxley hoped to use one to solve their four simultaneous differential equations, but it was under repair, so he solved the equations on a hand-cranked calculator, taking over three months of tedious cranking to plot the numerical results. They published their measurements and their model (known as the Hodgkin-Huxley model) in a series of six papers[24] in 1952 that led to an explosion of research in electrophysiology, for which Hodgkin and Huxley won the 1963 Nobel Prize in physiology or medicine.

The four-dimensional Hodgkin–Huxley model stands out as a classic example of the power of phenomenological modeling when combined with accurate experimental observation. Hodgkin and Huxley were able to ascertain not only the existence of ion channels in the cell membrane, but also their relative numbers, long before such molecular channels were ever directly observed using electron microscopes. The Hodgkin–Huxley model lent itself to simplifications that could capture the essential behavior of neurons while stripping off the details. For instance, in 1961 Richard FitzHugh (1922–2007), a neurophysiology researcher at the National Institute of Neurological Disease and Blindness (NINDB) of the National Institutes of Health (NIH), created a surprisingly simple model of the neuron that retained only a third-order nonlinearity, just like the third-order nonlinearity that Rayleigh had proposed and solved in 1883, and that van der Pol extended in 1926. Around the same time that FitzHugh proposed his mathematical model, the electronics engineer Jin-Ichi Nagumo in Japan created an electronic diode circuit with an equivalent circuit model that mimicked neural oscillations. Together, this work by FitzHugh and Nagumo led to the so-called FitzHugh–Nagumo model. The conceptual importance of this model is that it demonstrated that the neuron was a self-oscillator, just like a violin string or wheel shimmy or the pacemaker cells of the heart. Once again, self-oscillators showed themselves to be common elements of a complex world—and especially of life.

These advances in the understanding of natural neural systems were proceeding in parallel with, yet independently from, the development of electronic computers. The theoretical basis for artificial computation was closely allied with the field of mathematical logic. Alan Turing made this connection explicitly in 1936 in his famous paper on computability where he introduced the concept of what is today called a Turing Machine.[25] Turing called it an *a-machine* (automatic machine) and proved that it was "universal" in the sense that it could compute any operation on the natural numbers, no matter how complex. Turing tended to be visionary, and he alluded to the possibility of artificial intelligence, but he stopped short of making the analogy that the brain could be a Turing machine, or even that mathematical logic could be applied to neural functions. That bold step would be taken a few years later by an odd collaboration between a psychiatrist and a homeless genius.

In 1941, Warren S. McCulloch (1898–1969) arrived at the Department of Psychiatry at the University of Illinois at Chicago. He had an MD from Columbia University and had become interested in neural function while working at the neurology laboratory at Yale. As soon as he arrived in Chicago he made contact with the mathematical biology group at the University of Chicago led by Nicolas Rashevsky (1899–1972), widely acknowledged as the father of mathematical biophysics in the United States. An itinerant member of Rashevsky's group at the time was a brilliant, young and unusual mathematician Walter Pitts (1923–1969). Pitts had read Bertrand Russell's *Mathematica Principia* before he was twelve and began a correspondence with Russell.[26] He was not enrolled as a student at Chicago, but had simply "showed up" one day, while he was still a teenager, at Rashevsky's office door and so impressed him that he was invited to attend their group meetings. His peculiar interest was in the application of mathematical logic to biological information systems.

When McCulloch met Pitts, he realized that Pitts had the mathematical background that complemented his own deepening views of brain activity as computational processes. Pitts was homeless at the time, so McCulloch invited him to live with his family, giving the two men ample time to work together on their mutual obsession with the logic of brain activity. Their goal was to provide a logical basis for brain activity in the way that Turing had provided it for computation. To accomplish this goal, they simplified the operation of individual neurons to their most fundamental character, envisioning a neural computing unit with multiple inputs (received from upstream neurons) and a single on-off output (sent to downstream neurons) with the additional possibility of feedback loops as downstream neurons fed back onto upstream neurons. They also discretized the dynamics in time, using discrete logic and time-difference equations rather than the complicated differential and integral equations that were being used by others in Rashevsky's group. They succeeded in devising a logical structure with rules and equations for the general operation of their nets of neurons, publishing their results in 1943 in the paper titled "A logical calculus of the ideas immanent in nervous activity."[27] This paper introduced computational language and logic to neuroscience. Their simplified neural unit became the basis for discrete logic, picked up a few years later by von Neumann[28] as an elemental example of a logic gate upon which von Neumann began constructing the theory and design of the modern electronic computer.

The natural computation by biological neural networks is an extreme opposite to the artificial computation by electronic computers as information processing machines. Where biological intelligence is based on massively parallel analog networks, electronic computers are highly compartmentalized binary devices. Where biological neurons are adaptive in nature, electronic circuits are static. Where natural intelligence can easily generalize and associate seemingly unrelated facts, circuits to perform these tasks must be programmed with complicated deterministic algorithms. In the late 1950s, a small number of researchers began to seek a middle ground, developing ideas of artificial neurons. Artificial neurons could be implemented in software or hardware. As software, they represented a new programming paradigm, while as hardware, they represented an alternative to digital computation, and the elementary neural computing unit of McCulloch and Pitts was the natural starting point for both.

The town of Cheektowaga, New York, on the banks of lissome Ellicott Creek winding its way through the Buffalo suburbs, is home to the Cornell Aeronautical Laboratory (known then as CAL and today as Calspan). In 1955, it was a semi-commercial research laboratory associated with Cornell University much as Draper and Lincoln Labs were associated with MIT. CAL is credited with the invention of crash test dummies, the automotive seat belt and mobile weather Doppler radar systems. The lab subsisted on outside contracts, many of them military, such as a contract from the US Office of Naval Research to develop an analog computational architecture for rapid image recognition for target identification. Frank Rosenblatt (1928–1971), with a PhD in psychology from Cornell University, was in charge of the cognitive systems section tasked with fulfilling the contract. Drawing from the work of McCulloch and Pitts, the team constructed a software system and then constructed a hardware model that adaptively updated the strength of the inputs, called neural weights, as it was trained on test images. The machine was dubbed the Mark I Perceptron,[29] and its announcement in 1958 created a small media frenzy, such as a New York Times article that reported the perceptron to be "the embryo of an electronic computer that [the navy] expects will be able to walk, talk, see, write, reproduce itself and be conscious of its existence."[30]

The perceptron had a simple architecture, with two layers of neurons consisting of an input layer and a processing layer, and it was programmed by adjusting the synaptic weights to the inputs. This computing machine was the first to adaptively learn its functions, as opposed to following predetermined algorithms like digital computers. It seemed like a breakthrough in cognitive science and computing, as trumpeted by the New York Times, but within a decade, the development had stalled because the architecture was too restrictive. This changed when an algorithm was discovered that could train a type of neural network called a feed-forward network. A major resurgence in neural network research ensued, fueled in part by an alternative artificial neural network architecture called a Hopfield network that represented a major deviation from the feed-forward network models.

John Hopfield (1933–) received his PhD from Cornell University in 1958, advised by Al Overhauser[31] in solid state theory, and he continued to work on a broad range of topics in solid state physics as he wandered from appointment to appointment at Bell Labs, Berkeley, Princeton, and Cal Tech. In the 1970s Hopfield's interests broadened into the field of biophysics, where he used his expertise in quantum tunneling to study quantum effects in biomolecules, and expanded further to include information transfer processes in DNA and RNA. In the early 1980s, he became aware of aspects of neural network research and was struck by the similarities between McColloch and Pitts' idealized neuronal units and the physics of magnetism. For instance, there is a type of disordered magnetic material called a spin glass in which a large number of local regions of magnetism are randomly oriented. In the language of solid-state physics, one says that the potential energy function of a spin glass has a large number of local minima into which various magnetic configurations can be trapped. In the language of dynamics, one says that the dynamical system has a large number of basins of attraction.

A basin of attraction is a region of phase space within which any initial condition leads to a single fixed point or limit cycle. The phase space of a complex system can contain many basins of attraction for many fixed points (see Fig. 11.4). In the case of a neural network, the system can evolve in time to find exact memories when seeded with partial memories. This process is called associative memory. If you are shown the picture of half of a dog, you still recognize it as a dog—your mind

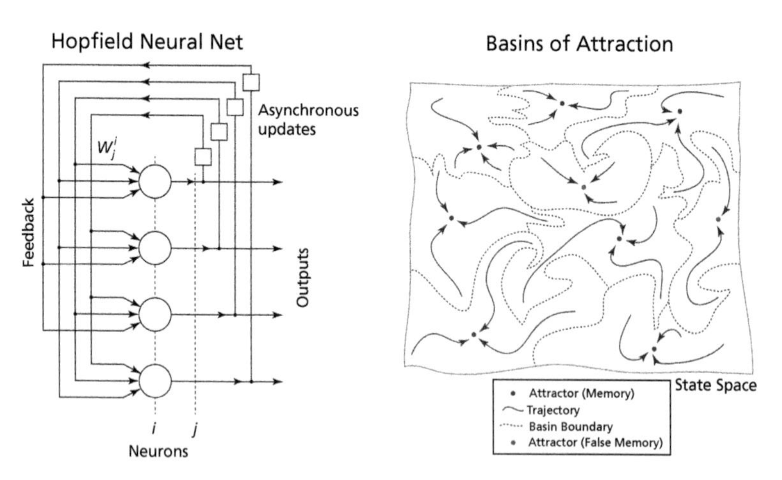

Fig. 11.4 The Hopfield neural network architecture and a schematic of the basins of attraction in state space. Any initial condition within a basin will attract to a fixed point of the dynamics. With many neurons, there can be many "false memories" when an initial condition crosses over a basin boundary to an adjacent basin

associates the partial input with an idealized representation of a dog in your mind. The amazing thing about associative memory is how little input is needed to recall perfect outputs. Even the tip of an elephant's trunk can let you recall the whole animal. The dynamical evolution of the neuron states from partial input to perfect recall is a dynamic trajectory caught within a basin of attraction. Hopfield published his recurrent neural network structure[32] in 1982. The recurrent neural network gave impetus, together with the rediscovery of error back propagation for perceptrons, to the field of artificial neural networks, which exploded in the late 1980s, pursued by a broad range of scientists drawn from neuroscience, computer science, electrical engineering and physics.

The state we call consciousness is an emergent phenomenon, constructed from billions of neural limit-cycle oscillators interacting through highly connected networks supporting billions of trajectories winding through a hyperspace of thought. The secret to consciousness is buried somewhere within the highly ramified structures of the human brain. The networks of the mind may yield a few of their secrets when examined using a theory developed initially to solve a problem of leisurely strollers in an Old Prussian city.

Seven Bridges of Königsberg

Leonhard Euler, who had been invited by the Tsarina Catherine I to join the Academy established by her husband, Peter the Great, at St. Petersburg, arrived in 1727 just in time to see her die and be replaced by the reactionary twelve-year-old Peter II. This second Peter lasted only three years, to be replaced in turn by Tsarina Anna Ivanovna whose reign has been characterized as a ship's captain asleep at the wheel. In this atmosphere of uncertainty, Euler learned to keep his head down, burying himself in his work and tending to his correspondences with a more enlightened Europe. In 1735, he received a letter from the mayor of Danzig in Prussia who posed a peculiar problem about the bridges of Königsberg, a neighboring city. The city was divided by the river Pregel into four regions that were connected by seven bridges: Blacksmith's bridge, Connecting bridge, High bridge, Green bridge, Honey bridge, Merchant's bridge, and Wooden bridge. As a game, the citizens of Königsberg, it is said, would stroll through their city on their casual Sunday afternoons trying to find a path that crossed each bridge

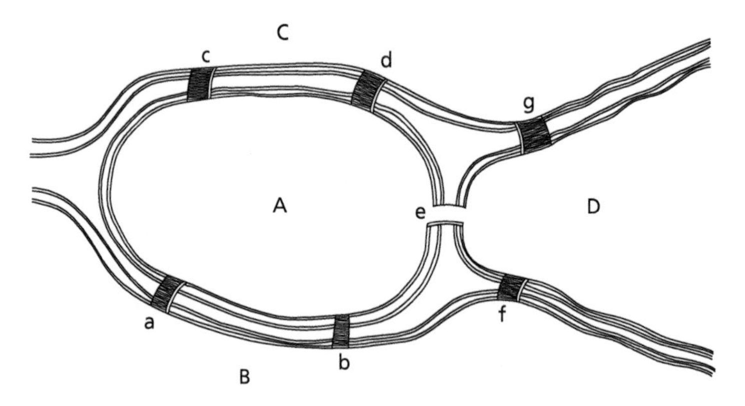

Fig. 11.5 Euler's drawing of the seven bridges of Königsberg. The Pregel River converges from the right, encircles the island of Kniephof and leaves to the left towards the Baltic Sea

once and only once (see Fig. 11.5). Many possible permutations existed, which is why the game was such fun, but no one had succeeded in finding a path, and many suspected that it did not exist. The mayor suggested that Euler could apply his considerable talent to the problem, as he was such an esteemed mathematician.

Euler's letter in reply was almost testy.[33] Why, he asked, would such a puzzle, that contained no numbers, be of interest to a mathematician? Yet the problem must have irked him, and intrigued him, because he set about to solve it. Euler thought that the solution might involve the famous calculus of position (*analysis situs*) that Leibniz first proposed in 1679 in a letter to Huygens. Leibniz had outlined a new field of study that would involve a geometry of position, without magnitudes, that would be as exact a science as mathematics. Today we would say that Leibniz was proposing to establish the field of *topology*. Despite Leibniz' voluminous output on so many topics, he was never able to construct such a theory. Euler was aware of Leibniz' hopes for an *analysis situs*, but he had never been sure what it would be good for, until now.

Euler's solution is as elegant as it is abstract, providing insight into why he was such a successful mathematician. He abstracted the problem to one of a succession of symbols representing the four regions, while ignoring the bridges, which formerly had been the center of attention. With the seven bridges of Königsberg, a successful path would have eight symbols. Furthermore, pairs of symbols of adjacent regions must occur as many times in the sequence as there are bridges between that pair. For example, one trajectory could be expressed as BACD-BACDA, although this is not a solution because the pair CD appears twice (bridge *g* is crossed twice). Such a pairing of symbols, in a symbolic trajectory, places a constraint on possible solutions, requiring most regions to have an even number of bridges attached to it, except for the starting and ending region. Because each region of Königsberg was connected by an odd number of bridges, a solution was impossible.[34] The game was over for the citizens of Königsberg.

More important than answering the bridge question in the negative, Euler went on to show how any such problem, with arbitrary numbers of regions and links, could be addressed to decide if an each-link-once path existed. Euler had succeeded in making the first concrete demonstration of Leibniz' long-sought topology. A successful path is called today an Eulerian path or an Euler walk, and if the starting and ending regions happen to be the same, it is an Eulerian circuit. Euler's solution to the Königsberg bridge problem is considered the first application of topology as well as graph theory.

Graph theory treats abstract sets of nodes (a node is also called a vertex) linked by edges. The edges, or links, can be directed by pointing from one node to another, or can be bidirectional, pointing both ways. Links can be viewed as paths of influence or information flow from one node to another. Graphs can have regular structure, such as complete graphs (where every node is connected to every other node) or lattices or regular trees (like a binary search tree). Or graphs can be irregular, such as random graphs with incomplete connections among the nodes (see Figure 11.6). Graphs can model a wide range of network structures and processes. The four-color map problem is a classic example of applied graph theory, as are questions about the distribution of electricity on national power grids, and the flow of information on the World Wide Web. Graph theory can pertain to social networks, or networks of scientific collaborators, and even to interaction networks among protein molecules associated with health and disease.[35]

Static graphs are not dynamic. The links may represent the dynamic flow of information from one node to another, yet their structure remains unaltered. Networks that change in time are more interesting, with connections added or removed in specified ways in a succession

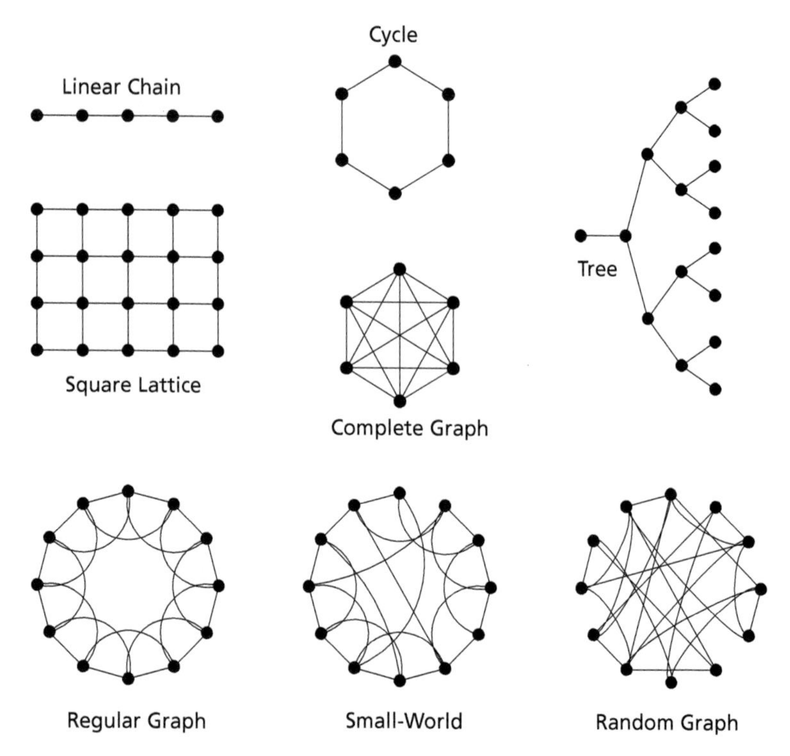

Fig. 11.6 Examples of graphs from graph theory. Regular graphs (linear chain, cycle, square lattice, complete graph, tree) have symmetry and patterns. A regular graph like a cycle with next-nearest-neighbor connections (bottom left) can be rewired with a small rewiring probability per link to generate a small world network (bottom middle) and a high probability per link to generate a random network (bottom right)

of steps. The graph evolves in time, performing a type of trajectory, the collective presence or absence of links representing the configuration or "state" of the network like the state of the binary hypercube discussed in Chapter 10.

Alternatively, a network trajectory consisting of a fixed number of nodes, but with an increasing density of links, displays a sudden transition from disconnected clusters of linked nodes to a large cluster that spans from one side of the network to the other. John Hammersley (1920–2004) and Simon Broadbent, who met by chance at a conference held at the Royal Society in London in 1954, were the first to discover this critical behavior. Hammersley was an assistant at Oxford, and Broadbent was a graduate student at Imperial College. Hammersley presented his ideas about using increasing computing power to perform a type of calculation called a Monte Carlo simulation (so called because one simulates many alternative patterns by using a random number generator like rolling dice at Monte Carlo), and in the discussion period afterwards Broadbent suggested the problem of gas infiltrating a porous medium that could be modeled as a lattice in two-dimensions (2D) or three-dimensions (3D) with random connections representing the connections among pores. He was thinking practically in terms of the design of gas masks, but the problem became an abstract mathematical one and the two joined forces. In 1957, they published a paper on the transition and called it *percolation*[36] because of the analogous problem of tea percolating through tea leaves. The paper proved that a critical percolation threshold existed in simple topological networks modeled as regular 2D and 3D lattices.

A few years later, in 1959, the Hungarian mathematician Paul Erdös (1913–1996), who has the distinction of having published more papers (over a thousand) than Euler, was collaborating with Alfred Rényi (1921–1970), another Hungarian, exploring the evolution of random graphs that were fundamentally unlike the lattices studied by Broadbent and Hammersley. Whereas a 2D lattice topology has an intrinsic 2D embedding dimension, a 3D lattice is embedded in 3D, and the binary genetic network of N bits is embedded in an N-dimensional space, random graphs have no analogous dimensionality. If the history of Galileo's trajectory is one of ever-expanding dimensionality, then this ultimate limit of Galileo's trajectory has transcended dimensionality entirely.

Erdös and Renyi discovered a phenomenon similar to Broadbent and Hammersley's percolation transition. If a graph has only a few links, there will be large numbers of small clusters of connected nodes that are all disconnected from each other. However, as the number of links is increased, the fragmented subclusters begin to join up, small clusters growing and merging with other small clusters to create larger clusters of nodes. As this process of adding links continues, there is a sudden coalescence, like a crystallization, as almost all of the disconnected clusters suddenly connect into a single giant cluster.[37] The sudden transition from disconnection to connection within the network, from disjoint behavior to common behavior, is a tantalizing metaphor for the appearance of new order, new awareness, that emerges with the right kind of network topology. Once networks are connected beyond the percolation threshold, they become surprisingly small.

Around the time that Broadbent and Hammersley were discovering the percolation transition in lattices, and Erdös was discovering the emergence of giant network components in random graphs, the American sociologist Stanley Milgram (1933–1984) was working on his PhD at Harvard. He graduated in 1960 and joined the faculty at Yale University where he began a series of groundbreaking and disturbing experiments on obedience. The experiments pretended to administer painful electric shocks to experimental subjects, who would scream and beg while the test subjects were ordered to increase the voltage. The experiment and its results became famous, first for the shocking number of ordinary people who complied with instructions to administer the shocks, and second for its unethical treatment of human test subjects. The outcry against this second factor contributed to the adoption of standards for the ethical treatment of human subjects and the establishment of institutional review boards (IRBs) for all human experiments. It also led, in part, to Milgram being denied tenure after he had moved from Yale to Harvard, forcing him to leave Harvard to join the faculty at the City University of New York. Despite his troubles at Harvard, it was during his last years there that Milgram succeeded in performing his most famous experiment on the problem of "six degrees of separation."

Since the early twentieth century, it was known that gossip could spread quickly through a large number of people in social contact, but this effect had never been quantified. In the mid-1960s, Milgram conceived of an ingenious experiment to measure it. He sent 160 packages to randomly chosen people in Omaha Nebraska and in Boston Massachusetts with instructions to relay the package to an identified individual living outside Boston, but only by sending the package to a person familiar to them on a first-name basis, and that person would do the same, and so on until the package was delivered to its target. Each person in the chain could only guess whether the next person to whom they shipped the package would know someone who knew someone, and so on. The astounding conclusion of the experiment was that the chain of acquaintances between the original person and the target was only about six links long despite there being over 100 million people in the United States that the package potentially could go through.

Milgram published his results, first in a trade magazine for general readers, and then in an archival journal with all the details. Only a handful of sociologists and network specialists knew of the study until the playwright John Guare wrote a play called "Six Degrees of Separation" performed on Broadway in 1990. The play touched on the interconnectedness of social networks while also striking a chord with playgoers. The success of the play was transformed into a Hollywood movie with the lead played by Will Smith in 1993. Not long after, the actor Kevin Bacon, who had nothing to do with either the play or the movie, quipped during an interview that he had either worked with everyone in Hollywood, or with someone who worked with them. This aside was picked up by a group of students at Albright College who started playing a game after watching Footloose (with Kevin Bacon as the lead) in which they began with some bizarre statement that would lead back, after a short number of links, to Kevin Bacon. They published a book about the game called "Six Degrees of Kevin Bacon" with Plume in 1996. This was all for fun, spoofing Milgram's societal experiment, but was also strangely prescient, because network science was about to be thrust into the forefront of everyone's thinking as the World Wide Web was exploding almost everywhere, not unlike a percolation transition, as billions and then trillions of dollars came up for grabs in the Dot-Com economy at the end of the twentieth century.

The rise of the Internet will stand as one of the major technological revolutions of modern history. Although it is premature to assess its ultimate impact on quality of human life, it is certainly as disruptive to daily behavior as electricity, gasoline engines and air flight were a century ago. The odd thing about the Internet is that it was never actually designed—it just grew on its own, organically. Therefore, no one knew what its architecture looked like, or what network properties it had. For something so important to human life, we were surprisingly ignorant of the structure and behavior of real-world networks. It was time, by the end of the 90s, to tackle the beast. Two pairs of names are associated with the rise of the new science of networks and with two seminal network archetypes. These are the Watts and Strogatz model for small-world networks, and the Barabasi and Albert model for scale-free networks.

Duncan Watts is a tall Australian who majored in physics as an undergraduate and entered graduate school in mathematics at Cornell in 1994. Steven Strogatz had just joined the faculty at Cornell, continuing his work on synchronization after his success with the fireflies, and was looking for a new student. Watts and Strogatz began by building a Rube–Goldbergesque contraption, reminiscent of Winfree's fly machine, to study the synchronization of crickets' chirps by letting them adjust the sound level heard by each cricket. As the experimental work progressed, Watts began to wonder what would happen if each cricket was not listening to every other cricket, but could hear only a few in the cohort. In that case, each cricket was like a node of a graph, and the few other crickets it could hear were connected to it by links. The links would be neither random like the Erdös–Renyi graph, nor regular like the ones studied by Broadbent and Hammersley. He asked Strogatz about this one day, and Strogatz realized that this type of network represented a middle ground that was relatively unexplored.

For mathematicians, the ideal would be to find a mathematical expression that bridged from the analytical results of the perfectly regular graph to the analytical results of a perfectly random graph by including increasing degrees of disorder. However, partial order presented theoretical difficulties, so they tackled the problem with the ever-increasing computer power of the mid 90s by using Monte Carlo simulations. To span from perfect order to perfect disorder, they created a process that began with a ring structure of N nodes with links among k nearest neighbors, creating a regular ring structure with many links. Then a random fraction p of the links is rewired at random. By adjusting p continuously from zero to unity, one goes smoothly from the perfectly regular network to the perfectly random network. As they studied this progression, they tracked two statistical network properties called the average path length and the clustering coefficient. The average path length measures the average distances between nodes inside the network, which is a global property, while the clustering coefficient measures the clumping of nodes into cliques, which is a local property. What they discovered was a surprise.

When only a few links had been randomly rewired, the average size of the network crashed from a large value around N (which was in the thousands) to values of only a handful of links that you literally could count on your fingers. On the other hand, the clustering of the graph, describing local cliques, was virtually immune to the rewiring. This means that only a few randomly rewired links make every node just a few links away from every other node, while preserving the local clusters—they had created a small world just like the social network that Milgram had discovered. As one important consequence, it meant that individuals in a group of friends would feel isolated and safe from the perils of the larger world, where deadly viruses spread rampantly, like Ebola or AIDS. Yet this feeling of isolation is an illusion, created by the indifference of the clustering coefficient to rewiring, even as the rewiring brings everyone in the world only a few links away from everyone else.

To test whether this small-world property was obeyed in real-world networks, Watts and Strogatz analyzed the network structure of the electric power grid, the neural network in a small worm, and the network connections among actors, including Kevin Bacon. Each of these highly disparate networks, governed by radically different underlying physical processes, showed the same small-world property. Watts and Strogatz published their discovery in *Nature* magazine with the title *Collective Dynamics of Small World Networks*. With a bit of tongue-in-cheek, they also analyzed the six degrees of Kevin Bacon and found that he was not, after all, the most linked person in Hollywood. That dubious honor went to a pair of actors known for acting in long strings of bad movies.[38]

When the *Nature* article of Watts and Strogatz appeared in 1998, it immediately caught the attention of physicist Laszlo-Albert Barabasi (1967–) then at Notre Dame University in Indiana who had been pursuing his own study of the structure of complex networks—networks that were neither purely regular nor random—with an interest in the structure of the World Wide Web. Barabasi was a solid-state theorist and a specialist in statistical mechanics, giving him a unique perspective that focused on statistical properties. He had his post-doc Hong Jeong build a web crawler robot to explore the set of web-page links on the WWW, and when Hong plotted the statistical distribution of the number of links for each node, he discovered that it had none of the properties expected for a random graph. Instead, the probability for a node to have a given number of links (known as its *degree*) was independent of scale, just as a fractal structure has no intrinsic scale. The degree distribution was a power law. Furthermore, as they studied computer chips, as well as the neural network of the same type of worm that Watts and Strogatz had analyzed, and the power grid data that they received from Watts, they found the same striking scale-free power-law degree distribution. When systems as far afield as the brain of a worm and the electrical power distribution grid show the same statistical structure, it begins to look like a universality. The question was how such robust properties could arise across so many different systems.

Barabasi tells the story of how he was attending a conference hosted at a seminary in Porto, Portugal, sitting half listening to the conference talks, when he took a walk to clear his head and to grapple with origins of scale-free networks.[39] As he walked, he had an epiphany, recognizing that power laws have long tails that arise when events compound, like the compounding of wealth (the rich get richer) that was well known from the Pareto distribution of wealth. He dashed off a fax to his graduate student Reza Albert asking her to run a computer simulation in which a network grows in time, adding nodes whose links are linked to other nodes according to their degree. In this way, nodes with high degree are more likely to be linked to, further raising their degree. In terms of the WWW, this means that new web pages tend to link to already popular web pages. Albert quickly confirmed that a growing network with preferential attachment to popular nodes produced a scale-free power law. As Barabasi flew home from Portugal, he wrote the paper *Emergence of Scaling in Random Networks* on the plane, finishing before landing, and submitting it to *Science* a few weeks later, where it was published in 1999.

In rapid succession, after the publication of the two seminal papers at the end of the century, and driven by the intensifying awareness of, and reliance on, the World Wide Web, small-world and scale-free network structures were identified across a widening range of disciplines, launching the modern field of Network Science. Network science collected many sub disciplines into a unifying framework—graph theory, percolation theory, social networks, communication grids, ecology, economics, signaling, and many more. What had been a diverse assortment of disconnected results and observations became understood as common elements of overarching principles shared by all these fields. The connectivity patterns linking the elements of complex systems had universal classes and properties that could be enlisted to understand and explain the bewildering assortment of behaviors that emerge from complex and dynamic systems—such as the inner workings of health and disease.

Network medicine is a new subfield of network science that seeks to understand the state of health of cells, tissues, organs and organisms. Life and consciousness are perhaps the ultimate complex systems—final frontiers of the science of inner space. At the extreme small scale, the elements of life are the molecules of life. The human genome, a macromolecule of DNA that sheds uncountable numbers of messenger

RNAs, codes for the synthesis of over 25,000 different types of proteins. These proteins are the cogs and the gears of the machinery of cellular life. But the cell is an ephemeral machine, built of thousands of transient associations among all its constituents. Network medicine provides insight into this chaotic dance by associating molecules with network nodes, and molecular interactions with links. Virtually all molecules in the cell interact with a host of others, combining into clusters of interactions, called modules, to perform semi-autonomous functions that partially overlap with other modules.

When disease strikes, the state of cellular dynamics is altered from its normal behavior, and the modules represent the functional level of the disease. One of the revelations from network medicine is that disease is rarely associated with a single gene mutation, or a single protein alteration. If such simple causes were the source of disease, then medicine would be easy. Instead, disease is often associated with changes in numerous interacting molecules, affecting nearly all the nodes and links of some of the semi-autonomous network modules, and in small-world spirit, a few links will span to distant modules, entraining the entire life of the cell into the disease state. In a hierarchical cascade, cells entrain tissues, tissues entrain organs, and organs entrain the life of the host. In the face of such complexity, it is understandable why diseases such as cancer are so difficult to address using single therapies targeting single molecules. If cellular interaction networks are like the meshwork of a tapestry, then disease is like a stain that seeps along the threads. Only a whole-scale approach can hope to eradicate the contamination.

Nonetheless, there is hope. The central tenant of network medicine is that organizing principles govern the structure and behavior of the complex nets of interactions. For example, many cellular networks have scale-free character, with central hubs possessing a huge numbers of links to other nodes in the network. Because hubs exert a disproportionate influence over the network dynamics, one might be concerned that hubs would be the origins of disease modifications, infiltrating broad reaches of the cellular network. Yet, these central hubs are so critical to the life of the organism that they tend to be robust against disease—fortified through eons of evolution to be stable centers of control. Instead, many of the foci of disease occur in the periphery of the interaction networks,[40] isolating them and making them more amenable to orchestrated intervention. Therefore, while the idea of a single therapeutic target for a single disease is too simplistic, network medicine gives hope that limited ranges of the complex cellular networks can be treated through structured approaches steered by the structure of the network.

The network trajectory has become supremely abstract, at the other end of the spectrum from Galileo's simple parabola. Unlike concrete ideas of mass and distance, the nodes of a network are purely representational, standing in for anything—neurons, money markets, species, molecules, power stations, words, circuits. Nodes can even stand in for other networks to describe networks of networks, and so on. This is the power of the network approach to complex systems. A single node in an abstract network can represent an entire complex system, condensed into a single active element. Networks become hierarchies of complexity, but simplified at each level of perspective to manageable sets of interacting nodes. This simplification is essential for understanding health and disease that depend, at their finest scale, on more than an Avogadro's number of molecules. Boltzmann confronted this incomprehensible complexity and simplified it by establishing equilibrium principles and applying them to trajectories in phase space. Similarly, network simplification draws from coarse-graining in state space, lumping smaller-scale processes into units that interact at larger scales, creating a new era of medicine that traces and gently adjusts our personal molecular trajectories—our life trajectories—as fitting descendants of Galileo's original trajectory, controlling our ultimate fates, but no longer completely out of our own control.

12

Epilogue: Galileo at Home in the Multiverse

Galileo achieved immortality despite his house arrest and despite the banning of his books that he feared had erased his legacy. Ironically, his house arrest gave him time to assemble and record his lifelong investigation into the science of motion—and banning his books made them immensely popular. He is often held up as a martyr for science, although his fall from grace was more by personal failings than by his beliefs. In Bertolt Brecht's 1943 play *Life of Galileo*, his former pupil Andrea chooses to see Galileo's abjuration as a heroic ploy to dupe the Church so that he can complete his *Two New Sciences* as an eternal gift to the world. But Brecht's Galileo demurs, confessing that his abjuration was made out of simple fear.

Regardless of his motives, Galileo's name is engraved in our scientific consciousness as well as in our popular culture. In physics, the law of the differences of squared times for uniformly accelerated motion is carved into the mind of every physics student, as is the application of Galilean relativity for inertial frames. Galilean thermometers are popular mantle pieces in homes. The Galileo spaceship, launched in 1989 from the Space Shuttle Atlantis became the first spacecraft to orbit Jupiter when it arrived there in 1995 after several gravitational assists that sent it careening about the inner solar system. The Galilaei Lunar Crater is a tribute to his important discovery of the geography of the moon, and, of course, the Galilean moons of Jupiter forever mark the nights in January of 1610 when he turned his telescope on the brilliant planet rising in the East.

In pop culture, the Galileo Affair is a cultural meme, serving as the archetypical conflict between faith and reason. Familiarity with his name was renewed for a new generation in the rock ballad *Bohemian Rhapsody* by Queen (1975), and in songs by Amy Grant (1991) and by the Indigo Girls (1992). Dan Brown's 2000 novel and 2009 film *Angels and Demons* credited Galileo with launching the supposed Illuminati

Galileo Unbound. David D. Nolte, Oxford University Press (2018).
 DOI: 10.1093/oso/9780198805847.001.0001

movement, and Galileo's dubious diary makes an appearance in Disney's 2017 film *Pirates of the Caribbean: Dead Men Tell no Tales*. But flying far above cultural fluency is his most important legacy—his trajectory.

Galileo's trajectory has come far from its humble birthplace in Padua on the outskirts of Venice, the shimmering jewel of the Adriatic coast. The simple curve of the parabola has become Ariadne's thread, winding its way, unbroken, through a labyrinth of dimensions and parallel universes. Now it has found a new home, somewhere far beyond Galileo's dreams, passing through many hands along its way.

Newton held it aloft as he stood on Galileo's shoulders, bragging to his nemesis Hooke. Maupertuis used it providentially, striving in vain to prove the existence of God. Lagrange freed it from the prison of our three-space, allowing it to fly, like the flight of a swallow, into abstract new dimensions. Riemann warped it, like the warps of a tapestry, bending it to his will. Boltzmann averaged it, creating little bins of probability to grapple with an Avogadro's number of dimensions that lay beyond his comprehension. Poincaré saved it at the last minute from an embarrassing publication, making it an icon of nonlinear dynamics immortalized in the three-body problem. Einstein refracted it, distilling its quintessence to create the most-subtle probe of gravity. Heisenberg tried to abolish it, relenting only to Bohr, his mentor, whom he would later betray. Feynman went crazy with it, letting it go anywhere and do anything, but always finding its resting place where all phases add up. Arnold turned it into the Cheshire cat, disappearing into translucent noise only to reconstitute itself again like magic. Lorenz released it gently as a butterfly, flapping its sensitive wings in response to the slightest breeze. Darwin grew it into a tree, branching into all the kingdoms on Earth. Rayleigh and van der Pol condensed it into a paradigm, providing the elementary unit for networks of synchronization. And finally, we all are holding the legacy of Galileo's trajectory tentatively in our own hands, hoping it will give us a glimpse into our futures.

Endnotes

Chapter 1: Flight of the Swallows

1 Heidegger, M. (1991). *The principle of reason*. Bloomington: Indiana University Press.
2 Gleick, J. (1987). *Chaos: Making a New Science*, New York: Viking.
3 Medio, A. (1992). *Chaotic Dynamics: Theory and Applications to Economics*. New York: Cambridge University Press.
4 Descriptions of configuration space, state space and phase space can be found in Nolte, D. D. (2015). *Introduction to Modern Dynamics: Chaos, Networks, Space and Time*. Oxford: Oxford University Press

Chapter 2: A New Scientist

1 Sobel, D. (2000). *Galileo's Daughter: A Historical Memoire of Science, Faith and Love*, New York: Penguin Books, p. 56.
2 An historical timeline of the events in this chapter can be found at the Berkeley Press website: https://works.bepress.com/ddnolte/.
3 Heilbron, J. (2010). Galileo. Oxford: OUP Oxford., p. 24.
4 Heilbron, p. 164.
5 Heilbron, p. 80.
6 Heilbron, p. 108.
7 Heilbron, p. 153.
8 The highest mountain on the moon, Mont Huygens, is about 3.3 miles high, with many others between 2 and 3 miles, so Galileo was high in his estimates by about a factor of two.
9 Io 1.8 days, Europa 3.6 days, Ganymede 7.2 days and Callisto 16.7 days.
10 Galileo's original intention for the title translation was The Starry *Message*, but Kepler and others translated it as The Starry *Messenger*, and their usage stuck.
11 Heilbron, p. 167.
12 Galileo, *Letter to the Grand Duchess Christina* (1615).
13 Sobel, p. 73.
14 Heilbron, p. 240.
15 Heilbron, p. 241.
16 Sobel, p. 251.
17 Sobel, p. 253.
18 Sobel, p. 253.

Chapter 3: Galileo's Trajectory

1 An historical timeline of the events in this chapter can be found at the Berkeley Press website: https://works.bepress.com/ddnolte/.
2 Clagett, Marshall (1968), *Nicole Oresme and the Medieval Geometry of Qualities and Motions; a treatise on the uniformity and difformity of intensities known as Tractatus de configurationibus qualitatum et motuum*, Madison: Univ. of Wisconsin Press, p. 537.
3 Drake (1989). *History of Free Fall* Toronto: Wall & Thompson, p. 25.
4 Drake, p. 28.
5 The bell tower leans to this day, pitching precariously towards the old canal.
6 Drake, p. 27.
7 Drake, p. 25.
8 Drake, p. 32.
9 Drake, p. 49.
10 Naylor, R. H. (1980). "Galileo theory of projectile motion." *Isis* **71**(259): 550–70.
11 Redrawn from Naylor, R. H. (1976). "GALILEO - search for parabolic-trajectory." *Annals of Science* **33**(2): 153–72.
12 Naess, A. (2005). *Galileo Galilei: When the World Stood Still*, Berlin: Springer, p. 185.
13 Naess, p. 186.

Chapter 4: On the Shoulders of Giants

1 An historical timeline of the events in this chapter can be found at the Berkeley Press website: https://works.bepress.com/ddnolte/.
2 1669 was a banner year for Newton. He also became the Lucasian Professor of Mathematics at Cambridge and invented the first reflecting telescope.
3 Voltaire, *Lettres philosophiques* (1734). "Lettre XIV. Sur Descartes et Newton." In: Voltaire, Amsterdam: E. Lucas au Livre d'or, p. 54.
4 Dugas, R. and J. R. V. Maddox (1988). *A history of mechanics*. New York: Dover Publications, p. 257.
5 Maupertuis, (1746). "Loix du mouvement et du repos," in *Histoire de l'Adadémie Royale des Sciences et des Belles Lettres*, Berlin: Haude, pp. 267–94.
6 Dugas, p. 268.
7 *Euler Biography, Dictionary of Scientific Biography* (2008), Detroit: Charles Scribner's & Sons. pp. 467–84.
8 Terrall, M. (2002). *The Man Who Flattened the Earth: Maupertuis and the Sciences in the Enlightenment*, Chicago: University of Chicago Press, p. 304.
9 Boudri, J. C. (2002). *What was Mechanical about Mechanics: The Concept of Force between Metaphysics and Mechanics from Newton to Lagrange*. Dordrecht: Kluwer Academic Publishers, p. 208.

10 Boudri, p. 209.
11 Lagrange, J.-L. (1811). *Mécanique Analytique*, Courcier.

Chapter 5: Geometry on my Mind

1 MacTutor History of Mathematics, http://www-history.mcs.st-andrews.ac.uk/index.html.
2 Descartes, R. (1637). *La Géométrie* (Geometry), Amsterdam: Elzevir.
3 An historical timeline of the events in this chapter can be found at the Berkeley Press website: https://works.bepress.com/ddnolte/.
4 Gauss, C. F. (1827). *Disquisitiones generales circa superficies curvas* (General Investigations of Curved Surfaces)," Göttingen: Typis Dieterichianis.
5 $\alpha + \beta + \gamma = \pi + K \cdot A$, where the three angles are α, β and γ and the total curvature is $K = 1/R^2$.
6 $ds^2 = Edu^2 + 2F\, dudv + Gdv^2$.
7 Boyer, C. B. (1985). *A History of Mathematics*. Princeton: Princeton University Press.
8 Riemann, B. (1854). *Über die Hypothesen, welche der Geometrie zu Grunde liegen, Habilitationsvortrag*, Göttingen: Abhandlung König. Gesell. Wiss., vol. 13, 1867 1854.
9 Gauss, C. F. (1851). "Bestimmung des kleinsten Werts der Summe . . . fur gegebene U . . . gleichungen u > 0," in *Gauss Werke* vol. 10.1. Leipzig: Teubner Verlag (1917), 473–81.
10 Betti. E., (1871). "Sopra gli spazi di un numero qualunque di dimensioni," *Annali di Matematica*, 4: 140–57.
11 Quoted in Kolmogorov and Yushkevich, Eds., (1996). *Mathematics of the 19th Century*, p. 96.
12 Peano, G. (1888). *Calcolo geometrico secundo l'Ausdehnungslehre di H. Grassmann e precedutto dalle operazioni della logica deduttiva*. Turin: Fratelli Bocca Editori.
13 Hilbert, D. (1900). "Mathematical Problems," *Göttinger Nachrichten*, pp. 253–97.
14 Hilbert, D. (1912). *Grundzuge einer allgemeinen Theorie der linearen Integralgleichungen*. Leipzig and Berlin: Teubner.
15 Weierstrass, K. (1872). "Uber continuirliche Functionen eines reellen Arguments, die fur keinen Werth des letzteren einen bestimmten Differentialquotienten besitzen," *Gelesen in der Königl. Akademie der Wissenschaften*, II: 71–4.
16 The Weierstrass function is given by the infinite Fourier series $g(x) = \sum_{n=1}^{\infty} a^n \cos(b^n \pi x)$ where a and b are constants with 0 < a < 1, and b is an odd integer.
17 Cantor, G. (1874). "Ueber eine Eigenschaft des Inbegriffs aller reellen algebraischen Zahlen." *Journal für die reine und angewandte Mathematik*, 258–62.
18 Cantor, G. (1883). *Grundlagen einer allgemeinen Mannigfaltigkeitslehre*. Leipzig: B. G. Teubner.

19 This is an infinite set of points composed of all the numbers that are generated by the formula $z = \frac{c_1}{3} + \frac{c_2}{3^2} + \ldots + \frac{c_m}{3^m}$ for $\lim m \to \infty$, where the c_m take on all permutations of the two integers [0, 2].
20 Peano, G. (1890). "Sur une courbe, qui remplit toute une aire plane." *Mathematische Annalen*, 36: 157–60.
21 Hausdorff, F. (1914). *Grundzuge der Mengenlehre*. Leipzig: Veit.
22 Hausdorff, F. (1919). "Dimension und ausseres Mass," *Mathematische Annalen*, 79: 157–79.
23 Edgar, G. A. (2004). *Classics on fractals*. Boulder, Colo.: Westview Press.
24 The Cantor ternary set has the unusual property that it is dense nowhere, yet is uncountable.
25 Mandelbrot, B. (1967). "How Long Is the Coast of Britain? Statistical Self-Similarity and Fractional Dimension," *Science*, 156: 636–8.
26 Mandelbrot, B. (1983). *The Fractal Geometry of Nature*. San Francisco: W. H. Freeman.

Chapter 6: The Tangled Tale of Phase Space

1 Arnold, V. I. (1978). *Mathematical Methods in Classical Mechanics*. New York: Springer.
2 Portions of this chapter were published in, *Physics Today*, 63, 33–8 (2010).
3 Liouville, J. (1838). "Note sur la théorie de la variation des constantes arbitraires." *Liouville Journal* 3: 342–9.
4 Lutzen, J. (1990). *Joseph Liouville 1809–1882: Master of Pure and Applied Mathematics*. New York: Springer-Verlag.
5 Struik, D. (1948). "*A Consice History of Mathematics*," Fourth ed.: Dover, 1948; E. Segre, (1984). *From Falling Bodies to Radio Waves*. San Francisco: Freeman.
6 An historical timeline of the events in this chapter can be found at the Berkeley Press website: https://works.bepress.com/ddnolte/.
7 Liouville's Theorem of 1838 (1838), in *Note sur la théorie de la variation des constantes arbitraires. J. mathematique purés et appliquées*, 3: 342–9.
8 Hamilton, W. R. (1834). "On a general method in dynamics I," *Phil. Trans. Roy. Soc.*, pp. 247–308; W. R. Hamilton, (1835). "On a general method in dynamics II," *Phil. Trans. Roy. Soc.*, pp. 95–144.
9 Lutzen, p. 661.
10 Jacobi, C. G. J. (1866). *Vorlesungen über Dynamik*. Berlin: Georg Reimer.
11 Jacobi's rederivation of Liouville's Theorem, p. 92 in *Vorlesungen über Dynamik* (Reimer, 1866).
12 Cayley, A. (1843). "Chapters in the analytical geometry of n dimensions," *Cambridge mathematical Journal*, 4: 119–27.
13 Grassmann, H. and L. C. Kannenberg (1995). *A new branch of mathematics: The "Ausdehnungslehre" of 1844 and other works. Translated by Lloyd C. Kannenberg*. Chicago: Open Court.

14 Riemann, B. (1990). *Gesammelte mathematische Werke, wissenschaftlicher Nachlass und Nachtrage. Collected papers. Nach der Ausgabe von Heinrich Weber und Richard Dedekind neu hrsg. von Raghavan Narasimhan*. Berlin: Teubner.

15 Joule, J. P. (1845). "On the Mechanical Equivalent of Heat," *Brit. Assoc. Rep.*, trans. Chemical Sect, p. 31.

16 Clausius, R. (1857). "Über die Art der Bewegung, die wir Wärme nennen," *Annalen der Physik*, 100: 353–79.

17 Maxwell, J. C. (1860). "Illustrations of the dynamical theory of gases. Part I. On the motions and collisions of perfectly elastic spheres," *Philosophical Magazine*, 4th series, 19: 19–32.

18 Boltzmann, L. (1871). "Über das Wärmegleichgewicht zwischen mehratomigen Gasmolekülen," *Wien. Ber.*, 63: 397–416.; Boltzmann, L. (1871). "Einige allgemeine Sätze über Wärmegleichgewicht," *Wien. Ber.*, vol. 63, pp. 676.

19 Boltzmann, L. (1871). "Einige allgemeine Sätze über Wärmegleichgewicht," *Wien. Ber.*, 63: 676; Brush, S. (2003). *The Kinetic Theory of Gases*. London: Imperial College Press.

20 Boltzmann, L. (1884). "Über die Eigenschaften Monocyklischer und andere damit verwandter Systeme," *Crelles Journal*, 98, 68–94.

21 Boltzmann, L. (1872). "Weitere Studien über das Wärmegleichgewicht unter Gasmolekülen," *Wien. Ber.*, 66: 275.

22 Loschmidt, J. (1876). "Über den Zustand des Wärmegleichgewichts eines Systems von Körpern mit Rücksicht auf die Schwerkraft," *Wien. Ber.*, 73: 135, 366.

23 Boltzmann, L. (1877). "Über die Beziehung zwischen dem zweiten Hauptsatze der mechanischen Wärmetheorie und der Wahrscheinlickeitsrechnung respektive den Sätzen über das Wärmegleichgewicht," *Wien. Ber.*, 76: 373.

24 Euler, L. (1764). Nov. Comm. Acad. Imp. Petropolitanae, vol. 10, pp. 207–42.

25 The fact that the published paper was not the original submission to the prize committee was apparently not widely known until the 1990s. Barrow-Green, L. (1997). *Poincaré and the three body problem*. London: London Mathematical Society.

26 Reproduced with permission from Nolte, D. D. (2015). *Introduction to Modern Dynamics: Chaos, Networks, Space and Time*. Oxford: Oxford University Press, p. 116.

27 Poincaré, H. and D. L. Goroff (1993). *New methods of celestial mechanics … Edited and introduced by Daniel L. Goroff*. New York: American Institute of Physics, Section 397.

28 Poincare, H. (1890). "Sur les equations de la dynamique et le probleme des trois corps." *Acta Mathematica*, 13, 1–270.

29 The reason why the dynamical space has 2p-1 dimensions is two for the position and velocity of each particle, minus one dimension overall because of the constraint of the conservation of energy.

30 Boltzmann, L. (1964). *Lectures on Gas Theory*. New York: Dover Publications:
31 Lutzen, J., p. 665.
32 $dU = TdS - pdV + \sum_n \mu_n dN_n.$
33 Gibbs, J. W. (1902). *Elementary Principles in Statistical Mechanics*. New York: C. Scribners & Sons.
34 Minkowski, H. (1909). "Raum und Zeit", *Physikalische Zeitschrift*, 10: 104–11.
35 P. Ehrenfest and T. Ehrenfest, (1911). "Begriffliche Grundlagen der statischen Auffassung in der Mechanik," in *Encyklopädie der mathematischen Wissenschaften*, vol. IV, Part 32. Leipzig: B. G. Teubner.
36 Hertz, P. (1910). "Uber die mechanischen Grundlagen der Thermodynamik." *Annalen der Physik*, 33, 225–74 and 537–52.
37 Rosenthal, A. (1913). "Proof of the Impossibility of Ergodic Systems," *Annalen der Physik*, 42, 796–806.
38 Plancheral, M. (1913). "Proof of the Impossibility of Ergodic Mechanical Systems," *Annalen der Physik*, 42, 1061–3.
39 Epstein, P. S. (1918). "Structure of the phase-space of a partially periodic system," *Abhandlungen der Preussischen Akademieder Wissenschaften*, 23: 435–46.

Chapter 7: The Lens of Gravity

1 Thorne, K. S. (1994). *Black holes and time warps: Einstein's outrageous legacy*. New York: W.W. Norton, p. 256.
2 Minkowski biography. (2008). *Complete Dictionary of Scientific Biography*. Detroit: Charles Scribner's Sons.
3 Poincaré, H. (1906). "Sur la dynamique de l'électron." *Rendiconti del circolo matematico di Palermo*, 21: 129–76.
4 Minkowski, H. (1909). "Raum und Zeit." *Jahresbericht der Deutschen Mathematikier-Vereinigung*, pp. 75–88.
5 Walter, S. (2008). "Minkowski's Modern World," in V. Petkov (ed.), *Minkowski Spacetime: A Hundred Years Later,* New York: Springer, p. 12.
6 Einstein, A. (1907). Jahrb. Rad. Elektr. vol. 4, 411.
7 Pais, A. (2005) *Subtle is the Lord: The Science and the Life of Albert Einstein*: Oxford: Oxford University Press, p. 178.
8 Pais, p. 178.
9 Reprinted with permission from Nolte, D. D. (2015). *Introduction to Modern Dynamics: Chaos, Networks, Space and Time.* (Oxford: Oxford University Press, p. 28.
10 Pais, p. 199.
11 Pais, p. 203.
12 Pais, p. 201.
13 Pais, p. 213.
14 Pais, p. 216.

15 Attributed to John Wheeler.
16 Dyson, F. W., A. S. Eddington and C. Davidson (1920). "A Determination of the Deflection of Light by the Sun's Gravitational Field, from Observations Made at the Total Eclipse of May 29, 1919." *Philosophical Transactions of the Royal Society of London. Series A, Containing Papers of a Mathematical or Physical Character*, **220**(291): 571–81.
17 Thorne, p. 212.
18 Thorne, p. 239.
19 Reprinted with permission from Nolte, D. D. (2015). *Introduction to Modern Dynamics: Chaos, Networks, Space and Time.* Oxford: Oxford University Press, p. 403.
20 Hawking, S. W. (1974). "Black Hole Explosions." *Nature*, **248**(5443): 30–1.

Chapter 8: On the Quantum Footpath

1 http://www-history.mcs.st-and.ac.uk/Biographies/Heisenberg.html.
2 An historical timeline of the events in this chapter can be found at the Berkeley Press website: https://works.bepress.com/ddnolte/.
3 Cassidy, D. C. (2010). *Beyond Uncertainty: Heisenberg, Quantum Physics, and The Bomb.* New York: Bellevue Literary Press, p. 89.
4 Cassidy, *Beyond Uncertainty*, p. 11.
5 Cassidy, *Beyond Uncertainty*, p. 89.
6 Cassidy, *Beyond Uncertainty*, p. 100.
7 Lindley, D. (2007). *Uncertainty: Einstein, Heisenberg, Bohr, and the struggle for the soul of science*. New York: Doubleday, p. 89.
8 Lindley, *Uncertainty*, p. 116.
9 Ehrenfest, P. (1917). "Adiabatic invariants and the theory of quanta." *Philosophical Magazine*, **33**(193–98): 500–13. An interesting correspondence between classical and quantum mechanics is encountered in the adiabatic invariants. If a classical quantity is an adiabatic invariant, then the associated quantum value is also an invariant, known as a quantum number. Quantum numbers are not altered by adiabatic variation of physical parameters, i.e., the adiabatic variation induces no transitions between stationary states. This correspondence was first demonstrated out by Paul Eherenfest in 1913, pursuing a suggestion made by Einstein at the 1911 Solvay Congress. The full theory of quantum adiabatic invariants was completed by Dirac, in 1925 just prior to the advent of Heisenberg's quantum mechanics.
10 Cassidy, *Beyond Uncertainty*, p. 119.
11 Lindley, Uncertainty, g. 140.
12 Born, M. and P. Jordan (1925). "The quantum theory of aperiodic processes." *Zeitschrift Fur Physik*, **33**: 479–505.

13 Born, M., W. Heisenberg and P. Jordan (1926). "Quantum mechanics II." *Zeitschrift Fur Physik*, **35**, (8/9): 557–615.

14 Dirac, P. A. M. (1926). "Quantum mechanics and a preliminary investigation of the hydrogen atom." *Proceedings of the Royal Society of London Series A*, **110**(755): 561–79.; Pauli, W. (1926). "The hydrogen spectrum from the view point of the new quantal mechanics." *Zeitschrift Fur Physik*, **36**(5): 336–63.

15 Baggott, J. E. (2011). *The quantum story: a history in 40 moments*. Oxford, New York: Oxford University Press, p. 64.

16 Schrodinger, E. (1926). "Quantisation as an eigen value problem." *Annalen Der Physik*, **79**(4): 361–8.

17 Born, M. (1926). "Quantum mechanics in impact processes." *Zeitschrift Fur Physik*, **38**(11/12): 803–40.

18 E. Schrodinger, (1926). "On the connection of Heisenberg-Born-Jordan's quantum mechanics with mine," *Annalen Der Physik*, 79: 734–56.

19 Cassidy, *Beyond Uncertainty*, p. 157.

20 Heisenberg, W. (1927). "'Über den anschaulichen Inhalt der quantentheoretischen Kinematik und Mechanik'." *Zeitschrift für Physik* **43**(3–4): 172–198.

21 Cassidy, *Beyond Uncertainty*, p. 169.

22 Cassidy, *Beyond Uncertainty*, p. 165.

23 Whitaker, A. (2006). *Einstein, Bohr, and the quantum dilemma: from quantum theory to quantum information*. New York: Cambridge University Press.

24 Einstein, A., B. Podolsky and N. Rosen (1935). "Can quantum-mechanical description of physical reality be considered complete?" *Physical Review*, **47**(10): 777–80.

25 Bohr, N. (1935). "Can quantum-mechanical description of physical reality be considered complete?" *Physical Review* **48**(8): 696–702.

26 von Neumann, J. (1932). *Mathematical Foundations of Quantum Mechanics*. Princeton: Princeton University Press.

27 Schrodinger, E. (1935). "The current situation in quantum mechanics." *Naturwissenschaften* **23**: 807–12.

28 Dirac, P. A. M. (1928). "The quantum theory of the electron." *Proceedings of the Royal Society of London Series A*, **117**(778): 610–24; Dirac, P. A. M. (1928). "The quantum theory of the electron - Part II." *Proceedings of the Royal Society of London Series A*, **118**(779): 351–61.

29 A famous Feynman story places him in Brazil playing percussion in a samba band in 1952. The vibration modes of the head of a drum make a direct analogy to quantum eigenfunctions on closed domains, although that is probably not why he played them.

30 Gleick, J. (1993). *Genius: The life and science of Richard Feynman*. New York, Vintage Books, p. 96.

31 Dirac, P. A. M. (1933). "The Lagrangian in Quantum Mechanics." *Physikalische Zeitschrift der Sowjetunion* **3**(1): 64–72.

32 Schweber, S. S. (1994). *QED and the Men Who Made it*. Princeton: Princeton University Press, p. 412.
33 Feynman, R. P. (1948). "Relativistic cut-off for quantum electrodynamics." *Physical Review*, **74**(10): 1430–8. 34 Feynman, R. P. (1949). "Space-time approach to quantum electrodynamics." *Physical Review*, **76**(6): 769–89.
35 Dyson, F. J. (1949). "The radiation theories of Tomonaga, Schwinger and Feynman." *Physical Review*, **75**(3): 486–502.
36 Feynman, R. P. (1949). "The theory of positrons." *Physical Review* **76**(6): 749–59.; Feynman, R. P. (1949). "Space-time approach to quantum electrodynamics" *Physical Review*, **76**(6): 769–89.
37 Kaiser, D., K. Ito and K. Hall (2004). "Spreading the tools of theory: Feynman diagrams in the USA, Japan, and the Soviet Union." *Social Studies of Science*, **34**(6): 879–922.

Chapter 9: From Butterflies to Hurricanes

1 Nolte, D. D. (2001). *Mind at light speed: A new kind of intelligence*. New York: Free Press.
2 An historical timeline of the events in this chapter can be found at the Berkeley Press website: https://works.bepress.com/ddnolte/.
3 Kolmogorov, A. N. (1954). "On conservation of conditionally periodic motions for a small change in Hamilton's function.," *Dokl. Akad. Nauk SSSR (N.S.)*, 98: 527–30.
4 All real positive numbers can be expressed as a continued fraction

$$\alpha = a_0 + \cfrac{1}{a_1 + \cfrac{1}{a_2 + \cfrac{1}{a_3 + \ldots}}} = [a_0; a_1, a_2, a_3, \ldots]$$

The irrational number with the slowest convergence is the golden mean

$$\phi = \frac{1+\sqrt{5}}{2} = [1; 1, 1, 1, 1, \ldots]$$

5 Siegel, C. L. (1942). "Iteration of Analytic Functions," *Ann. Math.*, 43: 607–12.
6 Arnold, V. I. (1997). "From superpositions to KAM theory," *Vladimir Igorevich Arnold. Selected*, 60: 727–40.
7 Action variables are constructed from action integrals (like Maupertuis and Euler's). Karl Schwarzschild showed that action-angle coordinates enabled the separation of the Hamilton-Jacobi equations. Schwarzschild's paper was published the same day that he died on the Eastern Front.
8 Moser, J. (1962). "On Invariant Curves of Area-Preserving Mappings of an Annulus.," *Nachr. Akad. Wiss. Göttingen Math.-Phys*, Kl. II, 1–20.
9 Arnold, V. I. (1963). "Small denominators and problems of the stability of motion in classical and celestial mechanics (in Russian)," *Usp. Mat. Nauk.*, 18:

91–192,; Arnold, V. I. (1964). "Instability of Dynamical Systems with Many Degrees of Freedom." *Doklady Akademii Nauk Sssr* **156**(1): 9.

10 Chirikov, B. V. (1969). *Research concerning the theory of nonlinear resonance and stochasticity*. Institute of Nuclear Physics, Novosibirsk. **4**.
The Standard Map

$$J_{n+1} = J_n + \varepsilon sin\theta_n$$
$$\theta_{n+1} = \theta_n + J_{n+1}$$

is plotted in Fig. 3.31 in Nolte, *Introduction to Modern Dynamics* (2015) on p. 139. For small perturbation ε, two fixed points appear along the line J = 0 corresponding to p/q = 1: one is an elliptical point (with surrounding small orbits) and the other is a hyperbolic point where chaotic behavior is first observed. With increasing perturbation, q elliptical points and q hyperbolic points emerge for orbits with winding numbers p/q with small denominators (1/2, 1/3, 2/3 etc.). Other orbits with larger q are warped by the increasing perturbation but are not chaotic. These orbits reside on invariant tori, known as the KAM tori, that do not disintegrate into chaos at small perturbation. The set of KAM tori is a Cantor-like set with non-zero measure, ensuring that stable behavior can survive in the presence of perturbations, such as perturbation of the Earth's orbit around the Sun by Jupiter. However, with increasing perturbation, orbits with successively larger values of q disintegrate into chaos. The last orbits to survive in the Standard Map are the golden mean orbits with p/q = ϕ–1 and p/q = 2–ϕ. The critical value of the perturbation required for the golden mean orbits to disintegrate into chaos is surprisingly large at $\varepsilon_c = 0.97$.

11 Several years later Smale won the Field Medal for proving Poincarés conjecture on the topology of spheres in at least five dimensions.

12 Structurally stable systems retain their global properties in the presence of small perturbations. Structural stability applies to the global phase-space pattern and not to individual trajectories. Structural stability was investigated by Andronov and Pontryagin in 1937. Andronov, A. and L. Pontryagin (1937). "Coarse systems." *Comptes Rendus De L Academie Des Sciences De L Urss*, **14**: 247–50.

13 Levinson, N. (1949). "A 2nd order Differential Equation with Singular Solutions." *Annals of Mathematics*, **50**(1): 127–53.

14 Cartwright, M. L. and J. E. Littlewood (1945). "On the non-linear differential equation of the second order. I. The equation y′′ − k(1 − y^2)y′ + y = bλk cos(λt + a), k large." *Journal of the London Mathematical Society* **20**: 180–9. Discussed in Aubin, D. and A. D. Dalmedico (2002). "Writing the History of Dynamical Systems and Chaos: Longue DurÈe and Revolution, Disciplines and Cultures." *Historia Mathematica*, **29**: 273.

15 Emanuel, K. (2011). *Edward Norton Lorenz*. National Academy of Sciences, National Academy Press.

16 Birkhoff, G. D. (1913). "Proof of Poincare's geometric theorem." *Transactions of the American Mathematical Society*, **14**(1–4): 14–22.
17 The first MacIntosh computer, sold in 1984, used a 32-bit processor operating at 6 MHz and had 128 kB of memory.
18 Lorenz, E. N. (1993). *The essence of chaos.* Seattle: University of Washington Press, p. 136.
19 Lorenz, E. N. (1963). "Deterministic Nonperiodic Flow." *Journal of the Atmospheric Sciences* **20**(2): 130–41.
20 Reproduced with permission from D. D. Nolte, (2015). *Introduction to Modern Dynamics: Chaos, Networks, Space and Time* (Oxford: Oxford University Press, p. 129.
21 Widom B. (1965). "Equation of state in the neighborhood of the critical point." *J. Chem Phys*, 43: 3898–905.
22 Kadanoff L. (1966). Scaling laws for Ising models near Tc. *Physics*, 2: 263.
23 Wilson K. G. (1971). Renormalization group and critical phenomena. I. Renormalization group and the Kadanoff scaling picture. *Phys Rev B*, 4: 3174–83; Wilson K. G. (1971). Renormalization group and critical phenomena. II. Phase-space cell analysis of critical behavior. *Phys Rev B*, 4: 3184–205.
24 Ruelle, D. and F. Takens (1971). "On the Nature of Turbulence." *Communications in Mathematical Physics* **20**(3): 167–92.
25 Gollub, J. P. and H. L. Swinney (1975). "Onset of Turbulence in a Rotating Fluid." *Physical Review Letters*, **35**(14): 927–30.
26 May, R. M. (1976). "Simple Mathematical-Models with very complicated Dynamics." *Nature*, **261**(5560): 459–67.
27 Reproduced with permission from D. D. Nolte, (2015). *Introduction to Modern Dynamics: Chaos, Networks, Space and Time* (Oxford: Oxford University Press, p. 120.
28 Reproduced with permission from D. D. Nolte, (2015). *Introduction to Modern Dynamics: Chaos, Networks, Space and Time.* Oxford: Oxford University Press, p. 121.
29 Gleick, J. (1987). *Chaos: Making a New Science*, New York: Viking. p. 180.
30 Gleick, p. 183.
31 Aleksandr Lyapunov (1857–1918) and George David Birkhoff (1884–1944) were mathematicians who contributed substantial theorems to dynamical theory.

Chapter 10: Darwin in the Clockworks

1 Darwin, Charles. (2008). *Voyage of the Beagle*, Aukland: The Floating Press, p. 9.
2 Thomson, K. (2009). *The Young Charles Darwin: Influences and Ideas.* New Haven, US, Yale University Press., p. 180.
3 Thomson, p. 214.
4 Gale (2008). *Charles Darwin, Complete Dictionary of Scientific Biography*, p. 571.

5 Thomson, p. 238.
6 Darwin, *Origin.* pp. 154–6.
7 The golden mean has the slowest convergence, among the real numbers, for its continued fraction representation, making it "farthest" away from any rational number.
8 A recent analysis of Leonardo's travels, and the unique applicability of the Fibonacci sequence to the ancestry of bees, has opened the possibility that the rabbits were surrogates for the bees of Bejaia. T. Scot, *Mactutor History of Mathematics* (2014) [online].
9 Bacaer, N. (2011). *A Short History of Mathematical Population Dynamics*, Springer, Bacaer, p. 37.
10 Bacaer, p. 71.
11 Bacaer, p. 74.
12 The American mathematician John Nash (1928–2015), the winner of the 1994 Nobel Prize in economics and the subject of the book and movie *A Beautiful Mind*, wrote several papers in the early 1950s that extended von Neumann's results for zero-sum games to the situation for non-zero-sum games.
13 Wright, S. (1932). "The roles of mutation, inbreeding, crossbreeding and selection in evolution." *Proc Sixth Internat Congr Genetics.* Ithaca New York **1**: 356–66.
14 Reprinted with permission from D. D. Nolte, (2015). *Introduction to Modern Dynamics: Chaos, Networks, Space and Time.* Oxford: Oxford University Press, p. 257.
15 Kauffman, S. (1993). *The Origins of Order: Self-Organization and Selection in Evolution.* Oxford: Oxford University Press.
16 Wright, S. "Evolution in Mendelian Populations." *Genetics*, **16**(1931): 97–159.
17 pg. 164, Morgan, G. J. (1998). "Emile Zuckerkandl, Linus Pauling, and the molecular evolutionary clock, 1959–1965." *Journal of the History of Biology* **31**(2): 155–78.
18 Zuckerkandl, E. and L. B. Pauling (1962). "Molecular disease, evolution, and genic heterogeneity," in M. Kasha and B. Pullman eds. *Horizons in Biochemistry.* New York: Academic Press: 189–225.
19 Kimura, M. (1968). "Evolutionary rate at molecular level." *Nature*, **217**(5129): 624.
20 Gavrilets, S. (1997). "Evolution and speciation on holey adaptive landscapes." *Trends in Ecology & Evolution*, **12**(8): 307–12.
21 von Neumann, J. (1932). *Mathematical Foundations of Quantum Mechanics.* Princeton: Princeton University Press.
22 von Neumann, J. (1928). "The theory of parlour games." *Mathematische Annalen* **100**: 295–320.

23 R. C. Lewontin suggested that game theory could provide a useful tool for evolution in a 1961 paper. Lewontin, R. C. (1961). "Evolution and the theory of games." *Journal of Theoretical Biology*, **1**(3): 382–403.
24 Taylor, P. D. and L. B. Jonker (1978). "Evolutionarily stable strategies and game dynamics." *Mathematical Biosciences*, **40**(1–2): 145–56.
25 Schuster, P. and K. Sigmund (1983). "Replicator dynamics." *Journal of Theoretical Biology*, **100**(3): 533–8.
26 Page, K. M. and M. A. Nowak (2002). "Unifying evolutionary dynamics." *Journal of Theoretical Biology*, **219**(1): 93–8.
27 Eigen, M. and P. Schuster (1977). "Hypercycle - principle of natural self-organization. Emergence of hypercycle." *Naturwissenschaften*, **64**(11): 541–65.
28 Mustonen, V. and M. Laessig (2009). "From fitness landscapes to seascapes: non-equilibrium dynamics of selection and adaptation." *Trends in Genetics*, **25**(3): 111–19.

Chapter 11: The Measure of Life

1 Winfree, A. T. (1967). "Biological rythms and behavior of populations of coupled oscillators." *Journal of Theoretical Biology* **16**(1): 15–42.
2 Strogatz, S. H. (2015). *SYNC: how order emerges from chaos in the universe, nature, and daily life*. New York: Hachette Books. p. 31.
3 Bedini, S. A. (1991). *The pulse of time: Galileo Galilei, the determination of longitude, and the pendulum clock*. [Firenze]: L.S. Olschki.
4 Pikovsky, A. S., M. G. Rosenblum and J. Kurths (2003). *Synchronization: A Universal concept in nonlinear science*. Cambridge: Cambridge University Press. p. 3.
5 J. W. Strutt, Baron Rayleigh, On maintained vibrations, *Philos. Mag*., (ser. 5) **15**(1883): 229–35.
6 Jenkins, A. (2013). "Self-oscillation." *Physics Reports-Review Section of Physics* Letters, **525**(2): 167–222.
7 Rayleigh, L. (1907). "Acoustical notes. - VII." *Philosophical Magazine*, **13**(73–78): 316–33.
8 van der Pol, B. (1926). "On 'relaxation oscillations.'." *Philosophical Magazine*, **2**(11): 978–92.
9 van der Pol's equation is a variation on the equation Rayleigh published in 1883.
10 Ginoux, J. M. and C. Letellier (2012). "Van der Pol and the history of relaxation oscillations: Toward the emergence of a concept." Chaos **22**(2): 023120–1 to 023120–15.
11 van der Pol, B. (1927). "Forced oscillations in a circuit with non-linear resistance. (Reception with reactive triode.)." *Philosophical Magazine*, **3**(13): 65–80.

12 Reproduced with permission from Oxford University Press. D. D. (2015). Nolte, *Introduction to Modern Dynamics: Chaos, Networks, Space and Time.* Oxford: Oxford University Press p. 110.
13 This same type of frequency pulling occurs in externally pumped lasers and is called mode pulling.
14 Cartwright, M. L. and J. E. Littlewood (1945). "On the non-linear differential equation of the second order. I.." *Journal of the London Mathematical Society*, **20**: 180–9; Aubin, D. and A. D. Dalmedico (2002). "Writing the History of Dynamical Systems and Chaos: Longue DurÈe and Revolution, Disciplines and Cultures." *Historia Mathematica*, **29**: 273.
15 Ginoux, J. M. and C. Letellier (2012). "Van der Pol and the history of relaxation oscillations: Toward the emergence of a concept." *Chaos*, **22**(2) 023120-1–15.
16 Winfree, A. T. (1984). "The prehistory of the Belousov-Zhabotinsky oscillator." *Journal of Chemical Education*, **61**(8): 661–3.
17 Winfree, A. T. (1967). "Biological rythms and behavior of populations of coupled oscillators." *Journal of Theoretical Biology*, **16**(1): 15–42.
18 Van der Pol and J. Van der Mark (1928). *L'Onde E'lectrique* 7, 365.
19 van der Pol, B. and J. van der Mark (1928). "The heartbeat considered as a relaxation oscilation, and an electrical model of the heart." *Philosophical Magazine*, **6**(38): 763–75; Van Der Pol, B. (1940). "Biological rhythms considered as relaxation oscillations." *Acta Medica Scandinavica*, 76–88.
20 Winfree, A. T. (1977). "Phase-control of neural pacemakers." *Science* **197**(4305): 761–3.
21 Mackey, M. C. and L. Glass (1977). "Oscillation and chaos in physiological control-systems." *Science*, **197**(4300): 287–8.
22 Guevara, M. R., L. Glass and A. Shrier (1981). "Phase locking, period-doubling bifurcations, and irregular dynamics in periodically stimulated cardiac-cells." *Science* **214**(4527): 1350–3.
23 Schwiening, C. J. (2012). "A brief historical perspective: Hodgkin and Huxley." *Journal of Physiology-London*, **590**(11): 2571–5.
24 Hodgkin, A. L. and A. F. Huxley (1952). "A quantitative description of membrane current and its applications to conduction and excitation in nerve." *Journal of Physiology-London*, **117**(4): 500–44. This paper has over 10 thousand citations on Web of Science.
25 Turing, A. M. (1937). "On computable numbers, with an application to the Entscheidungsproblem." *Proceedings of the London Mathematical Society*, **42**: 230–65.
26 Piccinini, G. (2004). "The first computational theory of mind and brain: A close look at McCulloch and Pitts's 'logical calculus of ideas immanent in nervous activity'." *Synthese*, **141**(2): 175–215.

27 McCulloch, W. S. and W. Pitts (1943). "A logical calculus of the ideas immanent in nervous activity." *Bull Math Biophys*, **5**(4): 115–33.

28 von Neumann, J. (1951). "The General and Logical Theory of Automata," in *Cerebral Mechanisms in Behavior*. L. A. Jeffress, ed. New York: Wiley, pp. 1–41.

29 Rosenblatt, F. (1958). "The perceptron – a probabilistic model for information-storage and organization in the brain." *Psychological Review*, **65**(6): 386–408.

30 Olazaran, M. (1996). "A sociological study of the official history of the perceptrons controversy." *Social Studies of Science*, **26**(3): 611–59.

31 Al Overhauser was a colleague of mine in the Physics Department at Purdue University. He is most famous for discovering the Overhauser Effect that plays an important role in MRI.

32 Hopfield, J. J. (1982). "Neural networks and physical systems with emergent collective computational abilities." *Proceedings of the National Academy of Sciences of the United States of America-Biological Sciences*, **79**(8): 2554–8.

33 Hopkins, B. and R. Wilson (2007). "The Truth about Konigsberg", in R. E. Bradley and C. E. Sandifer eds., *Leonhard Euler: Life, Work and Legacy*.409–20.

34 Euler, L. 1736.

35 Watts, D. J. and S. H. Strogatz (1998). "Collective dynamics of 'small-world' networks." *Nature* **393**(6684): 440–2.; Goh, K. I., M. E. Cusick, D. Valle, B. Childs, M. Vidal and A. L. Barabasi (2007). "The human disease network." *Proceedings of the National Academy of Sciences of the United States of America*, **104**(21): 8685–90.

36 Broadbent, S. R. and J. M. Hammersley (2008). "Percolation processes: I. Crystals and mazes." *Mathematical Proceedings of the Cambridge Philosophical Society*, **53**(3): 629–41.

37 There is no analogous "spanning" direction for such a giant cluster because random networks are dimensionless and have no "sides."

38 Strogatz, Strogatz, S. H. (2015). *SYNC: how order emerges from chaos in the universe, nature, and daily life*. New York: Hachette Books, p. 248.

39 Barabasi, A. L. (2016). *Network Science*. Cambridge: Cambridge University Press.

40 Barabasi, A. L., N. Gulbahce and J. Loscalzo (2011). "Network medicine: a network-based approach to human disease." *Nature Reviews Genetics*, **12**(1): 56–68.

Bibliography

Endnotes for each chapter contain specific citations to published works and page numbers. This Bibliography contains the general sources that were used for background material. Extensive use was made of biographical sources, such as

Complete Dictionary of Scientific Biography. Detroit, Charles Scribner's & Sons. (2008)

Helpful online sites were

MacTutor History of Mathematics, http://www-history.mcs.st-andrews.ac.uk/index.html

Gale Virtual Reference Library, Gale Group, Thomson Learning
Detroit, MI: 2002-

The Galileo Project, http://galileo.rice.edu/chron/galileo.html

The Stanford Encyclopedia of Philosopy, https://plato.stanford.edu/

Wikipedia, https://en.wikipedia.org/wiki/Main_Page

Crowd-sourced online biographical information, such as from *Wikipedia*, was checked for accuracy by corroborating information with alternate sources.

Published Sources:

Abraham, R. and Y. Ueda (2000). *The chaos avant-garde: Memories of the early days of chaos theory*. River Edge, NJ: World Scientific.

Acebron, J. A., L. L. Bonilla, C. J. P. Vicente, F. Ritort and R. Spigler (2005). "The Kuramoto model: A simple paradigm for synchronization phenomena." *Reviews of Modern Physics* 77(1): 137–85.

Andronov, A.A., A. A. Vitt and S.E. Khaikin (1937). *Theory of Oscillators* (tr. F. Immerzi from the Russian, first edition Moscow). London: Pergamon, 1966.

Arnold, V. I. (1989). *Mathematical methods of classical mechanics*. New York: Springer-Verlag.

Aubin A. and Dahan Dalmedico, D. (2002). "Writing the History of Dynamical Systems and Chaos: Longue Durée and Revolution, Disciplines and Cultures." *Historia Mathematica*, 29: 273–339.

Bacaer, N. (2011). *A Short History of Mathematical Population Dynamics*. Springer.

Baggott, J. E. (2011) *The quantum story: A history in 40 moments*. New York: Oxford University Press.

Barabási, A.-L. s. and M. r. Pósfai (2016). *Network science*. Cambridge: Cambridge University Press.

Barrow-Green, J. (1997). *Poincaré and the three body problem*. London: London Mathematical Society.

Bernkopf, M. (1966). "The development of function spaces with particular reference to their origins in integral equation theory." *Archive for History of Exact Sciences*, vol. 3, pp. 1–96.

Birkhoff, G. D. (1927). *Dynamical Systems* (reprinted with an introduction by J. Moser and a preface by M. Morse, 1966).Providence, RI: American Mathematical Society.

Bocaletti, D. (2016). *Galileo and the Equations of Motion*, New York: Springer.

Boudri, J. C. (2002). *What was Mechanical about Mechanics: The Concept of Force between Metaphysics and Mechanics from Newton to Lagrange* vol. 224. Dordrecht: Kluwer Academic Publishers.

Boyer, C. B. (1985). *A History of Mathematics*. Princeton, NJ: Princeton University Press.

Brown, Laurie M. (2005). *Feynman's Thesis-a New Approach to Quantum Theory*. River Edge, NJ: World Scientific Publishing Co.

Brush, S. (2003). *The Kinetic Theory of Gases*. London: Imperial College Press.

Brush, S. G. (1967). "Foundations of statistical mechanics 1845-1915." *Archive for History of Exact Sciences*, vol. 4, p. 145.

Brush, S. G. (1975). *The kind of motion we call heat: A history of the kinetic theory of gases in the 19th century*. Amsterdam: Elsevier.

Calcott, B. (2008). "Assessing the fitness landscape revolution." *Biology & Philosophy*, **23**(5): 639–57.

Cassidy, D. C. (1992). *Uncertainty: The life and science of Werner Heisenberg*. New York: W. H. Freeman.

Cassidy, David C. (2010). *Beyond Uncertainty: Heisenberg, Quantum Physics, and The Bomb*. New York, Bellevue Literary Press.

Cercignani, C. (1998). *Ludwig Boltzmann: the man who trusted atoms*. Oxford: Oxford University Press.

Chabert, J. L. (1990). "Half a Century of Fractals 1870-1920." *Historia Mathematica*, vol. 17, pp. 339–65.

Christianson, J. R. (2003). *On Tycho's island: Tycho Brahe, science, and culture in the sixteenth century*. New York: Cambridge University Press.

Clagett, M. (1959). *The science of mechanics in the Middle Ages*. Madison: University of Wisconsin Press.

Clagett, Marshall (1968). *Nicole Oresme and the Medieval Geometry of Qualities and Motions; a treatise on the uniformity and difformity of intensities known as Tractatus de configurationibus qualitatum et motuum*. Madison: University of Wisconsin Press.

Coopersmith, J. (2010). *Energy, the Subtle Concept: The Discovery of Feynman's Blocks from Leibniz to Einstein*. Oxford: Oxford University Press.

Crowe, M. J. (1967). *A history of vector analysis: the evolution of the idea of a vectorial system*. Notre Dame, IN: University of Notre Dame Press.

Crowe, M. J. (2007). *Mechanics from Aristotle to Einstein*. Santa Fe, NM: Green Lion Press.

D'Abro, A. (1951). *Rise of the new physics: its mathematical and physical theories* (formerly titled "Decline of mechanism"). New York: Dover.

Darwin, C. (2008). *Voyage of the Beagle*. Auckland, NZ: The Floating Press.

Darwin, Charles. (2007). On the Origin of Species, edited by Gillian Beer, Oxford: Oxford University Press.

Darwin, Charles Robert. (2008). *Complete Dictionary of Scientific Biography*. Detroit: Charles Scribner's Sons, Vol. 3, 565–77.

Dauben, J. W. (1979). *Georg Cantor: His Mathematics and Philosophy of the Infinite*. Cambridge, MA: Cambridge University Press.

Dawkins, R. and D. C. Dennett (1999). *The extended phenotype: The long reach of the gene.* Oxford: Oxford University Press.

Descartes, R., D. E. Smith and M. L. Latham (1925). *The geometry of René Descartes*. Chicago: Open Court Pub. Co.

Devreese, J. T. and en Vanden Berghe, G. (2007). *Magic is no magic. The wonderful World of Simon Stevin 1548–1620*. Southampton: WIT Press.

Diacu, F. and P. Holmes (1996). *Celestial encounters: The origins of chaos and stability*. Princeton, NJ: Princeton University Press.

Drake, S. (1989). *History of free fall: Aristotle to Galileo*. Toronto: Wall & Thompson.

Dugas, R. (1988). *A history of mechanics*. New York: Dover Publications.

Dumas, H. S. (2014). *The KAM Story: A friendly introduction to the content, history and significance of Classical Kolmogorov-Arnold-Moser Theory*, River Edge, NJ: World Scientific.

Edgar, G. A. (2004). *Classics on fractals*. Boulder, CO: Westview Press.

Ehrenfest, P. and T. Ehrenfest (1911). "Begriffliche Grundlagen der statischen Auffassung in der Mechanik," in *Encyklopädie der mathematischen Wissenschaften*. Leipzig: B. G. Teubner. IV, Part 32.

Eigen, M. and P. Schuster (1979). *The Hypercycle*. NewYork: Springer.

Einstein, A. (2005). *Einstein's Annalen papers: the complete collection 1901–1922*. Weinheim, Great Britain: Wiley-VCH.

Ekelund, R. B. and R. F. Hebert (2014). *A History of Economic Theory and Method*. Long Grove, IL: Waveland Press.

Euler, L. (1736). Mechanica sive motus scientia analytice exposita.

Euler, L. (1744). Methodus inveniendi lineas curvas maximi minimive proprietate gaudentes, sive solutio problematis isoperimetrici lattissimo sensu accepti.

Fisher, R. A. (1930). *The Genetical Theory of Natural Selection*. Oxford: The Clarendon Press.

Freely, J. (2014). *Before Galileo: the birth of modern science in medieval Europe*. London: Duckworth.

Galileo and S. Drake (1957). *Discoveries and opinions of Galileo*. New York: Doubleday.

Galilei, G. and S. Drake (1974). *Two new sciences: including centers of gravity & force of percussion*. Madison, WI: University of Wisconsin Press.

Galilei, G. and A. Van Helden (2015). *Sidereus nuncius or, The sidereal messenger*. Chicago: The University of Chicago Press.

Gamow, G. (1966). *Thirty years that shook physics: The story of quantum theory*. Garden City, NY: Anchor Books.

Gingerich, O. (1993). *The Eye of Heaven: Ptolemy, Copernicus, Kepler*. New York: American Institute of Physics.

Gingerich, O. (2010). *The book nobody read: Chasing the revolutions of Nicolaus Copernicus*. New York: Bloomsbury.

Ginoux, J. M. and C. Letellier (2012). "Van der Pol and the history of relaxation oscillations: Toward the emergence of a concept." *Chaos* **22**(2): 023120–1–15.

Glass, L. (2015). "Dynamical disease: Challenges for nonlinear dynamics and medicine." *Chaos* **25**(9) 097603–1–11.

Glass, L. and M. C. Mackey (1988). *From Clocks to Chaos*. Princeton: Princeton University Press.

Gleick, J. (1987). *Chaos: Making a New Science*. New York: Viking.

Gleick, J. (1993). *Genius: The life and science of Richard Feynman*. New York: Vintage Books.

Goldstine, Herman H. (1972). *The Computer from Pascal to von Neumann*. Princeton: Princeton University Press.

Gravilet, S. (2004). *Fitness Landscapes and the Origins of Species*. Princeton: Princeton University Press.

Gray, J. (1989). *Ideas of Space: Euclidean, Non-Euclidean and Relativistic*. Oxford: Oxford University Press.

Gray, J. (1999). *The symbolic universe: geometry and physics 1890–1930*. Oxford, New York: Oxford University Press.

Gutzwiller, M. (1990). *Chaos in classical and quantum mechanics* (Interdisciplinary applied mathematics, v. 1). New York: Springer-Verlag.

Gutzwiller, M. (1998). "Moon-Earth-Sun: The oldest three-body problem," in *Reviews of Modern Physics*, vol. 70, No. 2

Haldane, J. B. S. (1932). *The Causes of Evolution*. New York: Longmans.

Heidegger, M. (1991). *The principle of reason*. Bloomington, Indiana University Press.

Heilbron, J. (2010). *Galileo*. Oxford: Oxford University Press.

Hilborn, R. (2001). *Chaos and Nonlinear Dynamics: An Introduction for Scientists and Engineers*. Oxford: Oxford University Press.

Hill, D. K. (1988). "Dissecting Trajectories – Galileo Early Experiments on Projectile Motion and the Law of Fall." *Isis* **79**(299): 646–68.

Hoffman, P. (1998). *The man who loved only numbers: the story of Paul Erdos and the search for mathematical truth*. New York: Hyperion.

Hopkins, B. and R. Wilson (2007). The Truth about Konigsberg, in *Leonhard Euler: Life, Work and Legacy*. Ed. R. E. Bradley and C. E. Sandifer. Oxford: Elsevier Science & Technology pp. 409–20.

Huggett, N. (1999). *Space from Zeno to Einstein: classic readings with a contemporary commentary*. Cambridge, MA: MIT Press.

Huxley, J. (1942). *Evolution: The Modern Synthesis*. London: Allen & Unwin.

Izhikevich, E. M. (2007). *Dynamical Systems in Neuroscience: The Geometry of Excitability and Bursting*. Cambridge: MIT University Press.

Jammer, M. (1989). *The conceptual development of quantum mechanics*. Woodbury, NY: American Institute of Physics.

Jenkins, A. (2013). "Self-oscillation." *Physics Reports-Review Section of Physics Letters* **525**(2): 167–222.

Kaiser, D. (2005). *Drawing Theories Apart: The Dispersion of Feynman Diagrams in Postwar Physics*. Chicago IL: University of Chicago Press.

Kauffman, S. (1993). *The Origins of Order: Self-Organization and Selection in Evolution*. Oxford: Oxford University Press.

Kimura, M. (1968). *The Neutral Theory of Molecular Evolution*. Cambridge: Cambridge University Press.

Klein, M. J. (1982). *Paul Ehrenfest: The Making of a Theoretical Physicist*. Amsterdam, North-Holland: Springer.

Klima, Gyula (2008). *John Buridan*. New York: Oxford University Press.

Kuhn, H. W. and Tucker, A. W. (1958). "John von Neumann's work in the theory of games and mathematical economics." *Bull. Amer. Math. Soc.* 64 (Part 2) (3): 100–22.

Kurz, H. (2013). *Economic Thought: A Brief History*. New York: Columbia University Press.

Lacki, J. A. N. (2000). "The Early Axiomatizations of Quantum Mechanics: Jordan, von Neumann and the Continuation of Hilbert's Program," *Archive for History of Exact Sciences*, vol. 54, pp. 279–318.

Lagrange, J. L. (1760–1761). "Application de la méthode exposée dans le mémoire précédent a la solution de différents problemes de dynamique," *Miscellanea Teurinensia*.

Lagrange, J. L. (1760–1761). "Essai d'une nouvelle méthod pour dEeterminer les maxima et lest minima des formules intégrales indéfinies," *Miscellanea Teurinensia*.

Lagrange, J. L., A. Boissonnade, and V. N. Vagliente (1997). *Analytical mechanics. Translated from the "Mecanique analytique, nouvelle Èdition" of 1811. Translated and edited by Auguste Boissonnade and Victor N. Vagliente*. Dordrecht: Kluwer Academic.

Lanczos, C. (1970). *Space through the ages: The evolution of geometrical ideas from Pythagoras to Hilbert and Einstein*. New York: Academic Press.

Lattis, J. M. (2010). *Between Copernicus and Galileo: Christoph Clavius and the Collapse of Ptolemaic Cosmology*. Chicago, US: University of Chicago Press.

Lindley, D. (2001). *Boltzmann's atom: The great debate that launched a revolution in physics*. New York: Free Press.

Lindley, D. (2007). *Uncertainty: Einstein, Heisenberg, Bohr, and the struggle for the soul of science*. New York: Doubleday.

Lorenz, E. N. (1993). *The essence of chaos*. Seattle: University of Washington Press.

Lutzen, J. (1990). *Joseph Liouville 1809–1882: Master of Pure and Applied Mathematics*, New York: Springer-Verlag.

Malthus, T. R. (1798). *An Essay on the Principle of Population*.

May, R. M. (1995). "Necessity and chance- deterministic chaos in ecology and evolution." *Bulletin of the American Mathematical Society* **32**(3): 291–308.

Maynard Smith, J. and E. Szathmary (1997). *The Major Transitions in Evolution*. Oxford: Oxford University Press.

Medio, A. (1992). *Chaotic Dynamics: Theory and Applications to Economics*. New York: Cambridge University Press.

Minsky, M. and S. Papert (1988). *Perceptrons: an introduction to computational geometry*. Cambridge, MA: MIT Press.

Mirollo, R. E. and S. H. Strogatz (1990). "Synchronization of Pulse-Coupled Biological Oscillators." *Siam Journal on Applied Mathematics* **50**(6): 1645–62.

Moore, G. H. (1995) "The Axiomatization of Linear Algegra: 1875-1940," *Historia Mathematica*, vol. 22, pp. 262–303.

Morgan, G. J. (1998). "Emile Zuckerkandl, Linus Pauling, and the molecular evolutionary clock, 1959-1965." *Journal of the History of Biology* **31**(2): 155–78.

Naess, A. (2005). *Galileo Galilei: When the World Stood Still*. New York: Springer.

Naylor, R. H. (1976). "Galileo – Search for Parabolic Trajectory." *Annals of Science* **33**(2): 153–72.

Naylor, R. H. (1980). "Galileo Theory of Projectile Motion." *Isis* **71**(259): 550–70.

Newman, M. E. J. (2010). *Networks: An Introduction*. Oxford: Oxford University Press.

Newton, Roger G. (2009). *How Physics Confronts Reality: Einstein Was Correct, but Bohr Won the Game*. Singapore: World Scientific Publishing Co.

Nolte, D. D. (2001). *Mind at light speed: a new kind of intelligence*. New York: Free Press.

Nolte, D. D. (2015). *Introduction to Modern Dynamics: Chaos, Networks, Space and Time*. Oxford: Oxford University Press.

Nowak, M. A. (2006). *Evolutionary Dynamics: Exploring the Equations of Life*. Cambridge, MA: Harvard University Press.

Oliveira, A. R. E. (2014). *A history of the work concept: From physics to economics*. Dordrecht: Springer.

Pais, A. (2005). *Subtle is the Lord: The Science and the Life of Albert Einstein*: Oxford: Oxford University Press.

Pesic, P. (2014). *Music and the Making of Modern Science*, Cambridge, MA: MIT Press.

Peterson, I. (1993). *Newton's Clockworks: Chaos in the Solar System*. London: Macmillan.

Piccinini, G. (2004). "The first computational theory of mind and brain: A close look at McCulloch and Pitts's 'logical calculus of ideas immanent in nervous activity'." *Synthese* **141**(2): 175–215.

Pigliucci, M. (2008). “Sewall Wright’s adaptive landscapes: 1932 vs. 1988.” *Biology & Philosophy* **23**(5): 591–603.

Pikovsky, A. S., M. G. Rosenblum and J. Kurths (2003). *Synchronization: A Universal concept in nonlinear science*. Cambridge: Cambridge University Press.

Poincaré, H. and D. L. Goroff (1993). *New methods of celestial mechanics* . . . Edited and introduced by Daniel L. Goroff. New York, American Institute of Physics.

Provine, William B. (1986). Sewall Wright and Evolutionary Biology. Chicago: University of Chicago Press.

Rayleigh, J. W. S., Baron (1877). *The Theory of Sound*. London: Macmillan and Co.

Rowe, D. E., J. McCleary and E. Knobloch (1989). *The history of modern mathematics: Proceedings of the Symposium on the History of Modern Mathematics, Vassar College, Poughkeepsie, New York, June 20–24, 1989*. Boston: Academic Press.

Ruelle, D. (1991). *Chance and Chaos*, Princeton, NJ: Princeton University Press.

Sandelin, B., H.-M. Trautwein and R. Wundrak (2014). *A Short History of Economic Thought*. Florence, GB: Routledge.

Sanz, A. S. (2012). *A trajectory description of quantum processes*. Heidelberg: Springer.

Schweber, S. S. (1994). *QED and the men who made it: Dyson, Feynman, Schwinger, and Tomonaga*. Princeton: Princeton University Press.

Schwiening, C. J. (2012). “A brief historical perspective: Hodgkin and Huxley.” *Journal of Physiology-London* **590**(11): 2571–5.

Segrè, E. (1980) *From x-rays to quarks: modern physicists and their discoveries*. San Francisco: W. H. Freeman.

Shone, R. (2001). *An Introduction to Economic Dynamics*. Cambridge: Cambridge University Press.

Sklar, L. (2013). *From virtual work to Lagrange’s equation*. Cambridge: Cambridge University Press.

Smale, S. (1980). *The Mathematics of Time*. New York: Springer.

Sobel, D. (1999). Galileo’s daughter: a historical memoir of science, faith, and love. New York: Walker & Co.

Straffin, P. D. (1993). *Game Theory and Strategy*, MAA New Mathematical Library.

Strogatz, S. H. (1994). *Nonlinear Dynamics and Chaos*, Boulder, CO: Westview Press.

Strogatz, S. H. (2015). *SYNC: How order emerges from chaos in the universe, nature, and daily life*. New York: Hachette Books.

Strogatz, S. H. and I. Stewart (1993). “Coupled oscillators and biological synchronization.” *Scientific American* **269**(6): 102–9.

Struik, D. J. (1981). *The land of Stevin and Huygens: a sketch of science and technology in the Dutch Republic during the Golden Century*. Boston: D. Reidel Pub. Co.

Suisky, D. (2009). *Euler as Physicist*, New York: Springer.

Tazzioli, R. (1994). Rudolf Lipschitz’s Work on Differential Geometry and Mechanics, in *The History of Modern mathematics*. D. E. Rowe, ed., Academic Press. **3:** 113–38.

Terrall, M. (2002). *The Man Who Flattened the Earth: Maupertuis and the Sciences in the Enlightenment*: Chicago: University of Chicago Press.

Thomson, K. (2009). *The Young Charles Darwin: Influences and Ideas*. New Haven: Yale University Press.

Thorne, K. S. (1994). *Black holes and time warps: Einstein's outrageous legacy*. New York: W.W. Norton.

von Neumann, J. (1928). "The theory of parlour games." *Mathematische Annalen* **100**: 295–320.

von Neumann, J. (1932). *Mathematical Foundations of Quantum Mechanics*, Princeton: Princeton University Press.

von Neumann, J. and O. Morgenstern (2007). *Princeton Classic Editions: Theory of Games and Economic Behavior*. Princeton: Princeton University Press.

Wheeler, J. A. (1994). *At home in the universe*. Woodbury, NY: American Institute of Physics.

Whitaker, A. (2006). *Einstein, Bohr, and the quantum dilemma: from quantum theory to quantum information*. New York: Cambridge University Press.

Wiener, N. (1948). *Cybernetics, or Control and Communication in the Animal and the Machine*, Cambridge, MA: MIT Press/ John Wiley and Sons.

Winfree, A. T. (1980). *The Geometry of Biological Time*, New York: Springer.

Winfree, A. T. (1984). "The prehistory of the Belousov-Zhabotinsky oscillator." *Journal of Chemical Education* **61**(8): 661–3.

Wright, S. "Evolution in Mendelian Populations." *Genetics* 16 (1931): 97–159.

Wright, S. (1986). "Evolution: Selected papers."

Yourgrau, W. (1960). *Variational principles in dynamics and quantum theory*. New York: Pitman Pub. Corp.

Zuckerkandl, E. and L. Pauling (1965). Evolutionary Divergence and Convergence in Proteins. *Evolving Genes and Proteins*. V. Bryson and H. Vogel. New York, Academic Press: 97–166.

Index

A

Abbott, Edwin 103, 261
Académie Royale des Sciences 72
Acceleration, uniform 42, 152
Action, Least 74, 78–85, 87–8, 91, 93, 148, 155, 195–6, 198, 204–5
Action, Stationary 87, 93–4, 197–8
Action potential 284
Aether 67, 70, 72–3
Albert of Saxony 41–2, 48
Alexandria 20, 40, 74, 207
Algebra 41, 51, 94–5, 106, 177
Allele 250–2
Andronov, Aleksandr A. 212, 225
Apollonius of Perga 22
Arcetri 4, 60–1
Archimedes 13–14, 49, 51–2
Aristotle 3–4, 11, 15, 19, 26–7, 30, 39–40, 45, 49–50, 52, 56, 148
Arnold, Vladimir Igorevich 121, 209–11, 214, 302
Arnold diffusion 211
Assayer 31
Associative memory 288–9
Atomic bomb 263
Attractor 213, 219–21, 225, 227, 288
Autonomous oscillator 273, 275
Avicenna 40–2
Axon 282–3

B

Barabasi, Laszlo-Albert 295, 297–8
Barberini 31–2, 34–5, 37
Basin of attraction 289
Beagle, HMS 233
Bell Labs 287
Bellarmino, Robert 28–31, 36
Belousov, B. P. 277–9
Belousov–Zhabotinsky reaction (BZR) 278–9
Benedetti, Giovani Battista 50–1
Berkeley 163–4, 202, 212–14, 287
Bernini 34
Bernoulli, Johann 73, 80–1, 88
Bessel, Friedrich 33
Bethe, Hans 198–201, 222
Betti, Enrico 102
Bifurcation 221, 225–9
Binomial theorem 66
Birkhoff, George 164, 216, 225, 230
Black hole 166, 168–71
Bohr, Niels 165–6, 172, 174–5, 177–9, 182–91, 199, 203, 259, 302
Bohr-Sommerfeld model 175
Boltzmann, Ludwig 116, 128–32, 139–44, 230, 300
Bolyai 102
Born interpretation 183
Born, Max 149, 175–8, 181, 183, 187, 191, 259
Borro, Girolamo 12, 14–15
Bowditch, Nathanial 129
Buonamici, Francesco 12, 14–15, 52
Bradwardine, Thomas 40, 42, 45
Brahe, Tycho 20, 22, 65
Brain 271, 278, 281, 285–6, 289, 297
Brecht, Bertolt 301
Brescia, Italy 45–6
Broadbent, Simon 292–4, 296
Buoyancy 50, 52–3, 72
Buridan, Jean 40–2, 48, 244

C

Calculus 56, 65–7, 70, 80, 82, 86–7, 95, 106, 111, 114, 122, 156, 240, 286, 290
Cal Tech 119, 163–4, 222, 259–60, 287
Cambridge 66, 127, 158, 178, 233, 250, 271, 274, 277, 282
Cantor, Georg 113–16, 118–20, 208, 213–14
Cantor set 115, 119, 213–14
Capillary 282
Carnot, Sadi 125
Cartesian 70–3, 75, 93, 98, 101
Cartwright, Mary-Lucy 213, 277

Carrying capacity 225, 240–1
Cassini, Jacques 73–4, 79
Castelli, Benedetto 27
Cathodory, Konstantin 118
Cayley, Arthur 95, 125
Cell 267, 281–4, 299
Center of gravity 203
Central force 77
Centrifugal 67, 73, 90
Centripetal 67
Ceres 97
Cesi, Federico 25, 32, 34
Chandrasehkar, Subrahmanyan 163
Charlemagne 39
Charney, Jule 215–17
Chemical potential 141, 254
Cheyne–Stokes syndrome 281
Chirikov, B. V. 210–11
Chirikov map 210
Circle map 209–10
Clairaut, Alexis 73, 79, 81–2, 89, 91
Clapeyron, Émile 125
Clausius, Rudolf 125–8, 130–31, 144
Clifford, William Kingdon 103
Colgate, Stirling 167–8
Comet 29–30, 65–7, 70, 145
Complementarity 185–7, 190, 203
Compton scattering 185, 202
Configuration space 7, 94, 173, 203, 249, 254–5, 261
Congress of Vienna 111
Constructive interference 197–8
Cornell 167, 199, 201, 222–4, 226, 229, 268, 278, 287, 295
Copenhagen interpretation 182, 186–8, 190–2, 203
Copernican system 4, 18–20, 22, 24, 26, 28–30, 32–4, 36–7
Copernicus, Nicolaus 20–4, 29–33, 35–7, 54, 244
Cosimo II 20, 27, 31
Coulomb 200, 202
Cournot, Antoine Augustin 245–6
Covariance 157–8
Crelle's Journal 111
Cubic equation 13, 46
Curie, Pierre 279
Curvature 17, 96, 98–9, 101–2, 164–5, 180
Cybernetics 217

D

D'Alembert, Jean le Rond 68, 88–91, 94
D'Alembert's principle 70, 90–1, 150
Darboux, Gaston 107
Dawkins, Richard 262
Darwin, Charles Robert 231–7, 247, 250, 302
Darwinism 250, 252–3
Da Vinci, Leonardo 11, 224
de Broglie, Louis 179–80, 187
Debye, Peter 175, 179
Dedekind, Richard 102, 115, 120
De Forest, Lee 273–4
Delft 51–3
Dendrite 283
De Motu 12, 14, 52–3, 66–8
Descartes, René 63, 67–8, 70–3, 75–6, 79, 85, 95, 110, 120
Diderot, Denis 88–9
Differential equation 90, 120, 122, 124, 134–5, 179–181, 241, 272, 274, 284
Differential form 157
Differential geometry 96, 98, 122
Diophantus of Alexandria 207
Diophantine 207–10
Dirac, Paul 178, 180, 184, 185, 187, 190, 193, 195–6, 199
Diogenes of Sinope 3
Dirichlet, Peter 99, 133, 146, 208
Discrete map 135–6, 228
Dissipative 205
DNA 251, 254, 256, 259–60, 266, 288, 298
Doppler, Johann Christian 248, 287
du Chatelet, Emilie Marquise 81–2, 84, 91
Dumbleton, John 40
Duopoly 245–6
Dyson, Freeman 201–3, 222

E

East India Company 234
École Polytechnique 119, 132
Eclipse 154, 159–60
Economic dynamics 214, 244
Eddington, Arthur 74, 158–60, 169
Edison, Thomas 273

Ehrenfest, Paul 142, 144, 176, 187–8, 244
Eigen, Manfred 265
Einstein, Albert 3, 33, 43, 63, 70, 74, 76, 91, 94, 99, 103, 108, 140, 146, 148–58, 161–4, 169–72, 174, 179, 182–3, 186–91, 208, 244, 302
Electron 148, 173, 175, 177–8, 180–1, 183, 185, 191, 195–6, 199–202, 284
Elevator 151–2
Ellipse 20, 33, 68
Elliptical orbit 20, 22, 24, 68, 158
Energy, conservation of 126
ENIAC 215
Entanglement 187, 192–4
Entropy 122, 125–6, 130–1, 140–1, 143–4, 171, 277
Epicycle 22–4, 29, 267
EPR paradox 189–90
Equilibrium 10, 77, 79, 88, 91, 128, 132, 135, 141, 230, 246–7, 264–5, 300
Equipartition 129
Equivalence principle 91, 150–6
Erdös, Paul 293–4, 296
Erdös–Renyi graph 296
Ergodic 116, 129, 142–3, 207
ETH 155
Euclid 13, 43, 45, 48–50, 52, 95
Euclidean 98, 102, 119, 147, 149, 155–6
Euler, Leonhard 79–81, 83–8, 91, 95, 133–4, 195, 204, 227, 229, 235, 240, 247, 289–91, 293
Event horizon 170–1
Everett, Hugh 166
Evolutionary dynamics 1, 63, 232, 239, 253, 263, 266
Extension 104–6

F

Feedback 34, 212, 226, 230, 267, 275, 281, 284, 286, 288
Feigenbaum, Mitchel 222, 224, 226, 227, 228–9
Feigenbaum number 221, 228–9
Fermat, Pierre de 75–7, 195, 207
Feynman diagram 201
Feynman, Richard 93, 166, 194–204, 222, 302
Fibonacci 227, 238–40
Field theory 154–5, 223, 279–80
Finkelstein, David 168
Firefly 268
First-return map 136
Fisher, Michael 222–4
Fisher, Ronald Alymer 250–3
Fitness 251–8, 261–2, 264–5
FitzHugh, Richard 284–5
Fixed point 9–10, 135–6, 138, 210, 214, 223, 227, 242, 250–1, 288
Fleming, John 273–4
Flow line 9, 276
Force-free 4, 145, 150, 157, 161
Free fall 41, 47, 144, 149–50, 157, 165, 170
Frame 91, 132–3, 145, 148, 150–3, 157, 161–2, 168
Fractal 113, 119–20, 208, 213–14, 297
Frederick the Great 79–80, 84–5, 91
French Revolution 65, 234
Frequency locking 275, 281

G

Galilei, Galileo 11–63, 85, 132, 148–9, 161, 170, 203, 230, 268–70, 301
Galilei, Vincenzo 12–14, 268–9
Galilei, Virginia 17
Galapagos Islands 233–4
Galton Laboratory of Genetics 263
Gamba, Marina 17, 31
Game theory 246, 262–3
Gauss, Carl Friedrich 96–102, 108, 146, 184
Generalized coordinates 92–5, 143, 266
Genetics 237, 247, 249, 252–3, 260, 263
Genome 252–7, 265, 298
Genotype 251, 255, 261, 264–5
Geodesic 86–7, 94, 98–9, 103, 156, 161, 169
Geometric growth 235, 240, 247
Giant Axon 282–3
Gibbs, Josiah Willard 106, 112, 140–1, 143–4
Giordano, Bruno 15, 27, 29, 37
Glass lens 16–17, 75, 185
Glass, Leon 281
Gleick, James 8
Golden mean 208, 210–11, 229, 239
Göttingen University 97, 99–100, 105, 108, 146, 149, 162, 174–8, 192, 207–9
Gould, John 233
Graph theory 80, 291–2, 298

Grassmann, Hermann Günter 104–7, 112, 125, 141
Gravitational cutoff 167–8
Gravity 10, 14, 50, 54, 66, 70, 72, 90, 94, 121, 127, 133, 144–5, 150–2, 155–8, 160–1, 164–5, 168, 171, 203, 302
Grossmann, Marcel 155–7
Guidobaldo, Marchese del Monte 54–5, 57–8
Guinness beer 251

H

H bomb 167, 216, 263
Hahn, Otto 166
Haldane, J. B. S. 252–3
Halley, Edmund 65–6, 68
Hamilton, William Rowan 107, 110, 123–4, 143
Hamiltonian 110, 121, 207–11
Hammersley, John 292–4, 296
Hamming distance 255, 257
Hardy, Godfrey Harold 250–1
Harmonic oscillator 192–3, 242, 272, 277
Harriott, Thomas 18
Hausdorff, Felix 118–20
Hawking, Stephen 171
Heart 10, 205, 268, 271, 278, 280–1, 285
Heaviside 106
Heisenberg, Werner 110, 144, 172–88, 195, 199, 203–4, 224, 259, 262, 302
Heisenberg uncertainty principle 186, 188
Heliocentric 4, 27–8, 244
Helmholtz, Hermann von 126, 272–3
Hermite, Charles 107, 113, 132–3
Hero of Alexandria 74
Hertz, Heinrich 273
Hertz, Paul 142
Heytesbury, William 40, 45
Hilbert, David 103, 107–10, 116, 146, 149, 162, 175, 209–10
Hilbert space 103, 107, 109–10, 192
Hindu-Arabic numerals 238
Hodgkin, Alan Loyd 281–4
Homoclinic 135–8, 144, 212–3, 216
Hooke, Robert 64–8, 302
Hopf, Eberhard 221, 225
Hopfield, John 287–9
Huxley, Andrew Fielding 281–4
Huxley, T. H. 236–7
Huygens, Christiaan 24, 63, 67, 75, 79, 127, 195–6, 269–71, 273, 276, 278, 290
Hyperbolic 102, 147, 211, 214
Hypercube 255–6, 292
Hyperspace 103, 232, 267, 289

I

Ibn Sahl 75
Ibn al-Haytham 75
Ingoli, Francesco 29–30, 32–3, 35
Impetus theory 5, 41–3, 45, 48
Inertia 54, 67–8, 70, 90
Inertial frame 103, 144, 150–1, 157, 301
Inquisition 4, 12, 15–16, 26–7, 29, 35–8
Institute for Advanced Study 167, 201–3, 208, 215–7
Interference 93, 192–3, 197–8, 204
Interferometer 170
Invariant 3, 40, 108–9, 138, 144, 146–8, 153, 156, 163, 210–1, 223, 229, 247
Inverse square law 66–7
Irrational 114, 129, 207–10, 239
Irreversible 130–1
Isochrony 55, 269

J

Jacobi, Karl Gustav Jacob 94–6, 99, 123–5, 129, 138–9, 143–4, 155, 195
Jacobi's principle of stationary action 94
Jordan, Camille 102, 107
Jordan, Pascual 178, 184–5, 259
Joule, James Prescott 127
Julia, Gaston 119
Jupiter 4, 18–19, 21, 24, 39, 230, 301

K

Kadanoff, Leo 222–4
KAM 206, 210–1, 227
Kepler, Johannes 20, 24, 29, 33, 68
Kepler's laws 66, 67
Kimura, Motoo 260–1
Kinematics 5, 42, 56, 60
Kinetic energy 82, 92, 127–8, 180
Kinetic theory 127–30, 139–40
Klein, Felix 95, 102–3, 105, 107–8, 142, 146, 149, 175
Kolmogorov, Nikolaevich 206–10, 224–5
König, Johann 73, 82–5

Königsberg, Prussia 107–8, 123–4, 146, 289–91
Kramers, Hendrik 177–8, 182, 199
Kronig, August 127
Kuramoto, Yoshiki 279–80

L

Lagrange, Joseph-Louis 64–5, 86–8, 91, 93–5, 104, 106, 143, 150, 156, 195, 302
Landscape 205, 253–4, 256–8, 261, 265
Lamb shift 199–200, 202
Latitude of forms 40, 43–4, 56, 244
Law of fall 41–2, 50, 55, 57–8, 63, 145, 148, 169
Lawrence, E. O. 167
Least action, principle of 74, 78–85, 87–8, 91, 93, 148, 155, 195–6, 198, 204–5
Least time, principle of 76, 195
Lefschetz, Solomon 212
Lesbesgue measure 210
Legendre 99
Leibniz 44, 63, 67, 70–1, 73, 78, 80, 82–4, 91, 95, 104, 106, 110, 290–1
Length contraction 163
Levinson, Norman 213, 277
Lévy, Paul 119
Light cone 147
LIGO 170
Limit cycle 9–10, 136, 226–7, 242, 275–7, 288–9
Linear space 107
Line element 86, 98, 101–2, 109, 147, 155–6, 163
Linnaean Society 235–6
Liouville, Joseph 95, 114, 121–4, 129, 139–40, 143–4
Lissajou figure 129
Littlewood, John 213, 277
Lobachevsky 102
Lorentz transformations 146, 148–9, 152, 154–5, 183, 244
Lorenz, Edward 205, 215–21, 224–5, 230, 302
Lorenz attractor 144, 221
Lotka, Alfred James 225, 241–3, 264
Los Alamos 167, 201, 224, 226
Loschmidt, Joseph 130–2
Lyapunov 212, 230, 245
Lyncean Academy 25, 32

M

Mackey, Michael 281
Macroeconomics 245
Magnetization 279–80
Magnetron 276–7
Malthus, Robert 234–5, 240, 247
Mandelbrot, Benoit 119–20, 221, 229
Manifold 96, 101, 103, 125, 135–7, 147–8, 156, 230–1
Marconi, Guglielmo 273
Marine Biological Association 282
Matrix mechanics 173, 195, 199, 203–4
Maupertuis, Pierre Louis 71–4, 76–85, 87–8, 91, 148, 192, 195–6, 198, 204–5, 240, 302
Maxwell, James Clerk 127–31, 140, 152, 157, 196, 230
May, Robert 225
McCulloch, Warren S. 285–7
Mean-field theory 279–80
Measure theory 118, 142
Meitner, Lise 166
Meme 220, 262, 301
Mendel, Johann 248–52
Mercury 20–5, 28, 30, 158, 160
Merton College 40, 225
Meteorology 215–17, 225
Metric space 102
Metric tensor 147, 156–7, 163
Microscope 176, 185–6, 284
Milgram, Stanley 294–6
Military compass 16
Minimax 262
Mirollo, Renato 268
Miscellanea Taurinensea 87
Misner, Charles 166
MIT 175, 194, 209, 213, 215–18, 224, 287
Mittag-Leffler, Gösta 132, 135
Minkowski, Hermann 102, 107–9, 142, 146–50, 157, 162, 175
Molecular clock 258, 260
Momentum 7, 164, 172, 177–9, 183–5, 190, 192–3, 196, 203
Monge, Gaspard 96–7
Mongré, Paul 118

Monte Carlo 293, 296
Moon 18–20, 22, 24, 28, 30, 33, 72–3, 133, 159, 161
Moser, Juürgen 209–10

N

Nagumo, Jin-Ichi 285
Napoleon 97, 111
Natural motion 14, 41–2, 47–8, 54
Natural selection 234–6, 247, 250, 253, 261
Navier–Stokes 225
Neo-Darwinism 252–3
Network, neural 286–9, 297
Network science 1, 295, 298
Network, social 1, 271, 291, 295–6, 298
Nerve 205, 282–3
Neuron 205, 281, 283–9, 300
Neutral drift 260–1, 265
Neutron star 163, 166
Newcomen, Thomas 125
Newton, Isaac 11, 34, 53, 63–73, 78, 80, 85, 88, 91, 95, 110, 133, 148, 150, 160, 302
Nobel prize 170, 176, 181, 224, 229, 244, 259–60, 272, 284
Non-Euclidean 102, 147, 149, 155–6
Non-inertial 103, 151
Nonlinear dynamics 212, 215–17, 225, 232, 244, 276–7, 302
Nuclear 166–7, 198, 258–60
Null geodesic 169

O

Oldenburg 71
Operator 90, 179–80
Oppenheimer, J. Robert 163–8, 178, 198, 202
Orbit 22–3, 34, 66–7, 136, 145, 158, 160, 169–70, 183, 192, 206, 208, 210–11, 225, 245, 301
Oresme, Nicole 42–5, 244
Outer measure 118–19
Overhauser, Albert 287
Oxford 40, 45, 48, 225, 236–7, 252, 293

P

Pacemaker 267–8, 278, 281, 285
Padua 4, 15, 17, 24, 37, 53–4, 61, 302
Parabolic 1, 4, 42, 46, 48, 54, 57–9, 61, 63, 107, 145, 152, 225
Parallax 22, 23, 28, 29, 33, 159
Patagonia 234
Pauli, Wolfgang 173–5, 177, 178, 180, 184, 186, 187, 195
Pauling, Linus 199, 258–60
Payoff matrix 263–5
Peano, Guiseppe 103, 106–7, 109, 110, 115–18
Peixoto, Mauricio 212–13
Pendulum 55, 56, 61, 67, 205, 206, 212, 268–70, 273
Perceptron 287, 289
Percolation 293–5, 298
Perihelion 158
Periodic 129, 136, 209, 212, 213, 267, 269, 273, 277, 280, 281, 283
Phase portrait 9, 276
Phase space 7–9, 63, 120–25, 128, 129, 134–44, 173, 192, 193, 203, 207, 211, 212, 214, 230, 242, 244, 246, 254, 275, 281, 282, 284, 288, 298, 300
Phase transition 140, 222, 223, 229, 279
Philoponus 40
Piccolomini, Ascanio 38, 60
Pisa, Tower of 15, 52
Pitts, Walter 285–8
Plancheral, Michel 142–3
Planck, Max 157, 174, 176, 187
Planck's constant 171, 184, 188, 195
Poincare, Henri 63, 103, 107, 132–9, 142, 146, 148, 206–8, 211–13, 216, 225, 230, 242, 245–7, 275, 280, 281, 302
Poincare section 136–7
Podolski, Boris 189
Potential energy 77, 92, 128, 180, 205, 217, 254, 288
Power law 77, 222, 223, 297, 298
Prague 153, 157, 278
Precession 158, 160
Price, John 263
Princeton 93, 166, 167, 168, 192, 194, 201–3, 208, 212, 215, 216, 279, 287
Principia 65, 68, 70, 80, 82, 88, 91, 141, 285
Principle of least action 74, 78, 79, 81–5, 87, 88, 91, 93, 155, 195, 196, 198, 205
Probability 127–9, 131, 181, 182, 184, 192, 193, 196, 197, 201, 207, 230, 252, 253, 265, 292, 297, 302

Protestant Reformation 26
Prussian Academy 79
Ptolemaic 12, 20, 22, 24, 28, 29, 33
Ptolemy 13, 20–2, 27, 32, 267
Purdue 281
Pythagoras' theorem 101
Pythagorean 101, 102, 146

Q

QED 199–201, 203, 223
Quadratic form 101, 109, 146–8
Quantum computing 194
Quantum interference 93, 198, 204
Quantum mechanics 103, 110, 166, 171–3, 175, 178–83, 189, 191–2, 194–5, 197–9, 204, 224, 259
Quantum phase 196–8
Quark 222–3
Quasi-species 265

R

Radar 276–7, 283, 287
Rashevsky, Nicolas 285–96
Rational number 114, 208
Rayleigh 271–3, 276, 278, 284, 302
Refraction 74–5, 77–8, 118, 154
Relativity, Galilean 26, 33
Relativity, General 74, 99, 150, 153, 155, 157–8, 160–2, 168, 188
Relativity, Special 3, 108–10, 142, 146–52, 173, 183, 186, 188, 191, 301
Relaxation oscillator 275, 280
Renormalization 199–200, 222–4, 226, 229
Rényi, Alfred 293
Replicator 262–5
Resolving power 176, 186
Resonance 39, 207, 211, 275
Retrograde 19, 22, 33
Reversibility 126, 130–1
Ricci-Curbastro, Gregorio 102, 156, 158
Ricci, Ostilio 13, 52
Riemann, Georg Friedrich Bernhard 63, 99–102, 108, 117, 124, 146, 302
Riemannian geometry 102–3, 109, 146, 149, 156, 158, 216
Rosen, Nathan 189
Rosenblatt, Frank 287
Rosenthal, Artur 142–3
Royal Society 65–6, 71, 161, 274, 277, 292
Ruelle, David 220–1, 225–6

S

Saddle point 9, 10, 136, 245, 265
Sarpi, Paolo 15–17, 54–5, 58, 61
Saturn 21, 24, 127, 128
Scaling 223, 226, 229, 298
Scheiner, Christoph 25, 30, 61
Schrödinger, Erwin 110, 173, 179–83, 186, 187, 191–2, 195, 199, 203–4, 259, 262
Schrödinger's cat 191–2, 194
Schuster, Peter 265
Schwarzschild, Karl 149, 162, 164–5, 168
Schwarzschild radius 161–5, 168, 169
Schwarzschild solution 162–4
Schwinger, Julian 202
Self-energy 195, 199–200, 202
Self-oscillator 213, 267–8, 272–4, 276, 285
Sensitivity to initial conditions 138, 214, 220
Set theory 113–14, 118
Shelter Island 198–9
Sidereus Nuncius 19–20, 24
Siegel, Carl Ludwig 207–10
Siemens, Werner von 273
Simplex 254–65
Simultaneity 146, 188
Singularity 162–3, 168, 170–1
Six degrees of separation 294–5, 297
Smale, Stephen 212–14, 221, 225, 277
Smale horseshoe 213–14
Small divisors 208, 210
Small world 292, 296–7
Smith, John Maynard 263
Snel, Willebrord van Royen 75
Snell's law 75–6, 195–6
Snyder, Hartland 164–5, 167–8
Solvay 172, 186–9, 198
Sommerfeld, Arnold 149, 156, 173–6, 259
Space time 76, 94, 102–3, 109–10, 147–50, 156–7, 161, 164–5, 169–70, 183, 201
Speed of light 76, 78, 103, 147, 152–5, 162, 183, 189
Squid 282–3
Stability 127–8, 132, 134–5, 205–7, 209–12, 215, 227, 230, 244–5, 282

Statistical mechanics 116, 121–2, 127–8, 130, 141–2, 144, 261, 279–80, 297
Stettin, Prussia 104–5, 112, 125
Stevin, Simon 13, 51–3, 149
Standard map 210–11
St. Mark's square 17
St. Thomas Aquinas 22, 26
Strange attractor 96, 120, 220, 221, 225
Strogatz, Steven 268, 281, 295–7
Strutt, John William 271
Sun spot 11, 24–6, 33
Superposition 107, 191–2, 194
Swallow 6–8, 110, 254
Swineshead, Richard 40, 45
Synchronization 267–8, 271, 275–6, 278–80, 295–6
Szent-Györgyi, Albert 260

T

Taisner, Jean 50–1
Takens, Floris 225–6
Tartaglia, Niccolo 13, 46–50, 52
Telescope 15–19
Teller, Edward 15–19, 24–5, 54, 58, 61, 301
Temperature 128, 131, 219–20, 222, 229, 279
Tensor 147, 156–7, 163
Tesla, Nikola 273
Theorema Egregium 98
Thermionic 273
Thermodynamics 125–6, 128, 130, 141
Thompson, J. J. 274
Thomson and Tait 131
Thorne, Kip 166, 168, 170
Three-body problem 132–5, 137–8, 206–10, 230, 247, 302
Time dilation 163, 165
Topology 101, 122, 201, 206, 212–13, 290–1, 293–4
Transfinite 114
Trajectory, parabolic 1, 4, 42, 46, 48, 54, 57–9, 61, 63, 107, 152
Transmutation 233–5
Triode 213, 273–4
Tuning 13, 33, 272–3
Turbulence 206, 224–6
Turing, Alan 285–6
Twin paradox 3
Two New Sciences 4–5, 39, 53, 57, 61–2, 301
Tychonic system 20, 30–1, 33

U

Uncertainty 10, 37, 52–3, 97, 144, 182, 184–6, 188, 190, 203–4, 230, 251, 289
Uniformly difform 40, 44, 48
United Provenances 51
Uranium 166

V

Vacuum 72, 202, 205, 215, 219
van der Pol, Balthasar 225, 274–6, 278, 280, 284, 302
van der Pol oscillator 213, 274, 276, 277
Variational calculus 65, 80, 86, 87, 95, 156
Variational principle 83, 155, 156
Varignon, Pierre 88–9
Vector space 106, 107, 109, 115, 146
Venice 13–17, 20, 46, 49, 54, 60, 61, 302
Venus 4, 20–5, 28, 30
Verhulst, Pierre-François 225, 240–2
Violent motion 41, 42, 47, 48, 54
Virtual velocities 88, 91
Viviani, Vincenzo 53, 55, 269
Voltaire 72, 79, 81, 82, 84, 85, 88
Volterra, Vito 225, 242, 243, 264
von Koch, Helge 117
von Neumann, John 110, 186, 192, 199, 215–18, 262, 263, 286
Vortex 67, 70, 72, 73

W

Wallace, Alfred Russell 235–6
Walras, Marie-Esprit-Léon 246
Walras' law 246–7
Warp 76, 94, 103, 150, 154, 156, 157, 161, 169–71, 219, 302
Water Clock 55, 56, 63
Watts, Duncan 295–7
Wave function 109, 180, 181, 186, 189–91, 195, 203
Wave function collapse 186
Wave mechanics 173, 179, 181–3, 195, 199, 204, 259

Weierstrass, Karl Theodor Wilhelm 111–13, 117, 119, 120, 132, 135, 206, 208
Weiss, Pierre 279
Welser, Mark 25
Weyl, Hermann 109, 180
Wheatstone, Sir Charles 273
Wheeler, John Archibald 93, 145, 166–8, 194, 195, 198–200
Widom, Benjamin 222–3
Wien, Wilhelm 176, 186
Wiener, Norbert 110, 175, 217, 225
Wigner, Eugene 192–4
Wildcat Creek 2, 6, 8, 254, 268
Wilson, Kenneth 222–4, 226, 229
Winfree, Art 267, 278, 279, 281, 296
World line 147, 170
World war 118, 158, 160, 161, 166, 206–8, 216, 241, 252, 262, 276
World Wide Web 291, 298
Wren, Christopher 65–6
Wright, Sewall 253, 254, 256, 257, 261

X

X-ray 170, 259

Y

Yorke, James 220–1

Z

Zarlino, Gioseffo 13–14, 25
Zeeman effect 173–4, 244
Zeno 2, 3
Zero-sum 262, 264–5
Zhabotinsky, A. M. 278
Zuckerkandl, Emile 260